系统建模与仿真
（双语版）

钱慧芳　惠亚玲　卢　健　刘　薇　等编著

電子工業出版社
Publishing House of Electronics Industry
北京 · BEIJING

内 容 简 介

本书是中英文对照编写，可以作为自动化和控制类专业及相近专业的仿真课程双语教学用书。本书共4个主要模块。第一模块包括第1章和第2章，主要内容是MATLAB基础内容及其编程基本知识；第二模块包括第3章、第4章和第5章，主要内容是建立数学模型（基本传递函数模型建立和状态空间模型建立）；第三模块包括第6章、第7章和第8章，是系统稳定性分析和系统设计；第四模块包括第9章和第10章，是Simulink建模。除了中英文双语对照，本书另一大特色是实例丰富。每一章的知识点都以具体实例进行介绍，生动、灵活地呈现了运用MATLAB的具体操作步骤及其仿真过程。

本书结构清晰，突出实用性，以丰富的实例突出实践性。本书既可以作为自动控制等相关专业的教学参考书，也作为MATLAB语言入门参考书，亦可以作为专业英语教材。书中的实例为相关课程设计、毕业设计等提供了重要参考。

图书在版编目（CIP）数据

系统建模与仿真：双语版：汉英对照 / 钱慧芳等编著. —北京：电子工业出版社，2019.11

ISBN 978-7-121-35261-4

Ⅰ. ①系… Ⅱ. ①钱… Ⅲ. ①系统建模－高等学校－教材－汉、英②系统仿真－高等学校－教材－汉、英 Ⅳ. ①N945.12 ②TP391.92

中国版本图书馆 CIP 数据核字（2018）第 240983 号

责任编辑：窦　昊

印　　刷：三河市双峰印刷装订有限公司

装　　订：三河市双峰印刷装订有限公司

出版发行：电子工业出版社

北京市海淀区万寿路 173 信箱　　邮编：100036

开　　本：787×1 092　1/16　　印张：23　　字数：588.8 千字

版　　次：2019 年 11 月第 1 版

印　　次：2020 年 12 月第 2 次印刷

定　　价：69.00 元

凡所购买电子工业出版社图书有缺损问题，请向购买书店调换。若书店售缺，请与本社发行部联系，联系及邮购电话：(010) 88254888，88258888。

质量投诉请发邮件至 zlts@phei.com.cn，盗版侵权举报请发邮件至 dbqq@phei.com.cn。

本书咨询联系方式：(010) 88254466，douhao@phei.com.cn。

前　　言

本书采用中英文对照，是针对自动化和控制类专业以及相近专业高年级本科生编写的双语教科书。同时，本书中文部分自成一体，可以作为相关专业仿真课程的教学用书；英文部分可作为专业英语学习的参考内容。作者结合“控制系统的建模与仿真”的教学实践与研究成果，以 MATLAB R2014b 为系统仿真平台，采用简洁的风格和实用的准则编写了本书。

1．本书结构与内容安排

本书共 10 章，在结构上分成 4 个主要模块。第一模块是 MATLAB 基础内容及其编程基本知识，包括第 1 章和第 2 章；第二模块是建立数学模型（基本传递函数模型建立和状态空间模型建立），包括第 3 至第 5 章；第三模块是系统稳定性分析和系统设计，包括第 6 至第 8 章；第四模块是 Simulink 建模，包括第 9 章和第 10 章。

第一模块主要内容：以魔方例子介绍矩阵的相关操作；介绍 MATLAB 环境及其帮助系统的使用；MATLAB 二维和三维图形绘制及图像的简单操作；MATLAB 软件中的流程控制语句；脚本和函数具体使用，以实例分析编程技巧。

第二模块主要内容：重点介绍控制对象（包含传递函数和状态空间模型两种表达方式）的数学模型建立。其中包括三种传递函数（降幂排列的标准形式、零极点形式和部分分式展开式）及其三种连接方式（串联、并联和反馈）的 MATLAB 函数的表达，以及控制系统的三种模型［传递函数模型（tf）、零极点模型（zpk）和状态空间（ss）模型］形式及其相互转换的 MATLAB 函数的表达和实现。此外，通过两个实例建模及其仿真，分别说明简单系统的传递函数建立及其状态空间方程建立的具体过程。最后，以双容水箱液位系统为例，详细展示实际物理系统的模型建立过程，包括必要的假设、方程的推导和线性化等知识点，以及双容水箱液位系统的仿真。

第三模块主要内容：重点讨论控制系统的稳定性问题。首先，主要以两个实例对系统的稳定性和稳态误差进行详细的分析与仿真。其次，通过实例展示使用主导极点法的 PID 控制器设计思路，以及具体实现的步骤。

首先，以人控制的自行车系统为例，详细分析系统的稳定性因素及其稳定条件，包括模型的各种参数对其稳定性的影响、控制器参量对稳定性的影响，还讨论了自行车运行速度对稳定性的影响。

其次，以飞机飞行期间的波动控制为例，分析阶跃输入下的稳态误差和扰动情况下的稳态误差；重点研究飞机飞行的波动控制问题，在 P、PI 和 PID 三种控制器作用下，分别分析了系统在输入和扰动两种情况下的响应，在 PID 控制器中，发现有一个 PID 控制器满足所有的设计规范要求。

最后，通过比例与积分控制实例的比较、比例与微分控制实例的比较，对 PID 控制实例予以说明，分别得到相应的结论。以 PID 控制的下棋机器人为例，利用主导极点的概念，为下棋机器人设计了一个满足带宽和超调量指标的控制器，并具体展示该设计思想的实现过程。

第四模块主要内容：重点介绍 Simulink 下的模型建立，以简单的实例说明简单模型的建立过程；以房屋热系统为例，详细说明子系统及其封装技术，展示房屋热系统模型的实际搭建具体步骤及其仿真的实现。

2．本书特色

本书结构清晰，内容翔实，图文并茂，实例丰富，并具有以下特色：

（1）实例丰富，简明易学。本书中，所有介绍通过 MATLAB 实现控制系统理论仿真部分，均采用多种实例来具体说明其操作实现过程。只要跟随本书的步骤，就能完全学会并理解。很容易验证学过的理论，并深刻理解所学的知识。

（2）突出实践，简单易用。本书中有些章节本身就是个实际的例子，如第 5 章，就是双容水箱建模的实际例子。此例既可以作为课题知识讲述建立模型的过程和方法，也可以作为毕业设计和课程设计的基础内容进行相关拓展来使用。第 7 章、第 8 章和第 10 章都是以具体实例展开的，均突出了实践性，学生通过实际例子的建模仿真与系统的分析设计，能够把理论与实际相结合，做到融会贯通，把理论运用于实践中，更好地掌握知识。

（3）双语对照，学习专业英语易上手。本书英文部分均参照了英文原版的专业词汇及其表达，尽可能展现地道英语语言的表达方式，克服中文式的英语表达，把艰涩难懂的专业英语转换为清晰、简洁的英文表达，对学习专业英语的方式和方法进行了有益的探索。

3．本书编者

本书的主体结构及其章节的主要内容由钱慧芳负责，书中的内容简介与前言均由钱慧芳完成。第 1、4、7、8、9、10 章中文和英文内容均由钱慧芳编写，第 2、3、5、6 章的中文和第 2、5、6 章中的英文，由刘薇编写了初稿，钱慧芳修订并定稿，第 3 章英文内容由钱慧芳编写。全书的第 1 到第 10 章的中英文校对全部由惠亚玲完成。卢健参与书稿中部分例子的编写。

本书全部程序由研究生王戈在 MATLAB R2014b 平台上验证，书稿的文档格式及书中图的编辑也由王戈协助完成。游信勇在早期参与了书中部分例子的讨论与少部分编写工作。

4．致谢

对参与本书编写工作的所有人员表达谢意！感谢王戈和游信勇两名研究生的倾力协助！同时，对全力支持我书稿工作的女儿表达感谢！

由于作者水平有限，加之时间仓促，书中难免有不足与疏忽之处，敬请读者批评指正。

钱慧芳

目　　录

第1章 MATLAB基础知识

本章主要介绍 MATLAB 软件平台的基础知识。从一个有趣的魔方矩阵例子开始，介绍矩阵的基本操作及 MATLAB 的环境。对于 MATLAB 图形的绘制，重点描述二维绘图，对三维绘图和图像操作做了简单介绍。本章内容为后期进一步学习打下基础。

1.1 MATLAB 入门

MATLAB 入门的最好方法就是学会如何处理矩阵。在 MATLAB 中，矩阵是一个矩形的数字阵列。1×1 矩阵的特别意义为标量，仅有一列或一行的矩阵则为向量。通常，最好将一切都看成矩阵。不同于其他编程语言，一次处理一个数据，MATLAB 能够迅速而方便地处理整个矩阵。

在本课程中，使用一个很好的示例矩阵，它出现在文艺复兴时期一位德国艺术家兼业余数学爱好者阿布列西特·杜勒的雕刻作品《忧郁者 I》中。这幅图布满了数学符号，如果仔细观察，你会发现右上角有一个矩阵（如图 1.1.1 所示）。这个矩阵就是众所周知的魔方阵，早在杜勒时代，人们就相信它具有魔力，而它的确证明具有值得研究的神奇特性。

图 1.1.1 魔方阵

1.1.1 输入矩阵

可以用几种不同的方法将矩阵输入到 MATLAB 中：

1. 直接输入矩阵的元素列表。
2. 从外部数据文件加载矩阵。
3. 利用内置函数生成矩阵。
4. 在 M 文件中自定义函数创建矩阵。

需遵循的几个基本规则是：

1. 同一行内的元素用空格或逗号隔开。
2. 行与行之间，用分号“;”表示每行的结束。
3. 矩阵的全体元素使用方括号“[ ]”括住。

依据上述基本规则，要输入杜勒的矩阵，只需输入：

```
A=[ 16 3 2 13 ; 5 10 11 8; 9 6 7 12 ; 4 15 14 1 ]
```

按回车键，MATLAB 显示：

$$A = \begin{bmatrix} 16 & 3 & 2 & 13 \\ 5 & 10 & 11 & 8 \\ 9 & 6 & 7 & 12 \\ 4 & 15 & 14 & 1 \end{bmatrix}$$

它有何神奇之处呢？

1.1.2　求和、转置和对角线

你可能已经发现，魔方矩阵的特性与不同的求和方式有关。计算任一行或任一列，或两主对角线上元素的和，总会得到相同的结果。下面使用 MATLAB 来验证这一结论。

尝试的第一条语句

```
sum(A)
```

MATLAB 显示

```
ans =
    34    34    34    34
```

不指定输出变量时，MATLAB 默认用变量 ans（answer 的缩写）来储存最近的计算结果。已经计算出的结果是一个行向量，它包含了矩阵 A 所有列的所有元素之和。很明显，每一列的和不可思议地相同，都是 34。

那么，每一行的和会怎样呢？

先将矩阵转置，计算转置矩阵每一列之和，然后将结果再转置。转置操作符用一撇或单引号“'”表示。例如，输入

```
sum(A')'
```

```
ans =
    34
    34
    34
    34
```

此时得到的列向量为对矩阵 A 的每一行求和的结果。显然，各行的和仍为 34。

使用 diag 函数，很容易获得主对角线上所有元素之和。diag 函数的功能是提取矩阵对角线上的元素。

```
sum(diag(A))
```

```
ans =
    34
```

另外一个对角线，即所谓的反对角线，fliplr 函数的功能是将矩阵左右翻转，将反对角线转换成主对角线。

```
sum(diag(fliplr(A)))
```

```
ans =
    34
```

上述结果证明，矩阵 ***A*** 确实是一个魔方矩阵。在此过程中，列举了几个 MATLAB 矩阵运算。下面继续用此矩阵介绍 MATLAB 的其他功能。

1.1.3 下标

矩阵 ***A*** 的第 i 行、第 j 列的元素用 $\boldsymbol{A}(i,j)$表示。

那么，$\boldsymbol{A}(1,4)+\boldsymbol{A}(2,4)+\boldsymbol{A}(3,4)+\boldsymbol{A}(4,4)$的计算语句及其结果为

```
A(1,4)+A(2,4)+A(3,4)+A(4,4)
```

```
ans =
    34
```

除了上述双下标的元素表示方法，矩阵索引也可以用一个单下标 $\boldsymbol{A}(k)$表示矩阵元素。这种方法常用来表示行向量和列向量的元素。但它也可以用于二维矩阵，在这种情况下，二维矩阵被视为由其原矩阵各列依次构成的一个长的列向量。因此，对于魔方矩阵 ***A***，$\boldsymbol{A}(8)$与 $\boldsymbol{A}(4,2)$是对矩阵 ***A*** 中 15 这一元素的两种表示方法。

1.1.4 冒号运算符

冒号“:”是 MATLAB 中最重要的运算符之一，它有几种不同的表达形式。表达式 1:10 表示包含从 1 到 10 的整数行向量。

1 2 3 4 5 6 7 8 9 10

为了得到非单位的间距，可以指定一个增量，例如，100:−7:50 是

100 93 86 79 72 65 58 51

0:pi/4:pi 是

0 0.7854 1.5708 2.3562 3.1416

下标表达式中使用冒号可以表示矩阵的一部分。$\boldsymbol{A}(1:k,j)$指矩阵 ***A*** 的第 j 列前 k 个元素。sum(A(1:4,4))为计算 ***A*** 的第 4 列的和。但有一个更好的方式，用冒号表示矩阵中某一行或列的所有元素，用关键词 end 表示最后一行或列的所有元素。

因此，用命令

```
sum(A(:,end))
```

```
ans =
    34
```

来计算 ***A*** 的最后一列元素之和。

为什么 4×4 魔方矩阵的和等于 34？如果将 1 到 16 的整数之和等分为 4 份，那么每一份的和为 sum(1:16)/ 4，可得 34。

```
sum(1:16)/ 4
```

```
ans=
    34
```

1.1.5　魔方阵函数

实际上，MATLAB 提供了一个内嵌函数，它可以创建任意大小的魔方矩阵。毫不奇怪，这个函数就是 magic。

```
B = magic(4)
```

$$\boldsymbol{B}=\begin{bmatrix}16 & 2 & 3 & 13\\ 5 & 11 & 10 & 8\\ 9 & 7 & 6 & 12\\ 4 & 14 & 15 & 1\end{bmatrix}$$

这个矩阵与杜勒雕刻版上的矩阵几乎一样，具有所有相同的神奇性质。唯一的区别在于中间的两列调换了位置。为了将这个矩阵 $\boldsymbol{B}$ 转变成杜勒的矩阵 $\boldsymbol{A}$，调换 $\boldsymbol{B}$ 中间的两列

```
A = B(:,[1 3 2 4 ])
```

上式表示，对于矩阵 $\boldsymbol{B}$ 的每一行，将其列的顺序按照“1，3，2，4”重新编排。

1.2　表达式

像大多数其他编程语言一样，MATLAB 提供了数学表达式。但与大多数编程语言不同，MATLAB 的表达式针对整个矩阵，构成如下：

（1）变量

（2）数字

（3）运算符

（4）函数

1.2.1　变量

变量名由字母构成，后面跟字母、数字或下画线。MATLAB 具有区分大小写的特性，所以，“A”和“a”是两个不同的变量。

1.2.2　数字

MATLAB 使用传统的十进制记数法，带有一个可选的小数点和正负符号。科学记数法使用字母 e 表示以 10 为底的指数。虚数使用 i 或 j 作为后缀。

下面是一些合法数字的例子。

```
3            –99            0.000
9.6397238    1.60210e–20    6.02252e23
1i           –3.14159j      3e5i
```

1.2.3　运算符

使用常见算术符号的表达式和优先规则如表 1.1 所示。

表 1.1　使用常见算术符号的表达式

常用算术符号	含　义
+	加
-	减
*	乘
/	除
\	左除
^	幂
'	复共轭转置
()	指定计算顺序

1.2.4　函数

MATLAB 提供丰富的标准初等数学函数，包括 abs（绝对值）、sqrt（平方根）和 exp（指数函数）等。MATLAB 也提供很多高等数学函数，如 gamma（伽马）函数，这些函数中，大多数都接受复杂的参数。一些函数是 MATLAB 内置的，如 exp 和 cos。它们是 MATLAB 的部分核心构成，因此用起来比较高效，但内部的计算细节不能查询。

表 1.2 为一些特殊的函数提供了有用的常量值。

表 1.2　特殊函数的常量值

常　量	含　义
pi	3.14159265
i	虚数单位，-1
j	同 i
eps	浮点相对精度，2.2204e-16
real min	最小浮点数，2.2251e-308
real max	最大浮点数，1.7977e+308
Inf	无穷大
nan	不定数

我们已看到一些 MATLAB 表达式的例子，下面再举一些例子和运算结果。

```
rho = (1+sqrt(5))/2
rho =1. 6180
a = abs(3+4i)
a=5
huge = exp(log( realmax) )
huge =1. 7977e+308
toobig = pi *huge
toobig = Inf
```

1.3　矩阵操作

下面介绍生成矩阵的其他方法。

1.3.1　生成矩阵

MATLAB 提供 4 个生成基本矩阵的函数（见表 1.3）。

表 1.3　4 个基本函数

函　数	说　明
zeros	全 0 矩阵
ones	全 1 矩阵
rang	均匀分布的随机矩阵
rangn	正态分布的随机矩阵

1.3.2　M 文件

M 文件是包含 MATLAB 代码的文本文件，可以用 M 文件创建自己的矩阵。在 MATLAB 命令行键入一些命令，即可创建包含该命令的文件，将其文件名保存为.m 格式。

例如，创建一个文件包含

```
A = [16, 2, 3, 13; 5, 11, 10, 8; 9, 7, 6, 12; 4, 14, 15, 1]
```

将该文件名保存为 magik.m。magik 文件的文本内容，即上文的示例魔方矩阵。

1.3.3　连接

连接是一个将小矩阵合并成大矩阵的过程。事实上，你创建的第一个矩阵，就是将它的单个元素连接在一起获得的。方括号[ ]是连接运算符。例如，以 4×4 魔方矩阵 ***A*** 为基础，构造 ***B***

```
B = [A   A+ 32 ; A+ 48   A+16]
```

连接 4 个子矩阵，得到 8×8 矩阵 ***B***。

```
B =
    16   3    2    13   48   35   34   45
    5    10   11   8    37   42   43   40
    9    6    7    12   41   38   39   44
    4    15   14   1    36   47   46   33
    64   51   50   61   32   19   18   29
    53   58   59   56   21   26   27   24
    57   54   55   60   25   22   23   28
    52   63   62   49   20   31   30   17
```

这个矩阵仅是成为另一个魔方矩阵的过渡。它的元素是从 1～64 整数的重新排列。它的列向量之和正是一个 8×8 魔方矩阵的值。

```
sum(B)
```

```
ans = 260 260 260 260 260 260 260 260
```

但是，它的行向量之和 sum(B')'，并不总是相同的。为了使之成为真正 8×8 的魔方矩阵，还需要进一步的处理。

1.3.4 删除行和列

使用一对方括号，可删除一个矩阵的行和列。例如，***X*** =***A***，然后，删除 ***X*** 的第二列，用 ***X***(:,2)= [ ]表示。

```
X(:,2)=[]
```

结果是

```
X =
   16    2   13
    5   11    8
    9    7   12
    4   14    1
```

如果删除矩阵的单个元素，其结果不再是矩阵。所以，像这样的表达式 ***X***(1,2)= [ ]，结果会报错。因此，用一个单下标删除一个元素或一系列元素，并把剩下的元素再形成一个行向量。

```
X(2:2:10)=[]
```

结果是

```
X =
   16   9   2   7   13   12   1
```

1.4 命令窗口与常用函数

目前为止，已经使用了 MATLAB 命令行、键入命令和表达式，并且看到了命令窗口显示出的运行结果。

本节描述改变命令窗口外观的方法，并介绍几个常用函数。如果系统允许选择命令窗口的字体和字形，为了保证合适的间隔，建议采用固定宽度的字体，如 Fixedsys 或 Courier。

1.4.1 格式命令与长命令行

format 命令用于控制 MATLAB 显示的数据值格式。该命令只影响数据显示的方式，而不影响 MATLAB 对数据的计算和储存。

下面是向量 ***x*** 在不同格式和量级下的相应结果显示。

```
x=[4/3  1.2345e-6]
format short
1.3333   0.0000
format short e
1.3333e+000     1.2345e-006
format short g
1.3333   1.2345e-006
format long
1.333333333333333     0.000001234500000
format long e
1.333333333333333e+000 1.234500000000000e-006
format long   g
1.33333333333333   1.2345e-006
format rat
4/3   1/810045
format hex
3ff5555555555555   3eb4b6231abfd271
```

如果矩阵中最大的元素比 10^3 大，或者最小的元素比 10^{-3} 小，那么 MATLAB 会为 short 型和 long 型数据采用一个常用的比例因子。

除了上述命令格式，利用 format compact 命令能够消除输出中出现的空白行，让你能在屏幕或窗口查看更多的信息。

如果一条语句无法在一行输入完，则键入三个点“…”，然后按回车键，表示下一行继续该指令的输入。例如：

i=1+2+3+4+5+6+7+8+⋯+10+11;

1.4.2　隐藏输出

如果你简单输入一个语句，按回车键，MATLAB 会自动显示其运行结果。然而，如果该行命令是以“;”结尾的，MATLAB 仍会执行运算，但不显示任何输出。这一点对于生成大型矩阵时特别有用。

例如， A=magic(100);

（magic 函数是生成魔方矩阵的函数命令，后面会讲到。）

1.4.3　命令行编辑

键盘上的箭头键和控制键允许撤销、编辑或重新执行已输入的命令。

例如，假设错误地输入 rho = (1 + sqt(5)/2)，即错误拼写了 sqrt，MATLAB 会显示 Undefined function or variable 'sqt'（未定义的函数或变量 'sqt'）。

这时，不需要重新输入整行命令，只需按↑键，拼写错误的命令会重新显示。用←键，移动光标并添加遗漏的 r。也可重复使用↑键召回之前的命令行，键入几个字符，然后用↑键，找到前面以这些字符开头的命令行。

表 1.4 是常用的命令行编辑指令。

表 1.4 命令行编辑指令

↑	Ctrl+p	召回上一行
↓	Ctrl+n	召回下一行
←	Ctrl+b	后退一个字符
→	Ctrl+f	前进一个字符
Ctrl+→	Ctrl+r	右移一个单词
Ctrl+←	Ctrl+l	左移一个单词
Home	Ctrl+a	移至行始
End	Ctrl+e	移至行末
Esc	Ctrl+u	清除该行
Del	Ctrl+d	删除光标所在字符
Backspace	Ctrl+h	删除光标之前字符
	Ctrl+k	删除至行末

1.4.4 常用基本函数

1. sum 函数

求和函数，表示数组元素的和。

例如，一维数组 $\boldsymbol{A}=[1\quad 2\quad 3\quad 4\quad 5]$，对 $\boldsymbol{A}$ 求和，则 $\boldsymbol{S}=\text{sum}(\boldsymbol{A})$，计算结果为 $\boldsymbol{S}=15$。

又如，二维数组 $\boldsymbol{B}=\begin{bmatrix}1 & 4\\ 6 & 11\end{bmatrix}$，对 $\boldsymbol{B}$ 求和就是对 $\boldsymbol{B}$ 的列元素求和，则 $\boldsymbol{S}=\text{sum}(\boldsymbol{A})$，计算结果为 $\boldsymbol{S}=[7\quad 15]$。

2. transpose 函数

转置函数，对数组进行转置。假定矩阵 $\boldsymbol{B}=[1\quad 2\quad 3]$，对矩阵 $\boldsymbol{B}$ 使用转置函数，则 $\boldsymbol{C}=\text{transpose}(\boldsymbol{B})$，可以得到 $\boldsymbol{B}$ 的转置矩阵 $\boldsymbol{C}=\begin{bmatrix}1\\ 2\\ 3\end{bmatrix}$。

3. diag 函数

该函数创建对角矩阵或得到矩阵的对角元素。假定矩阵 $\boldsymbol{E}=[1\quad 2\quad 3]$，用 $\boldsymbol{D}=\text{diag}(\boldsymbol{E})$，对 $\boldsymbol{E}$ 使用 diag 函数，生成矩阵 $\boldsymbol{D}$ 的主对角线元素为 $\boldsymbol{E}$ 矩阵的元素，其余元素为 0，得到 $\boldsymbol{D}=\begin{bmatrix}1 & 0 & 0\\ 0 & 2 & 0\\ 0 & 0 & 3\end{bmatrix}$。

又如，矩阵 $\boldsymbol{F}=\begin{bmatrix}2 & 3 & 3\\ 3 & 2 & 3\\ 3 & 1 & 3\end{bmatrix}$，对 $\boldsymbol{F}$ 矩阵使用 diag 函数。输入 $\boldsymbol{G}=\text{diag}(\boldsymbol{F})$，提取 $\boldsymbol{F}$ 矩阵中的主对角线元素，从而构成了 $\boldsymbol{G}$ 矩阵的元素，即 $\boldsymbol{G}=[2\quad 2\quad 3]$。

4. fliplr 函数

翻转函数，使数组的列从左到右翻转。假定矩阵 $\boldsymbol{A}=[1\quad 2\quad 3\quad 4\quad 5]$，则使用

$\boldsymbol{H}=\text{fliplr}(\boldsymbol{A})$，得到矩阵 $\boldsymbol{A}$ 的翻转矩阵 $\boldsymbol{H}=[5\quad 4\quad 3\quad 2\quad 1]$。

又如，对矩阵 $\boldsymbol{F}=\begin{bmatrix}2&3&3\\4&2&3\\6&1&3\end{bmatrix}$，使用 $\boldsymbol{H}=\text{fliplr}(\boldsymbol{F})$，得到 $\boldsymbol{F}$ 的翻转矩阵 $\boldsymbol{H}$，$\boldsymbol{H}=\begin{bmatrix}3&3&2\\3&2&4\\3&1&6\end{bmatrix}$。

5. magic 函数

该函数创建一个魔方矩阵，使其具有相同的行数和列数，并在每行每列及其对角线上的和都相等的矩阵，且矩阵中的每个元素互不相同。例如，使用 $\boldsymbol{I}=\text{magic}(3)$ 这个命令函数可以得到 $\boldsymbol{I}=\begin{bmatrix}8&1&6\\3&5&7\\4&9&2\end{bmatrix}$。

1.4.5 常用进阶函数

学习了常用的基本函数之后，再介绍几种常用的进阶函数。

1. conv 函数

卷积和多项式乘法函数，创建包含多项式和向量系数的 $\boldsymbol{u}$ 与 $\boldsymbol{v}$。例如，$\boldsymbol{u}=[1\quad 0\quad 1]$，$\boldsymbol{v}=[2\quad 7]$，使用函数 $\boldsymbol{w}=\text{conv}(\boldsymbol{u},\boldsymbol{v})$，可以得到卷积后的向量 $\boldsymbol{w}=[2\quad 7\quad 2\quad 7]$。

2. find 函数

该函数查找非零元素的索引和值的函数。假定一个矩阵 $\boldsymbol{X}=\begin{bmatrix}1&0&2\\0&1&1\\0&0&4\end{bmatrix}$，使用函数 $\boldsymbol{Y}=\text{find}(\boldsymbol{X})$ 在矩阵 $\boldsymbol{X}$ 中找到非零元素为 $\boldsymbol{Y}=[1\quad 5\quad 7\quad 8\quad 9]$。

又如，有矩阵 $\boldsymbol{X}=\begin{bmatrix}8&1&6\\3&7&5\\4&9&2\end{bmatrix}$，使用函数 $\boldsymbol{Y}=\text{find}(\boldsymbol{X}>5)$，在矩阵 $\boldsymbol{X}$ 中找到元素值大于 5 的元素，将其单索引值输出到 $\boldsymbol{Y}$ 中，$\boldsymbol{Y}=[1\quad 5\quad 6\quad 7]$。

3. step 函数

该函数用于绘制连续系统的阶跃响应图，生成一个随机的稳定传递函数模型 $G(s)=\dfrac{\text{num}(s)}{\text{den}(s)}$，写成具体 s 函数形式为

$$G(s)=\frac{10s+25}{0.16s^3+1.96s^2+10s+25}$$

阶跃响应图实现的程序如下所示。

```
num=[10 25];
den=[0.16 1.96 10 25];
t=0:0.02:5;
step(num,den,t);
```

绘制出连续系统的阶跃响应如图 1.4.1 所示。

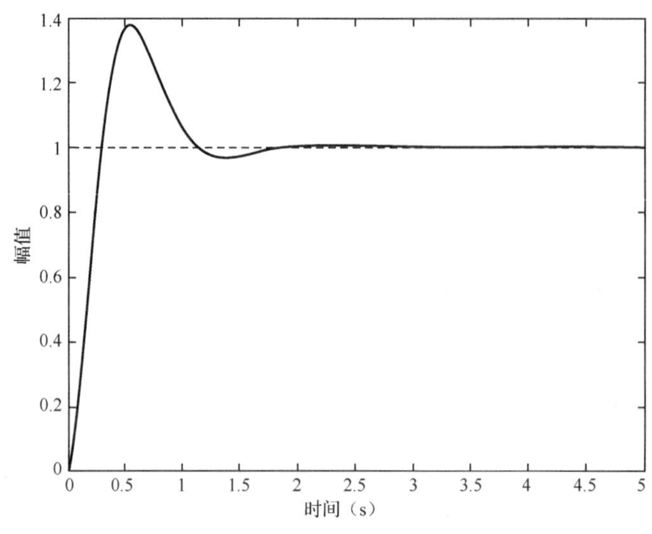

图 1.4.1 连续系统的阶跃响应

4. impulse 函数

该函数用于绘制连续系统的脉冲响应图，生成一个随机的稳定传递函数模型 $G(s)=\dfrac{\mathrm{num}(s)}{\mathrm{den}(s)}$，写成具体 s 函数形式为

$$G(s)=\frac{20}{s^4+8s^3+36s^2+40s+20}$$

实现的程序如下所示。

```
num=20;
den=[1 8 36 40 20];
t=1:0.1:10;
impulse(num,den,t);
```

绘制出连续系统的脉冲响应如图 1.4.2 所示。

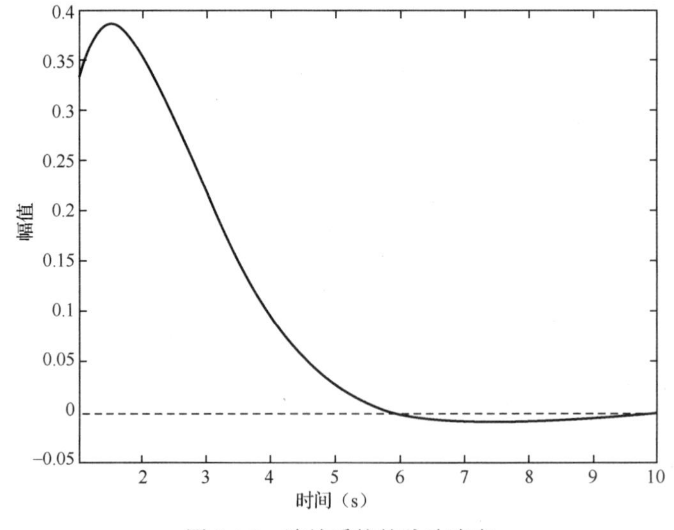

图 1.4.2 连续系统的脉冲响应

掌握以上几种常用的函数，将对后面课程的学习打下良好基础。

1.5　绘图

MATLAB 提供大量的工具用来图形化地显示向量和矩阵，同时也可以注释和打印这些图表。本节将介绍一些重要的图形化函数，及一些典型实例。

1.5.1　创建一个图

根据输入参数的不同，plot 函数有不同的形式。如果 y 是一个向量，则 plot(y)输出一个分段的线性图形，来显示 y 的各个元素与其对应下标的关系。如果指定两个向量作为参数，则 plot(x,y)表示 y 是 x 的函数而做出的图。

例如，绘制正弦函数从 0 到 2π 值的图，如图 1.5.1 所示，其中，图（a）是 plot(y)绘制出来的，图（b）是 plot(t,y)绘制出来的。两幅图的纵坐标是一样的，区别在于横坐标。前者是与 y 值相对应的下标，后者则是与 y 值相对应的 t 值。

```
t=0:pi/100:2*pi;
y=sin(t);
figure(1),plot(y);
figure(2),plot(t,y);
```

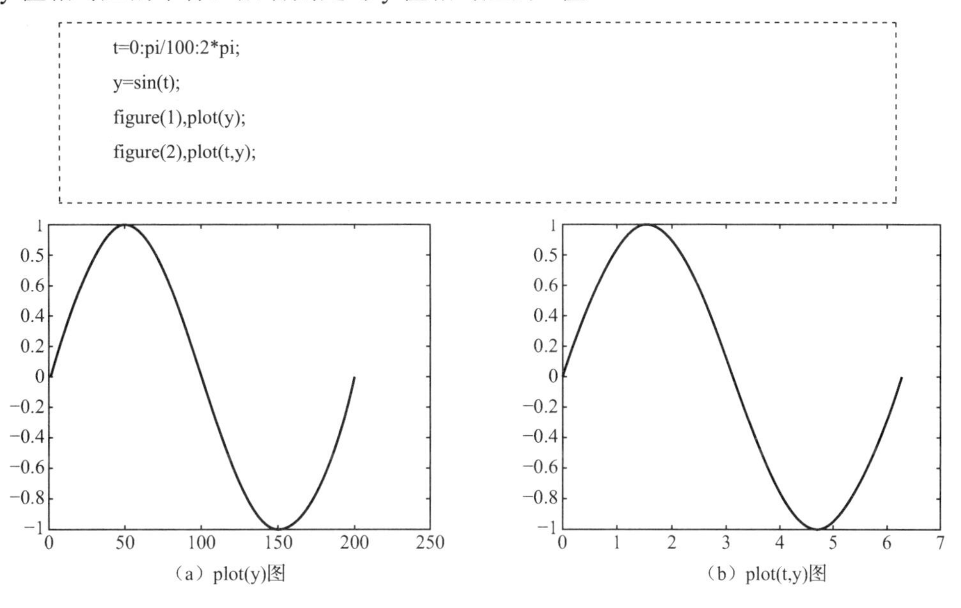

（a）plot(y)图　　（b）plot(t,y)图

图 1.5.1　正弦函数

通过一个 plot 函数，可对多个 x-y 组合绘制多重图。MATLAB 会自动循环使用预定义的颜色列表来区分每组数据（用户也可自定义）。

例如，下面程序是绘制三个 t 的相应正弦函数的图像（见图 1.5.2），每条曲线都用不同的颜色加以区分。

```
t=0:pi/100:2*pi;
y=sin(t);
y1=sin(t-0.25);
y2=sin(t-0.5);
plot(t,y,'r-',t,y1,'g--',t,y2,'b-.')
```

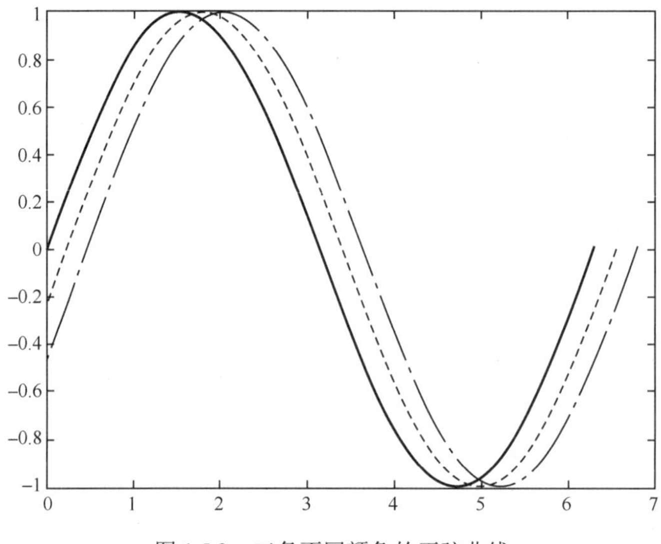

图 1.5.2　三条不同颜色的正弦曲线

曲线的颜色、线型和标记（如加号或圆圈），均可以自定义，即

```
plot(x，y，'colors tyle_marker')
```

color_style_marker 是字符串（由单引号括起来），由颜色、线型和标记构成，详见表 1.5。

表 1.5　曲线设置参数表

曲线颜色		曲线线型		数据点形	
选　项	意　义	选　项	意　义	选　项	意　义
b	蓝色（blue）	—	实线（默认）	·	实点
c	青色（cyan）	:	点线	+	十字形
g	绿色（green）	—·	点断线	O	圆圈
k	黑色（black）	——	虚线	*	星号
m	红紫色（magenta）			×	叉号
r	红色（red）			s	正方形
w	白色（white）			d	菱形
y	黄色（yellow）			h	六角形
				p	五角形
				∨	下三角
				∧	上三角
				<	左三角
				>	右三角

1.5.2　图形窗口

如果屏幕上无图形窗口，plot 函数自动打开一个新的图形窗口；若已有图形窗口存在，

plot 函数默认使用该窗口绘图。若要打开一个新的图形窗口，并使它成为当前图形窗口，则需要键入 figure 指令。

为了使已存在的图形窗口成为当前图形窗口，需要键入 figure(n)，*n* 为图形窗口的标题栏的序号。随后的图形命令结果均显示于该窗口（见图 1.5.3）。

图 1.5.3　figure(2)

1.5.3　添加图形至现有的图

Hold 命令可在已存在的图形窗口中添加图形。如果键入了 hold on 命令，MATLAB 不会移走现有图形，而是把新数据添加到当前图块，必要时会改变坐标轴的标尺。

1.5.4　绘制子图

subplot 函数可在同一窗口显示多幅图形或将其打印到同一张纸上。

输入 subplot(m,n,p)，将图形窗口分成 $m \times n$ 个子区域，并选择第 *p* 个区域作为当前图形。例如，在 figure 窗口的 4 个不同的子区域中分别作图，如图 1.5.4 所示。

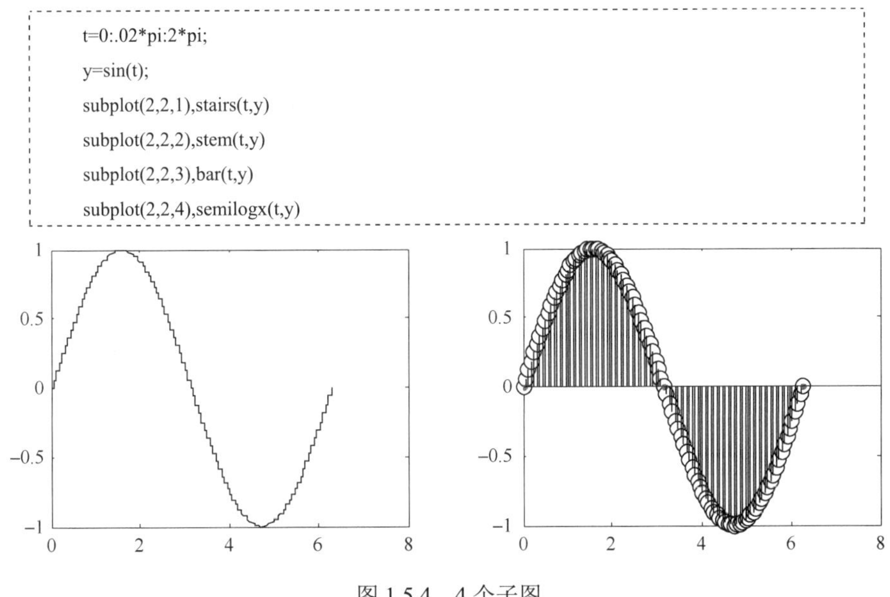

图 1.5.4　4 个子图

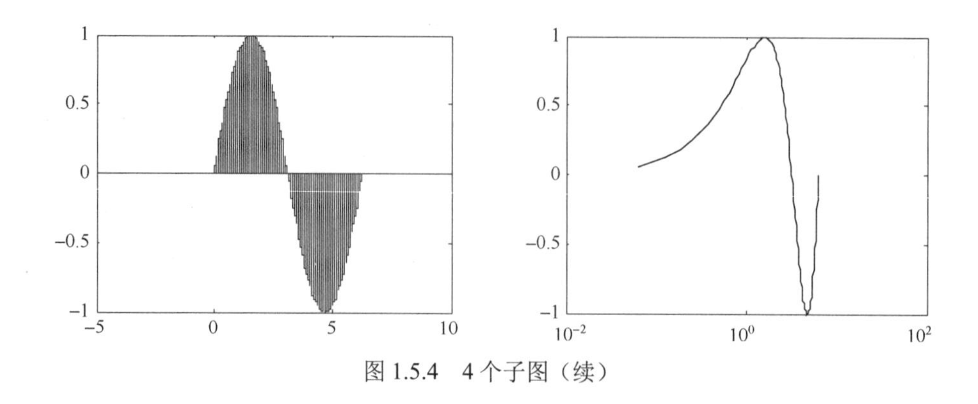

图 1.5.4　4 个子图（续）

1.5.5　隐函数绘图

ezplot(fun)在默认区间$-2\pi<x<2\pi$上绘制表达式 fun(x)的图像。

fun 是一个处理 M 文件函数句柄、一个匿名函数或一个字符串。

ezplot(fun[min,max])：指在区间 min$<x<$max 和 min$<y<$max 上绘制 fun(x,y)=0 的图像，如图 1.5.5 所示。

```
ezplot('x*y*sin(x^2+y^2)+(x+y)^2*exp(-(x+y))')
```

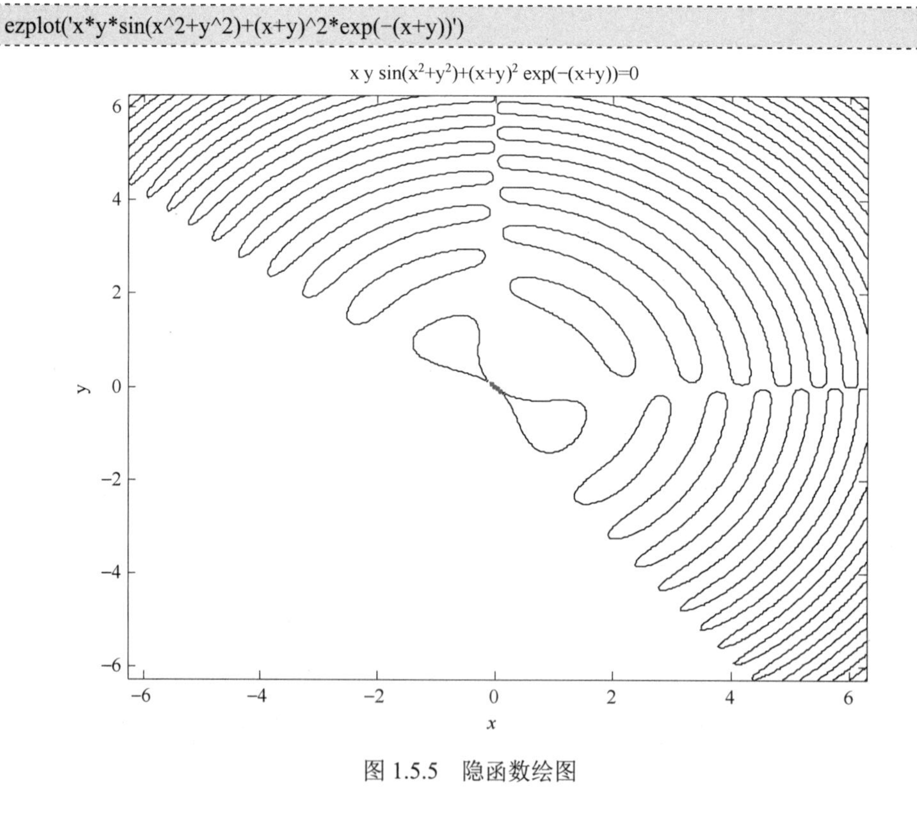

图 1.5.5　隐函数绘图

1.5.6　轴的控制

axis 函数包含多种选项，可自定义图像的缩放比例、方位和纵横比。

通常，MATLAB 可以根据数值的最大值和最小值来选择一个合适的坐标轴范围。axis

函数可以通过自定义坐标轴范围来覆盖默认设置。例如，

axis([xmin xmax ymin ymax])

axis 函数也接受一系列的轴控制关键词。

- axis auto 返回坐标轴的默认刻度，即自动模式
- grid off 关闭网格线
- grid on 显示网格线

1.5.7　轴标记和标题

用 xlabel、ylabel 和 zlabel 命令添加 x、y 和 z 轴标签。可用 title 函数在图形窗口的顶部添加标题，用 text 函数在图形内任意位置添加文本。tex 标记的子集则产生希腊字母、数学符号和替代字体。详见表 1.6。

表 1.6　图形标注所用特殊字符表

命　　令	所代表字符	命　　令	所代表字符
\alpha	α	geq	⩾
\beta	β	neq	≠
\gamma	γ	\equiv	=
\delta	δ	\approx	≈
\omega	ω	\leq	≦
\zeta	ζ	\leftarrow	←
\eta	η	\uparrow	↑
\lanbda	λ	\downarrow	↓
\xi	ξ	\rightarrow	→
\pi	π		

例子详见下面程序及图 1.5.6。

```
t=-pi:pi/100:pi;
y=sin(t);
plot(t,y)
axis([-pi pi -1 1])
xlabel('-\pi\leq\itt\leq\pi');
ylabel('sin(t)');
title('Graph of the sine function')
```

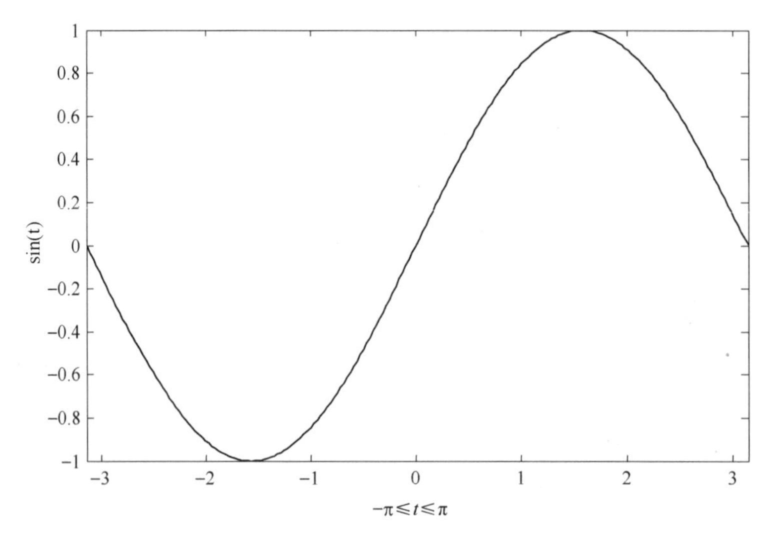

图 1.5.6　标记的正弦曲线图

1.5.8　plot3 和 mesh

plot3 函数用于显示一组数据点的三维图。例子详见下面程序及图 1.5.7。

```
t=0:.02*pi:2*pi;
x=t.^3.*sin(3*t).*exp(-t);
y= t.^3.*cos(3*t).*exp(-t);
z=t.^2;
plot3(x,y,z);grid on
stem3(x,y,z);hold on;
plot3(x,y,z),grid on
```

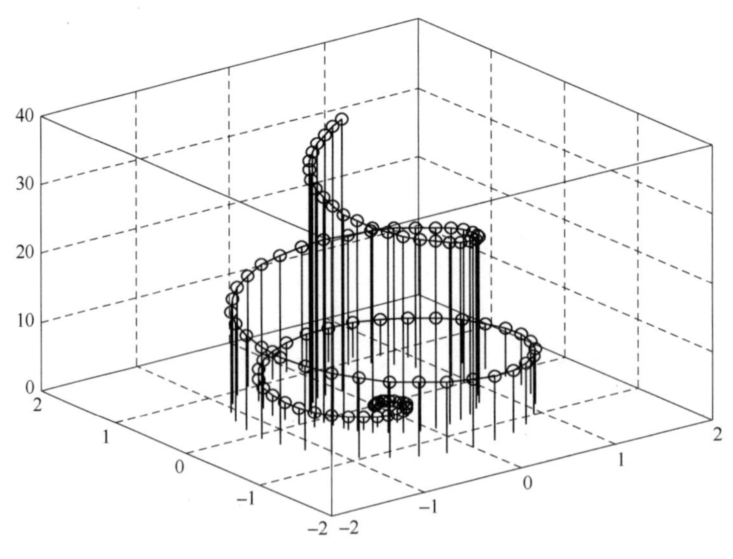

图 1.5.7　plot3 函数绘图

要显示两个变量的函数 $z=f(x,y)$图像，分别在函数的定义域，生成各自含有重复的行与

列的矩阵 ***X*** 和 ***Y***。然后用矩阵 ***X*** 和 ***Y*** 来计算和绘制函数。函数 meshgrid 会将两个向量 ***x***，***y*** 指定的定义域转换成矩阵 ***X*** 和 ***Y***，以用于计算双变量函数。***X*** 的行和 ***Y*** 的列分别是 ***x*** 和 ***y*** 向量的副本。

例子详见下面程序及图 1.5.8。

```
[x,y]=meshgrid(-3:0.1:3,-2:0.1:2);
z=(x.^2-2*x).*exp(-x.^2-y.^2-x.*y);
mesh(x,y,z)
```

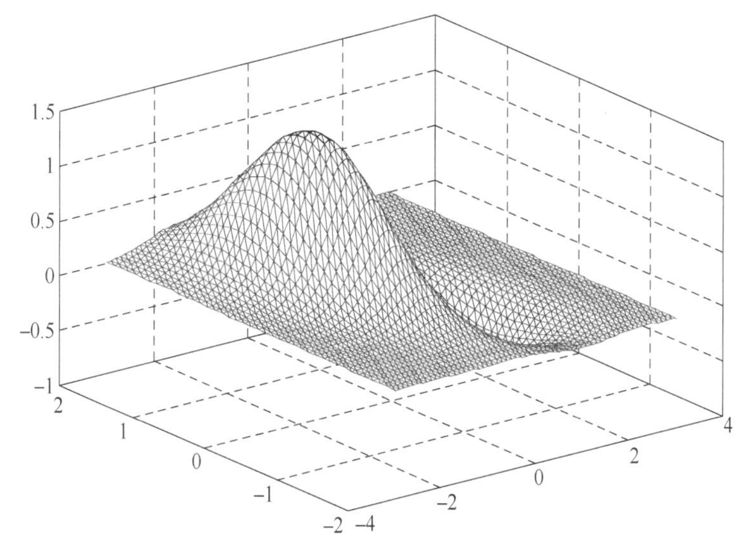

图 1.5.8　mesh 函数绘图

1.5.9　图像

从图形文件读取图像并显示图像。例子详见下面程序及图 1.5.9。

```
I=imread('d:\My Documents\My Pictures\5.jpg');
imshow(I);
whos I
```

图 1.5.9　I 图

whos I

Name	Size	Bytes	Class	Attributes
I	839×1024×3	2577408	uint8	

读取彩色图像的 RGB 三个通道并显示，例子详见下面程序及图 1.5.10。

```
I=imread('C:\Users\andy\Pictures\\I.jpg');
Rim=I(:,:,1);
Gim=I(:,:,2);
Bim=I(:,:,3);
subplot(2,2,1);
imshow(I),title('I');
subplot(2,2,2);
imshow(Rim),title('Rim')
subplot(2,2,3);
imshow(Gim),title('Gim')
subplot(2,2,4);
imshow(Bim), title('Bim')
```

图 1.5.10　RGB 三个通道图像

本章小结

1．本章开始以一个魔方矩阵为例，通过对魔方矩阵的验证，介绍了矩阵的相关知识：矩阵输入、矩阵运算、矩阵操作及 M 文件等。

2．介绍 MATLAB 环境及其常用函数。

3．介绍 MATLAB 二维和三维绘图，以及图像的操作。

Chapter 1
Basics of MATLAB

This chapter introduces the basics of the MATLAB software platform. Starting with an interesting magic matrix example, the basic operations of the matrix and a brief description of the MATLAB environment are introduced. For the drawing of MATLAB graphics, the focus is on the drawing of two-dimensional graphics and three-dimensional graphics as well as the operation of images. This chapter lays a foundation for further study.

1.1 Introduction to MATLAB

The best way for you to get started with MATLAB is to learn how to handle matrices. In MATLAB, a matrix is a rectangular array of numbers. Special meaning attached to 1-by-1 matrices is defined as scalars, and that to matrices with only one row or column, is defined as vectors. Usually, it is preferable to think of everything as a matrix. Unlike other programming languages working with numbers one at a time, MATLAB allows you to work with entire matrices quickly and easily.

A good example of matrix, applied throughout in this course, appears in the Renaissance engraving *Melancholia I* by the German artist and amateur mathematician *Albrecht Dürer*. It is filled with mathematical symbolism, and if you look carefully, you will see a matrix in the upper right corner (as shown in Fig.1.1.1). This matrix is known as a magic square and was believed by many in *Dürer*'s time to have genuinely magical properties. It does turn out to be possessed with fascinating characteristics worth further exploring.

Figure 1.1.1 Magic Squares

1.1.1 Entering Matrices

You can enter matrices into MATLAB in several different ways.

1. Enter an explicit list of elements.
2. Load matrices from external data files.
3. Generate matrices using built-in functions.
4. Create matrices with your own functions in M-files.

You have only to follow a few basic conventions:

1. Separate the elements of a row with blanks or commas.
2. Use a semicolon";" to indicate the end of each row.
3. Surround the entire list of elements with square brackets "[ ]" in a matrix.

According to the above basic rules, enter *Dürer*'s matrix, simply type:

```
A= [ 16 3 2 13 ; 5 10 11 8; 9 6 7 12 ; 4 15 14 1 ]
```

Press *Enter* and MATLAB will display

$$A=\begin{bmatrix}16 & 3 & 2 & 13\\ 5 & 10 & 11 & 8\\ 9 & 6 & 7 & 12\\ 4 & 15 & 14 & 1\end{bmatrix}$$

Why is it magic?

1.1.2 Sum, Transpose, and Diagonal

You're probably already aware that the special properties of a magic square have something to do with the various ways of summing its elements. If you take the sum along any row or column, or along either of the two main diagonals, you will always get the same. Let's verify the findings with MATLAB.

The first statement to try is

```
sum(A)
```

MATLAB show

```
ans =
     34   34   34   34
```

When you don't specify an output variable, MATLAB uses the variable *ans* (short for *answer*)to store the results of a calculation. You have computed a row vector containing the sums of the columns of ***A***. Sure enough, each of the columns has the same magic sum, 34.

How about the row sums?

To transpose the matrix, compute the column sums of the transpose, and then transpose the result. The transpose operation is denoted by an apostrophe or single quote ''. For example, input

```
sum(A')'
```

```
ans =
     34
     34
     34
     34
```

The column vector obtained at this time is the result of summing each row of the matrix ***A***. Obviously, the sum of the lines is still 34.

The sum of the elements on the main diagonal is easily obtained with the help of the *diag* function, which picks off that diagonal.

```
sum(diag(A))
```

```
ans =
```

```
        34
```

The other diagonal is the so-called antidiagonal. The function of the *fliplr* function is to flip the matrix left and right, converting the antidiagonal to the main diagonal.

```
sum(diag(fliplr(A)))
```

```
ans =
    34
```

The above results prove that matrix ***A*** is indeed a magic square and, a few MATLAB matrix operations are given in the process. The following sections continue to use this matrix to illustrate additional MATLAB capabilities.

1.1.3 Subscripts

The element in row *i* and column *j* of ***A*** is denoted by ***A***(*i, j*).

Then, the calculation statement of ***A***(1,4)+***A***(2,4)+***A***(3,4)+***A***(4,4) and the result is

```
A(1, 4)+ A(2, 4)+ A(3, 4)+ A(4, 4)
```

```
ans=
    34
```

Matrix index, in addition to the above double subscript element representation method, is also possible to refer to the elements of a matrix with a single subscript ***A***(*k*) . This is the usual way of referencing row and column vectors. But it can also apply to a fully two-dimensional matrix, in which case the array is regarded as one long column vector formed from the columns of the original matrix. So, for our magic square ***A***, ***A***(8) is another way of referring to the value 15 stored in ***A***(4, 2) .

1.1.4 The Colon Operator

The colon “:” is one of MATLAB’s most important operators. It occurs in several different forms. The expression 1:10 is a row vector containing the integers from 1 to 10

1 2 3 4 5 6 7 8 9 10

To obtain nonunit spacing, specify an increment. For example, 100: –7:50 is

100 93 86 79 72 65 58 51

and 0:pi/4:pi is

0 0.7854 1.5708 2.3562 3.1416

Subscript expressions involving colons refer to portions of a matrix. A(1:*k*,*j*) is the first *k* elements of the *j*-th column of ***A***. So *sum*(A(1:4, 4)), computes the sum of the fourth column. But there is a better way. The colon by itself refers to all the elements in a row or column of a matrix and the keyword *end* refers to the last row or column.

Therefore, use the command

```
sum(A(:,end))
```

```
ans =
    34
```

Compute the sum of the elements in the last column of ***A***.

Why is the magic sum for a 4-by-4 square equal to 34? If the integers from 1 to 16 are sorted into four groups with equal sums, that sum must be sum(1:16)/ 4, the result is 34.

```
sum(1:16)/4
```

```
ans =
    34
```

1.1.5 The Magic Function

MATLAB actually has a built-in function that creates magic squares of almost any size. Not surprisingly, this function is named *magic*.

```
B=magic(4)
```

$$\boldsymbol{B}=\begin{bmatrix}16 & 2 & 3 & 13\\ 5 & 11 & 10 & 8\\ 9 & 7 & 6 & 12\\ 4 & 14 & 15 & 1\end{bmatrix}$$

This matrix is almost the same as the one in the *Dürer* engraving and has all the same "magic" properties; the only difference is that the two middle columns are exchanged. To make this ***B*** into *Dürer*'s ***A***, swap the two middle columns.

```
A=B (:,[1 3 2 4])
```

This says "for each of the rows of matrix ***B***, reorder the elements in the order 1, 3, 2, 4."

1.2 Expressions

Like most other programming languages, MATLAB provides mathematical expressions, but unlike most programming languages, these expressions involve entire matrices. The building blocks of expressions are

(1) Variables

(2) Numbers

(3) Operators

(4) Functions

1.2.1 Variables

Variable names consist of a letter, followed by any number of letters, digits, or underscores. MATLAB is case sensitive; it distinguishes between uppercase and lowercase letters. "A" and "a" are not the same variable.

1.2.2 Numbers

MATLAB uses conventional decimal notation, with an optional decimal point and leading plus or minus sign, for numbers. Scientific notation uses the letter *e* to specify a power-of-ten scale factor. Imaginary numbers uses either *i* or *j* as a suffix.

Some examples of legal numbers are:

```
3            -99           0.000
9.6397238    1.60210e-20   6.02252e23
1i           -3.14159j     3e5i
```

1.2.3 Operators

Expressions using familiar arithmetic operators and precedence rules are shown in Table 1.1.

Table 1.1 Expressions using Familiar Arithmetic Operators

Familiar Arithmetic Operators	Familiar Arithmetic
+	addition
-	subtraction
*	multiplication
/	division
\	left division
^	power
'	complex conjugate transpose
()	specify evaluation order

1.2.4 Functions

MATLAB provides a large number of standard elementary mathematical functions, including *abs*(absolute value), *sqrt*(square root) and *exp*(exponential function). MATLAB also provides many more advanced mathematical functions, including *gamma* functions. Most of these functions accept complex arguments. Some of the functions, like *exp* and *cos*, are built-in. They are part of the MATLAB core so they are very efficient, but the computational details are not readily accessible.

Table 1.2 for several special functions provide values of useful constants.

Table 1.2 Constant Value of Special Function

Constants	Meaning
pi	3.14159265...
i	imaginary unit,-1
j	same as i
eps	floating-point relative precision, 2.2204e-16

续表

Constants	Meaning
real min	smallest floating-point number, 2.2251e–308
real max	largest floating-point number, 1.7977e+308
inf	Infinity
nan	Not-a-number

You have already seen several examples of MATLAB expressions. Here are a few more examples, and the resulting values.

```
rho = (1+sqrt(5))/2
rho =1. 6180
a = abs(3+4i)
a = 5
huge = exp(log( realmax) )
huge = 1.7977e+308
toobig = pi *huge
toobig = Inf
```

1.3 Operation with Matrices

This section introduces other ways to create matrices.

1.3.1 Generating Matrices

MATLAB provides four functions that generate basic matrices (See Table 1.3) .

Table 1.3 Four Basic Functions

Functions	Description
zeros	all zeros
ones	all ones
rang	uniformly distributed random matrices
rangn	normally distributed random matrices

1.3.2 M-Files

You can create your own matrices using M-files, which are text files containing MATLAB code. Just create a file containing the same statements you would type at the MATLAB command line. Save the file under a name that ends in *.m*.

For example, create a file containing these

```
A=[16, 2, 3, 13; 5, 11, 10, 8; 9, 7, 6, 12; 4, 14, 15, 1]
```

Store the file under the name *magik.m*. Then the statement *magik* reads the file containing our example magic.

1.3.3 Concatenation

Concatenation is the process of joining small matrices to make bigger ones. In fact, you made your first matrix by concatenating its individual elements. The pair of square brackets, [], is the concatenation operator. For an example, start with the 4-by-4 magic square ***A***, and form ***B***:

```
B = [A  A+ 32 ;  A+ 48  A+16]
```

The result is an 8-by-8 matrix, obtained by joining the four sub-matrices.

```
B=
     16  3   2   13  48  35  34  45
     5   10  11  8   37  42  43  40
     9   6   7   12  41  38  39  44
     4   15  14  1   36  47  46  33
     64  51  50  61  32  19  18  29
     53  58  59  56  21  26  27  24
     57  54  55  60  25  22  23  28
     52  63  62  49  20  31  30  17
```

This matrix is half way to being another magic square. Its elements are a rearrangement of the integers from1 to 64. Its column sums are the value for an 8-by-8 magic square.

```
sum(B)
```

```
ans =
   260  260  260  260  260  260  260  260
```

But its row sums, *sum*(***B***')' , are not all the same. Further manipulation is necessary to make this a valid 8-by-8 magic square.

1.3.4 Deleting Rows and Columns

You can delete rows and columns from a matrix using just a pair of square brackets. For example, ***X*** = ***A***; then, delete the second column of ***X***, expressed by X(:,2) = [ ].

```
X(:,2)=[]
```

We have:

```
X =
     16  2   13
     5   11  8
     9   7   12
     4   14  1
```

If you delete a single element from a matrix, the result isn't a matrix anymore. So, expressions like X(1,2) = [] result in an error. Therefore, using a single subscript deletes a single element, or sequence of elements, and reshapes the remaining elements into a row vector. So,

```
X(2:2:10) =[ ]
```

We have

```
X =
    16   9   2   7   13   12   1
```

1.4 The Command Window and Commonly Used Functions

So far, you have been using the MATLAB command line, typing commands and expressions, and seeing the results printed in the command window.

This section describes how to change the appearance of the command window, and introduces you to several commonly used functions. If your system allows you to select the command window font or typeface, we recommend you use a fixed width font, such as *Fixedsys* or *Courier*, to provide proper spacing.

1.4.1 The Format Command and Long Command Lines

The *format* command controls the numeric format of the values displayed by MATLAB. The command affects only how numbers are displayed, rather than how MATLAB computes or saves them.

Here are the corresponding results produced from a vector *X* in different formats and magnitudes.

```
x=[4/3   1.2345e-6]
format short
1.3333   0.0000
format short e
1.3333e+000     1.2345e-0.006
format short g
1.3333   1.2345e-006
format long
1.33333333333333 0.000001234500000
format long e
1.333333333333333e+000 1.234500000000000e-006
Format long g
1.33333333333333   1.2345e-006
format rat
4/3   1/810045
format hex
3ff5555555555555   3eb4b6231abfd271
```

If the largest element of a matrix is larger than 10^3 or smaller than 10^{-3}, MATLAB applies a common scale factor for the *short* and *long* formats.

In addition to the format commands shown above, format *compact* suppresses many of the blank lines that appear in the output. This lets you view more information on a screen or window.

If a statement does not fit on one line, use three periods, ···, followed by *Return* or *Enter* to indicate that the statement continues on the next line. For example:

i = 1+2+3+4+5+6+7+8+…+10+11;

1.4.2 Suppressing Output

If you simply type a statement and press *Return* or *Enter*, MATLAB automatically displays the results on screen. However, if you end the line with a semicolon, MATLAB performs the computation but does not display any output. This is particularly useful when you generate large matrices. For example

A = magic(100);

(The *magic* function is a function command that generates a cube matrix, which will be discussed later.)

1.4.3 Command Line Editing

Various arrow and control keys on your keyboard allow you to recall, edit, and reuse commands you have typed earlier.

For example, suppose you mistakenly enter $rho = (1 + sqt(5)/2)$, that is, you have misspelled *sqrt*. MATLAB responds with *Undefined function or variable* '*sqt*'.

Instead of retyping the entire line, simply press the key ↑. The misspelled command is redisplayed. Use the key← to move the cursor over and insert the missing r. Repeated use of the key ↑ recalls earlier lines. Type a few characters and then the key ↑ finds a previous line that begins with those characters.

Table 1.4 is some commonly used command line editing commands.

Table 1.4 Command Line Editing Commands

↑	Ctrl+p	Recall the previous line
↓	Ctrl+n	Recall the next line
←	Ctrl+b	Back one character
→	Ctrl+f	Advance by one character
Ctrl+→	Ctrl+r	Move right one word
Ctrl+←	Ctrl+l	Move left one word
Home	Ctrl+a	Move to the beginning of the line
End	Ctrl+e	Move to the end of the line
Esc	Ctrl+u	Clear the line
Del	Ctrl+d	Delete the cursor character
Backspace	Ctrl+h	Delete the cursor before the character
	Ctrl+k	Delete to the end of the line

1.4.4 Commonly Used Basic Functions

1. *sum* Function

It indicates the sum of array elements.

For example, find the sum of a one-dimensional array $\boldsymbol{A}=[1 \quad 2 \quad 3 \quad 4 \quad 5]$.

Key: Using $\boldsymbol{S}=sum(\boldsymbol{A})$, we have $\boldsymbol{S}=15$.

Example 2. Find the sum of a two-dimensional array $\boldsymbol{B}=\begin{bmatrix} 1 & 4 \\ 6 & 11 \end{bmatrix}$.

Key: To sum ***B*** is to sum all the elements in the column of ***B***. Therefore, using $\boldsymbol{S}=sum(\boldsymbol{A})$, we get $\boldsymbol{S}=[7 \quad 15]$.

2. *transpose* Function

It functions to transpose the array.

Assume Matrix $\boldsymbol{B}=[1 \quad 2 \quad 3]$. Try to apply *transpose* to Matrix ***B***.

Key: Using $\boldsymbol{C}=transpose(\boldsymbol{B})$ (representing transpose Matrix ***B***).

Then we get:

$$\boldsymbol{C}=\begin{bmatrix} 1 \\ 2 \\ 3 \end{bmatrix}.$$

3. *diag* Function

It indicates creating a diagonal matrix or getting the diagonal elements of a matrix. Assume matrix $\boldsymbol{E}=[1 \quad 2 \quad 3]$. Then, use $\boldsymbol{D}=diag(\boldsymbol{E})$, representing applying *diag* function to Matrix ***E*** and creating the main diagonal elements of Matrix ***D***, which are equivalent to the elements of Matrix ***E***, and the rest of Matrix ***D*** is zero. The result is as follows.

$$\boldsymbol{D}=\begin{bmatrix} 1 & 0 & 0 \\ 0 & 2 & 0 \\ 0 & 0 & 3 \end{bmatrix}.$$

Example 2. Apply *diag* function to matrix $\boldsymbol{F}=\begin{bmatrix} 2 & 3 & 3 \\ 3 & 2 & 3 \\ 3 & 1 & 3 \end{bmatrix}$ and ***F***. Enter $\boldsymbol{G}=diag(\boldsymbol{F})$. We get the diagonal elements of Matrix ***F***, which form the elements of Matrix ***G***, i.e. $\boldsymbol{G}=[2 \quad 2 \quad 3]$.

4. *fliplr* Function

It functions to flip the column of an array from left to right. Suppose the matrix $\boldsymbol{A}=[1 \quad 2 \quad 3 \quad 4 \quad 5]$. We use $\boldsymbol{H}=fliplr(\boldsymbol{A})$ to flip the elements of ***A*** and get its flip matrix $\boldsymbol{H}=[5 \quad 4 \quad 3 \quad 2 \quad 1]$.

For another example, for a matrix $\boldsymbol{F}=\begin{bmatrix} 2 & 3 & 3 \\ 4 & 2 & 3 \\ 6 & 1 & 3 \end{bmatrix}$, with $\boldsymbol{H}=fliplr(\boldsymbol{F})$, we can get the flip matrix ***H*** from ***F***. The result is as follows,

$$\boldsymbol{H}=\begin{bmatrix} 3 & 3 & 2 \\ 3 & 2 & 4 \\ 3 & 1 & 6 \end{bmatrix}$$

5. *magic* Function

It indicates creating a magic square with the same number of rows and columns, and the same sum of all the elements in each row, column and diagonal, but different in each element. For example, using the command function $I = magic(3)$, we get $I = \begin{bmatrix} 8 & 1 & 6 \\ 3 & 5 & 7 \\ 4 & 9 & 2 \end{bmatrix}$.

1.4.5 Commonly Used Advanced Functions

After a brief introduction to basic functions, let us move on to some commonly used advanced functions.

1. *conv* Function

This is the convolution and polynomial multiplication function, creating vectors $\boldsymbol{u}$ and $\boldsymbol{v}$ containing the coefficients of the polynomials. For example, suppose $\boldsymbol{u} = [1 \quad 0 \quad 1], \boldsymbol{v} = [2 \quad 7]$, using the function $\boldsymbol{w} = conv(u, v)$, we can get the convolved vector $\boldsymbol{w} = [2 \quad 7 \quad 2 \quad 7]$.

2. *find* Function

This is the function to find indices and values of nonzero elements.

Given a matrix $\boldsymbol{X} = \begin{bmatrix} 1 & 0 & 2 \\ 0 & 1 & 1 \\ 0 & 0 & 4 \end{bmatrix}$, we can use $\boldsymbol{Y} = find(\boldsymbol{X})$ to find the nonzero elements $\boldsymbol{Y} = [1 \quad 5 \quad 7 \quad 8 \quad 9]$ in Matrix $\boldsymbol{X}$.

Take another example.

For a matrix $\boldsymbol{X} = \begin{bmatrix} 8 & 1 & 6 \\ 3 & 5 & 7 \\ 4 & 9 & 2 \end{bmatrix}$, using $\boldsymbol{Y} = find(\boldsymbol{X} > 5)$ to find the elements in Matrix $\boldsymbol{X}$ whose values are larger than 5 and output their single index value to $\boldsymbol{Y}$, we get $\boldsymbol{Y} = [1 \quad 5 \quad 6 \quad 7]$.

3. *step* Function

This is to draw the step response graph of the continuous system and generate a stochastic stable transfer function model $G(s) = \dfrac{\text{num}(s)}{\text{den}(s)}$.

The format of the function is

$$G(s) = \frac{10s + 25}{0.16s^3 + 1.96s^2 + 10s + 25}$$

The script is generated as follows,

```
num=[10 25];
den=[0.16 1.96 10 25];
t=0:0.02:5;
step(num,den,t);
```

Then the step response of the continuous system can be drawn in Figure 1.4.1.

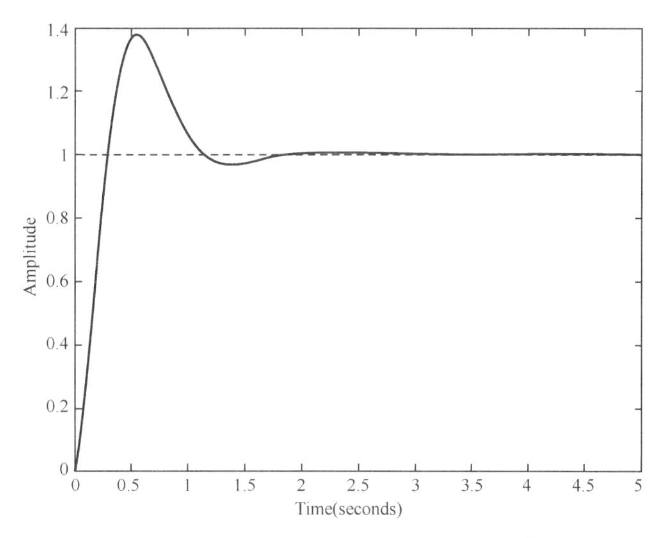

Figure 1.4.1 Step Response of Continuous System

4. *impulse* function

This is to draw the impulse response graph of the continuous system and generate a stochastic stable transfer function model $G(s)=\dfrac{\text{num}(s)}{\text{den}(s)}$,

The format of the function is

$$G(s)=\frac{20}{s^4+8s^3+36s^2+40s+20}$$

The script is generated as follows,

```
num=20;
den=[1 8 36 40 20];
t=1:0.1:10;
impulse(num,den,t);
```

Then the impulse response of the continuous system can be drawn in Figure 1.4.2.

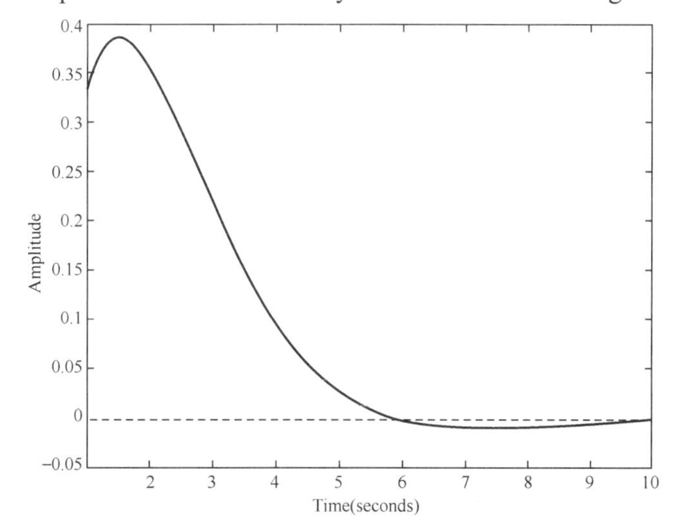

Figure 1.4.2 Impulse Response of Continuous System

Mastering the above commonly used functions will lay a good foundation for the study of following courses.

1.5 Graphics

MATLAB provides a number of tools for graphically displaying vectors and matrices, as well as annotating and printing these graphs. This section introduces some important graphical functions and gives typical examples.

1.5.1 Creating a Plot

The plot function takes different forms, depending on the input arguments. If *y* is a vector, plot (y) outputs a piecewise linear graph to show the relation ship between various elements of y and its corresponding subscript. If we specify two vectors as parameters, plot (x,y) represents a plot made by *y* as afunction of x.

For example, to plot the value of the sine function from 0 to 2π. The graph is shown in Figure 1.5.1 as follows, where (a) is plotted by plot(y), (b) is plotted by plot(t, y). The ordinates of the two figures are the same and the difference lies in the abscissa. The former is the subscript corresponding to y, and the latter is the t value corresponding to y.

```
t=0:pi/100:2*pi;
y=sin(t);
figure(1), plot(y);
figure(2), plot(t,y);
```

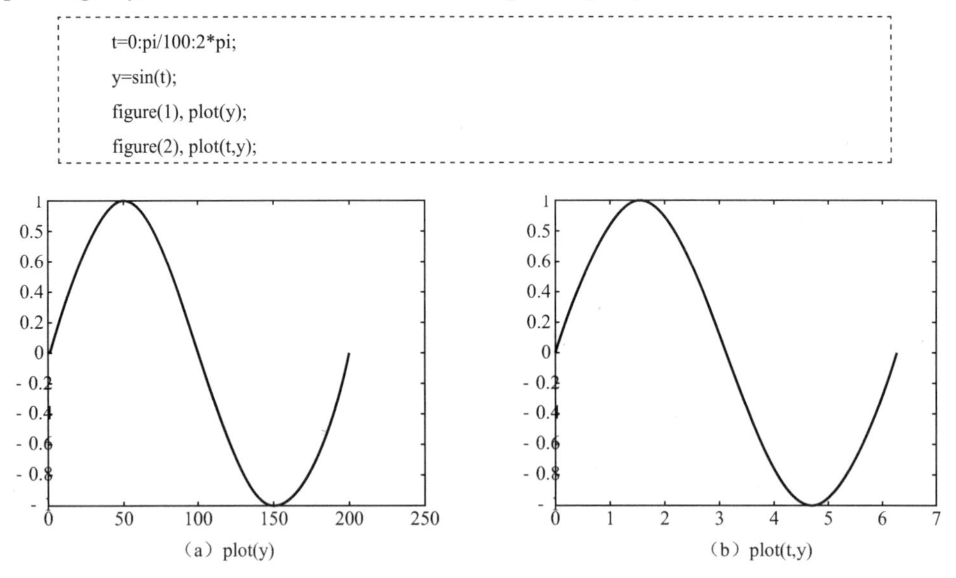

（a） plot(y)　　（b） plot(t,y)

Figure 1.5.1　Graph of *sine* Function

We can draw multiple plots against multiple x-y combinations using a single plot function. MATLAB automatically loop through a predefined list of colors to distinguish each set of data (users can also customize it).

For example, the following script is to plot the three sine functions corresponding to three t (see Figure 1.5.2). Each curve is distinguished by a different color.

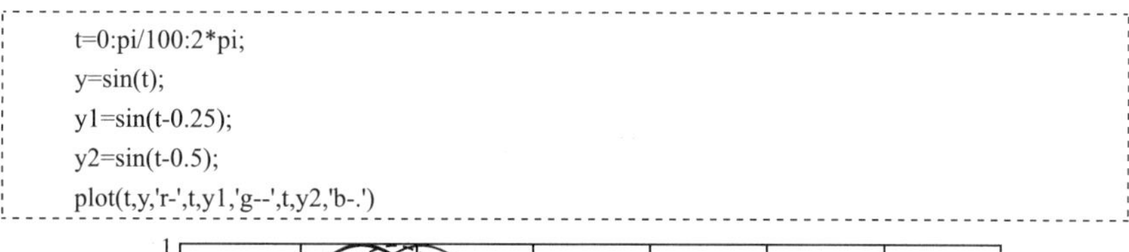

```
t=0:pi/100:2*pi;
y=sin(t);
y1=sin(t-0.25);
y2=sin(t-0.5);
plot(t,y,'r-',t,y1,'g--',t,y2,'b-.')
```

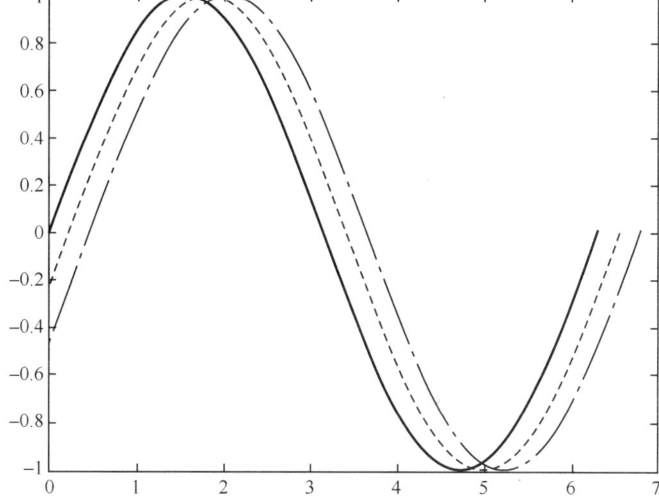

Figure 1.5.2 Sine Curve of 3 Colors

It is possible to customerize color, linestyle, and markers (such as plus signs or circles), with

```
plot ( x, y, ' Color_style_marker ' )
```

Color_style_marker is a character string (delineated by single quotation marks) constructed from a color, a linestyle, and a marker, as shown in Table 1.5.

Table 1.5 Curve Setting Parameter Table

Curve Color		Curve Type		Data Point Pattern	
option	sense	option	sense	option	sense
b	blue	—	Full line(default)	·	Real point
c	cyan	:	Point line	+	Cross
g	green	—·	Point break line	O	Circle
k	black	——	Imaginary line	*	Asterisk
m	magenta			X	X mark
r	red			s	Square
w	white			d	Rhombus
y	yellow			h	Hexagon
				p	Pentagon
				V	Suboccipital triangle
				Λ	Upper triangle
				<	left triangle
				>	Right triangle

1.5.2 Figure Windows

The *plot* function automatically opens a new figure window if there are no figure windows already on the screen. If a figure window exists, *plot* function uses that window by default. To open a new figure window and make it the current figure, type *figure*.

To make an existing figure window the current figure, type *figure* (*n*),where *n* is the number of the titlebar in the figure window. The results of subsequent graphics commands are displayed in this window (as shown in Figure 1.5.3).

Figure 1.5.3 figure(2)

1.5.3 Adding Plots to an Existing Graph

The *hold* command allows you to add plots to an existing graph. When you type *hold on*, MATLAB does not remove the existing graph; instead, it adds the new data to the current graph, changing the scale of the axes if necessary.

1.5.4 Subplots

The *subplot* function allows you to display multiple plots in the same window or print them on the same piece of paper.

Typing *subplot* (*m*, *n*, *p*) breaks the figure window into an *m*-by-*n* subplots and selects the *p*-th subplot for the current plot. For example, to plot data in four different subplots of the figure window (as shown in Figure 1.5.4).

```
t=0:.02*pi:2*pi;
y=sin(t);
subplot(2,2,1), stairs(t,y)
subplot(2,2,2), stem(t,y)
subplot(2,2,3), bar(t,y)
subplot(2,2,4), semilogx(t,y)
```

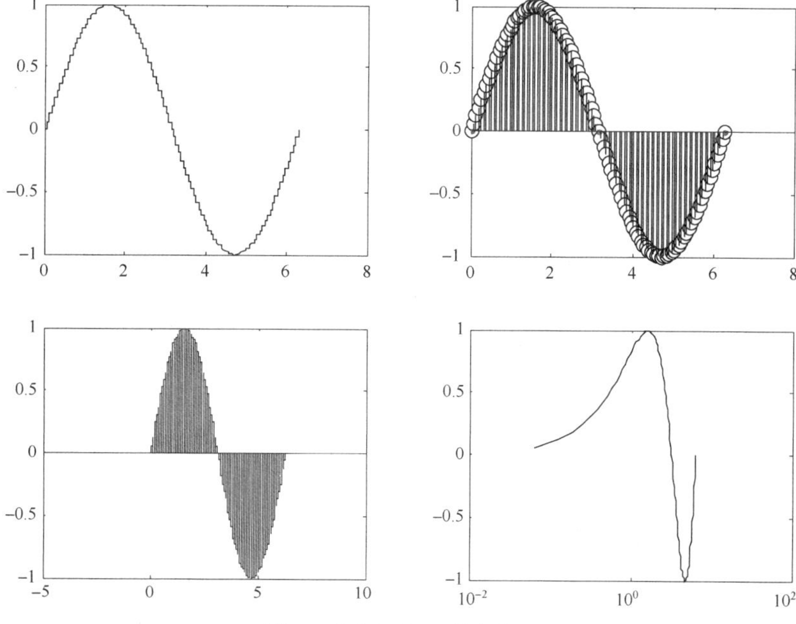

Figure 1.5.4 Four Subplots

1.5.5 ezplot

ezplot (*fun*) refers to plotting the expression *fun* (*x*) over the default domain $-2\pi < x < 2\pi$.

fun can be a function handle for an M-file function or an anonymous function or a string.

ezplot (*fun,[min,max]*) : to plot *fun* (*x,y*) = *0* over *min* < *x* < *max* and *min* < *y* < *max*, as shown in Figure 1.5.5.

```
ezplot('x*y*sin(x^2+y^2)+(x+y)^2*exp(-(x+y))')
```

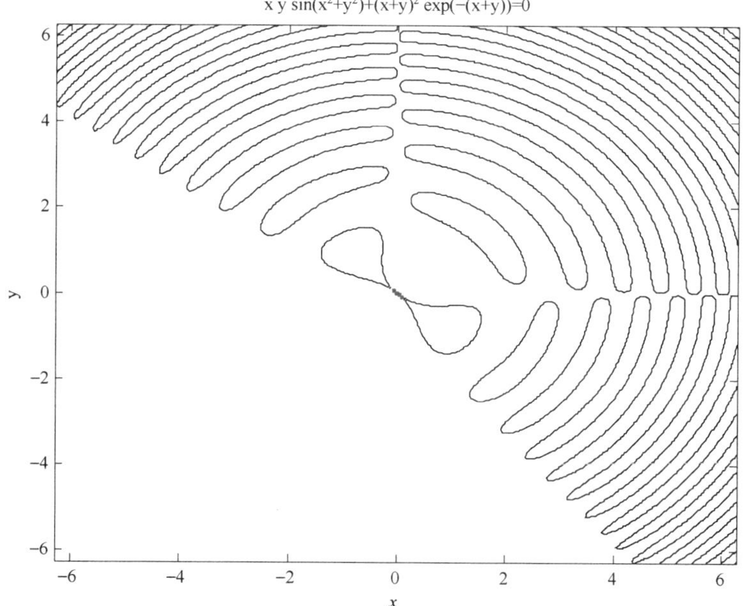

Figure 1.5.5 Graph of *ezplot*

1.5.6 Axes Control

The *axis* function has a number of options for customizing the scaling, orientation, and aspect ratio of plots.

Ordinarily, MATLAB finds the maxima and minima of the data and chooses an appropriate plot box and axes labeling. The *axis* function overrides the default by setting custom axis limits,

axis([*xmin xmax ymin ymax*])

axis also accepts a number of keywords for axes control.

axis auto: returns the axis scaling to its default, automatic mode.

grid off: turns the grid lines off.

grid on: turns them back on again.

1.5.7 Axis Labels and Titles

The *x* label, *y* label, and *z* label functions add *x-, y-*, and *z*-axis labels. The *title* function adds a title at the top of the figure and the *text* function inserts text anywhere in the figure. A subset of *Tex* notation produces Greek letters, mathematical symbols, and alternate fonts, as shown in Table 1.6.

Table 1.6 Special Character Table for Graphic Annotation

Option	Character	Option	Character
\alpha	α	geq	≥
\beta	β	neq	≠
\gamma	γ	\equiv	═
\delta	δ	\approx	≈
\omega	ω	\leq	≦
\zeta	ζ	\leftarrow	←
\eta	η	\uparrow	↑
\lanbda	λ	\downarrow	↓
\xi	ξ	\rightarrow	→
\pi	π		

The example is shown in the following script and Figure 1.5.6.

```
t=-pi:pi/100:pi;
y=sin(t);
plot(t,y)
axis([-pi pi -1 1])
xlabel('-\pi\leq\itt\leq\pi');
ylabel('sin(t)');
title('Graph of the sine function')
```

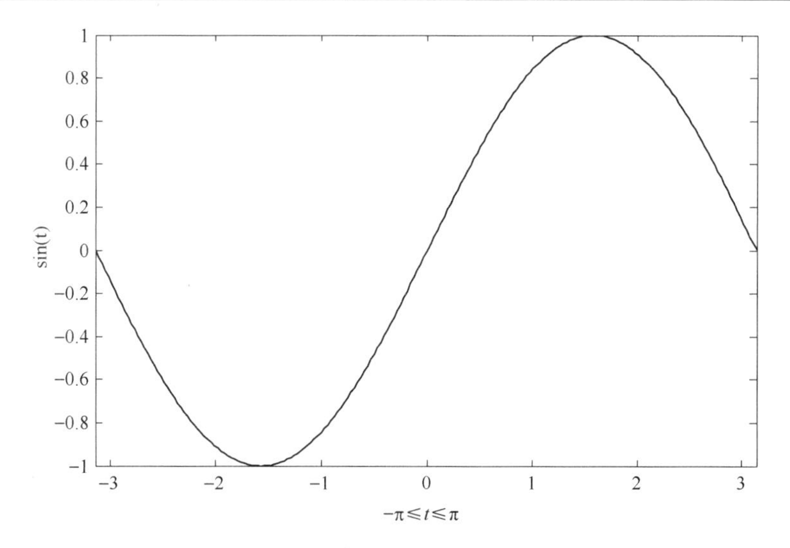

Figure 1.5.6 Graph of *sine* Function of Labels

1.5.8 *plot*3 and *mesh*

The *plot*3 function displays a three-dimensional plot of a set of data points. The example is shown in the following script and Figure 1.5.7.

```
t=0:.02*pi:2*pi;
x=t.^3.*sin(3*t).*exp(-t);
y= t.^3.*cos(3*t).*exp(-t);
z=t.^2;
plot3(x,y,z);grid on
stem3(x,y,z);hold on;
plot3(x,y,z),grid on
```

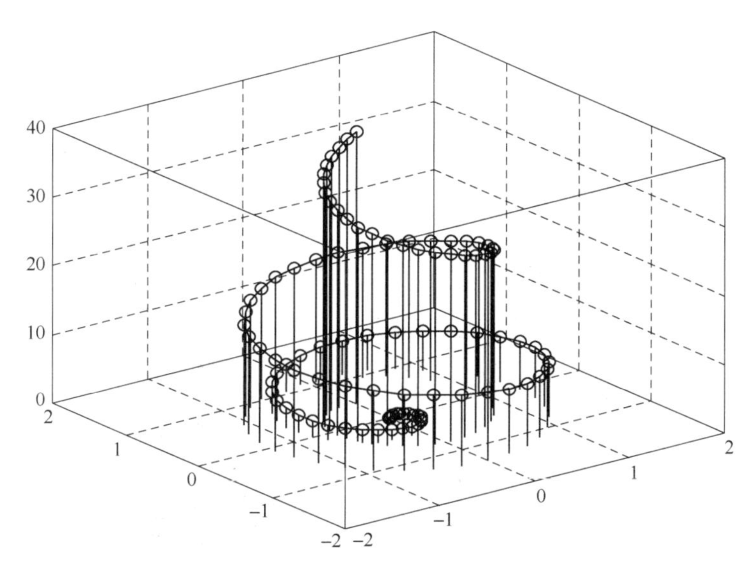

Figure 1.5.7 Graph of *Plot*3 Function

To display a function of two variables, $z=f(x,y)$, generate $\boldsymbol{X}$ and $\boldsymbol{Y}$ matrices consisting of repeated rows and columns, respectively, over the domain of the function. Then use these matrices to evaluate and graph the function. The *meshgrid* function transforms the domain specified by two vectors $\boldsymbol{x}$ and $\boldsymbol{y}$ into matrices $\boldsymbol{X}$ and $\boldsymbol{Y}$ for use in evaluating functions of two variables. The rows of $\boldsymbol{X}$ are copies of the vector $\boldsymbol{x}$ and the columns of $\boldsymbol{Y}$ are copies of the vector $\boldsymbol{y}$. The example is shown in the following script and Figure 1.5.8.

```
[x,y]=meshgrid(-3:0.1:3,-2:0.1:2);
z=(x.^2-2*x).*exp(-x.^2-y.^2-x.*y);
mesh(x,y,z)
```

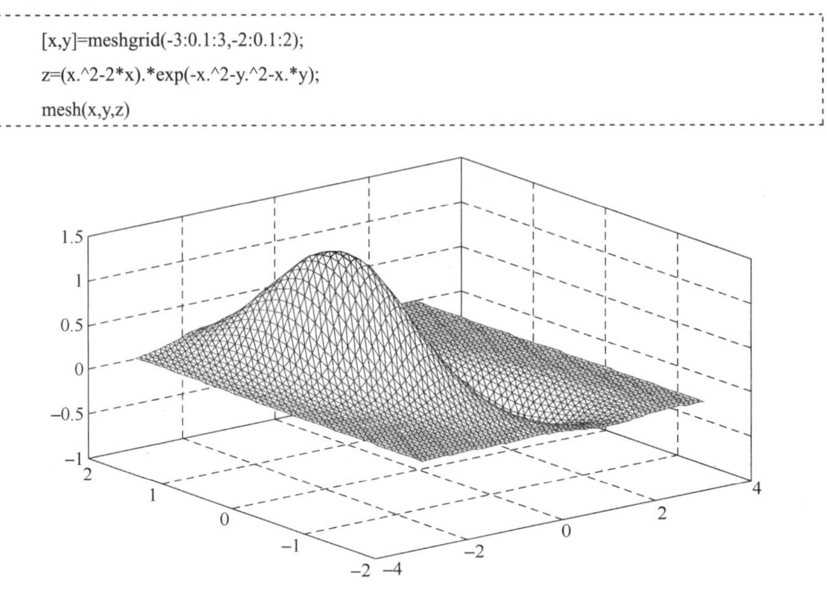

Figure 1.5.8 Graph of *mesh* Function

1.5.9 Images

Read image from graphics file and display the image. The example is shown in the following script and Figure 1.5.9.

```
I=imread('d:\My Documents\My Pictures\5.jpg');
imshow(I);
whos I
```

Figure 1.5.9 Image I

```
whos I
  Name        Size              Bytes    Class    Attributes
  I           839x1024x3        2577408  uint8
```

Read three RGB channels of color image. The example is shown in the following script and Figure 1.5.10.

```
I=imread('C:\Users\andy\Pictures \1.jpg');
Rim=I(:,:,1);
Gim=I(:,:,2);
Bim=I(:,:,3);
subplot(2,2,1);
imshow(I),title('I');
subplot(2,2,2);
imshow(Rim),title('Rim')
subplot(2,2,3);
imshow(Gim),title('Gim')
subplot(2,2,4);
imshow(Bim), title('Bim')
```

Figure 1.5.10 RGB Channel of Color Image

Summary

1. This chapter gives, an interesting example of the magic square to illustrate the basic knowledge of matrix including its input, computation, operation and M-file, etc.

2. MATLAB environment and its commonly used functions are also introduced.

3. MATLAB 2D/3D plotting and the operation of the image are introduced as well.

生词注解

matrix n. 矩阵

explicit	adj. 直接的；清楚的；明确的
magic square	魔方阵
properties	n. 性能
engraving	n. 雕刻
apostrophe	n. 撇号
single quote	n. 单引号
decimal	n. 小数
suffix	n. 后缀
rearrangement	n. 重新整理
concatenating	adj. 连接的
manipulation	n. 操作
reshapes	vt. 改造
font	n. 字体
coefficients	n. 系数
extensive	adj. 广泛的
rhombus	n. 菱形
hexagon	n. 六角形
pentagon	n. 五角形
meshgrid	n. 用于三维曲面的分格线坐标；产生“格点”矩阵
function	n. 函数

第 2 章 MATLAB 编程

本章的主要内容是 MATLAB 编程的基础知识，介绍 MATLAB 环境下的程序设计，包括循环的控制、程序脚本文件的生成和使用、函数的编写和调用。最后，通过程序实例分析编程技巧。

2.1 流控制

MATLAB 有 5 个流控制结构：

1．if 语句

2．for 循环

3．while 循环

4．break 语句

5．switch 语句

2.1.1 if 语句

if 语句判断逻辑表达式，并在表达式为真时执行一组语句。当表达式为假时，可选择 else 或 elseif 下的语句组执行。与 if 搭配的关键字 end 终止最后一组语句。所有语句都用这 4 个关键字描述，且不必用括号。

```
if 条件表达式 1
      语句段 1
elseif 条件表达式 2
      语句段 2
…
else
     语句段 n
end
```

上述判断流程为：如果满足条件表达式 1，则执行语句段 1；否则，判断是否满足条件表达式 2，满足，则执行语句段 2，不满足，则继续判断后面的条件。如果所有条件均不满足，则执行语句段 n。条件转移结构以 end 标识结束。

2.1.2 for 和 while 语句

for 循环依照一个预先设定的次数来重复执行一组语句。与之匹配的 end 语句会描述该语句组。

while 循环在一个逻辑条件的控制下，可无限次地重复执行一组语句。与之匹配的 end 语句会描述该语句组。

```
for 循环变量= w
   执行语句
end
```

```
while 条件表达式
    执行语句
end
```

例 2.1

```
s=0;
  for  i=1:100
        s=s+1;
  end
s
```

例 2.2

```
s=0;
  while (s<=100)
        s=s+1;
  end
s
```

```
输出：                              输出：
s =                                 s=
  100                                 101
```

在例 2.1 中，用 for 进行指定次数的循环控制，共循环 100 次，每进入一次循环，s 加 1，输出 s=100。在例 2.2 中，用 while 进行指定条件下的循环控制，共循环了 101 次，每进入一次循环 s 加 1。当 s=100 时，仍然满足条件，故而继续进入循环，执行赋值语句，得到 s=101。之后再进行条件判断，不满足条件，则循环结束。最终输出为 s=101。

```
例 2.3
H=[1,2,3;4,5,6;7,8,9];
 for i=1:3
     for j=1:3
        H(i,j)=H(i+j)+1;
     end
 end
H
输出：
H =
     2     3     4
     5     6     7
     8     9    10
```

```
例 2.4
 H=[1,2,3;4,5,6;7,8,9];
for k=1:9
     H(k)=H(k)+1;
end
H
输出：
H =
     2     3     4
     5     6     7
     8     9    10
```

```
例 2.5
H=[1,2,3;4,5,6;7,8,9];
 H=H+1;
 H
输出：
H =
     2     3     4
     5     6     7
     8     9    10
```

例 2.3、例 2.4 和例 2.5 为一组实验，完成同样的功能。在例 2.3 中，用两层循环嵌套来控制矩阵的行和列；MATLAB 中规定，按列顺序标记下标。故而，即使是二维矩阵，仍可以用单下标来控制对元素的操作，如在例 2.4 中。例 2.5 中，由循环控制的程序在 MATLAB 中可用一条矩阵赋值语句实现，MATLAB 强大的矩阵处理能力可见一斑。

2.1.3　break 语句

在 for 和 while 的循环中，可以加入 break 语句和 continue 语句对循环和运行予以终止。

break 的作用是终止循环，并跳出距它最近的循环结构；而 continue 的作用是略过本次循环，继续下次的判断与循环执行。

```
例 2.6
clc;
clear;
close all;
s=[];
num=0;
w=0;
for i=1:100
    w=w+i;
    if w>50
        break;
    end
    s=[s w];
```

```
输出：s =
     1     3     6    10    15    21    28    36    45
  num =
  9
```

```
    num=num+1;
end
s
num
```

例 2.6 中，定义空矩阵 s 用来存放每次计算的数据 w，并对每次计算的 w 进行判断。若满足 w>50，则 break 语句终止循环，输出 s 和 num。

2.1.4 switch 和 case

switch 语句是执行基于变量或表达式数值的语句组。关键字 case 和 otherwise 描述该语句组。程序只执行第一个与 case 匹配的语句。必须要有一个 end 与 switch 相匹配。

```
switch  开关表达式
    case  表达式 1
        语句段 1
    case  {表达式 2,表达式 3}
        语句段 2
    …
    otherwise
        语句段 n
end
```

例 2.7 要求按照考试成绩的等级来输出百分制分数段。

```
g=input('please input grade:','s');          %由用户交互输入成绩等级
     switch(g)                               %判断用户输入
          case{'A', 'a'}
               disp('85~100');               %用户输入 A，a 时，输出 85～100
          case{'B', 'b'}
               disp('70~84');                %用户输入 B，b 时，输出 70～84
case{'C', 'c'}
     disp('60~69');                          %用户输入 C，c 时，输出 60～69
case{'D', 'd'}
     disp('<60');                            %用户输入 D，d 时，输出<60
otherwise
     disp('输入错误' )                        %用户输入其他信息时，输出“输入错误”
end
```

2.2 脚本

MATLAB 是一种功能强大的编程语言，也是一种交互式计算环境。包含 MATLAB 语言代码的文件称为 M 文件。用户可以使用文本编辑器创建 M 文件，然后像使用其他 MATLAB 函数或命令一样使用它们。

MATLAB 软件有两种类型的 M 文件：

（1）脚本，不接受输入参数或返回输出参数。可以在工作空间运行数据。

（2）函数，可以接受输入参数和返回输出参数。内部变量是函数的局部变量。

当调用一个脚本时，MATLAB 仅执行文件中的命令。脚本可以对工作空间中已有的数据进行操作，或者创建新数据来运行。虽然脚本不能返回输出参数，但其创建的任何变量都保留在工作区间中，供后面的运算使用。另外，脚本可以提供图形输出，就像使用 plot 函数一样。

在脚本文件中，能够进行语句的修改，断点调试，便于进行 MATLAB 程序设计。因为 command window 中的命令不能进行断点调试，执行后的语句便不能被修改，所以，掌握脚本文件的使用非常重要。

例 2.8　创建一个名为 sin.m 的文件，其包含以下这些 MATLAB 的命令。其输出结果如图 2.2.1 所示。

```
clc;
close all;
clear;
t=0:pi/100:2*pi;
y=sin(t);
y1=sin(t-0.25);
y2=sin(t-0.5);
plot(t,y,'-',t,y1,'--',t,y2,':')
```

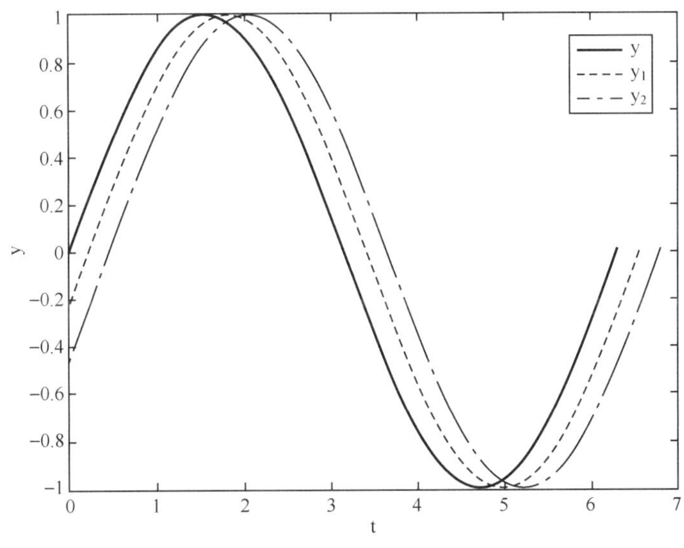

图 2.2.1　例 2.8 输出结果

2.3　函数

MATLAB 提供了丰富的标准初等数学函数，包括 abs（绝对值）、sqrt（平方根）、exp（指数函数）和 sin（正弦函数）等。对负数取平方根或取对数不会出错，此情况下 MATLAB 将自动生成相应的复数结果。MATLAB 也提供很多高等数学函数，如 bessel（贝塞尔）函数和 gamma（伽马）函数。这些函数绝大部分支持复数参数。

若要得到初等数学函数列表，键入 help elfun，则输出为

```
Elementary math functions.
  Trigonometric.
    sin         - Sine.
    sind        - Sine of argument in degrees.
    sinh        - Hyperbolic sine.
    asin        - Inverse sine.
    asind       - Inverse sine, result in degrees.
    asinh       - Inverse hyperbolic sine.
    cos              - Cosine.
    cosd             - Cosine of argument in degrees.
    cosh             - Hyperbolic cosine.
    acos        - Inverse cosine.
    acosd       - Inverse cosine, result in degrees.
    acosh       - Inverse hyperbolic cosine.
    tan         - Tangent.
    tand        - Tangent of argument in degrees.
    tanh        - Hyperbolic tangent.
    atan        - Inverse tangent.
    atand       - Inverse tangent, result in degrees.
    atan2       - Four quadrant inverse tangent.
    atanh       - Inverse hyperbolic tangent.
    sec         - Secant.
    secd        - Secant of argument in degrees.
    sech        - Hyperbolic secant.
    asec        - Inverse secant.
    asecd       - Inverse secant, result in degrees.
    asech       - Inverse hyperbolic secant.
    csc         - Cosecant.
    cscd        - Cosecant of argument in degrees.
    csch        - Hyperbolic cosecant.
    acsc        - Inverse cosecant.
    acscd       - Inverse cosecant, result in degrees.
    acsch       - Inverse hyperbolic cosecant.
    cot         - Cotangent.
    cotd        - Cotangent of argument in degrees.
    coth        - Hyperbolic cotangent.
    acot        - Inverse cotangent.
    acotd       - Inverse cotangent, result in degrees.
    acoth       - Inverse hyperbolic cotangent.
    hypot       - Square root of sum of squares.
  Exponential.
    exp              - Exponential.
    expm1            - Compute exp(x)-1 accurately.
    log              - Natural logarithm.
    log1p            - Compute log(1+x) accurately.
    log10            - Common (base 10) logarithm.
    log2             - Base 2 logarithm and dissect floating point number.
```

```
    pow2          - Base 2 power and scale floating point number.
    realpow       - Power that will error out on complex result.
    reallog       - Natural logarithm of real number.
    realsqrt      - Square root of number greater than or equal to zero.
    sqrt          - Square root.
    nthroot       - Real n-th root of real numbers.
    nextpow2      - Next higher power of 2.

  Complex.
    abs           - Absolute value.
    angle         - Phase angle.
    complex       - Construct complex data from real and imaginary parts.
    conj          - Complex conjugate.
    imag          - Complex imaginary part.
    real          - Complex real part.
    unwrap        - Unwrap phase angle.
    isreal        - True for real array.
    cplxpair      - Sort numbers into complex conjugate pairs.
Rounding and remainder.
    fix           - Round towards zero.
    floor         - Round towards minus infinity.
    ceil          - Round towards plus infinity.
    round         - Round towards nearest integer.
    mod           - Modulus (signed remainder after division).
    rem           - Remainder after division.
    sign          - Signum.
```

同理，若要得到高等数学函数和矩阵函数列表，则键入 help specfun 和 help elmat。

MATLAB 具有内置函数，如 sqrt 和 sin。它们是 MATLAB 的核心部分，使用效率很高，但计算细节不易获取。其他函数，如 gamma 和 sinh 是在 M 文件里执行的，可以看到其代码，如有必要，甚至可以修改。

除了使用 MATLAB 的内置函数，通常需要自己编写一些函数来完成需要的功能。函数的定义有特定的要求。

函数定义行：

```
function [返回变量列表]=函数名（输入变量列表）
```

函数主体：函数语句段，及必要的注释说明（由%引导）。

在函数定义行，当有多个返回变量时，用逗号分隔。当有多个输入变量时，也用逗号分隔。通常的要求是，函数名与对应的 M 文件名要保持一致。

在函数的编写过程中，每次修改后都需要先保存该函数，再对其进行调用和执行，否则执行的是未修改的函数。此外，函数只执行自己工作空间内的变量，函数的工作空间与通过 MATLAB 命令提示符访问的工作空间是分开的。

下面简单介绍本书后续要用到的 ode45( )函数的使用方法。

MATLAB 提供了常微分方程求解的函数 ode45( )。该函数采用了四阶五级的 RKF 方法，并采用自适应变步长的求解方法。函数调用格式为

```
[t , x] = ode45（方程函数名，tspan，x0，选项，附加参数）
```

其中，“选项”可以通过两个函数 odeget()和 odeset()来设置，具体的常用选项如下：

- RelTol 是相对误差容许上限，默认值为 0.001（即 0.1%的相对误差）。在一些特殊的微分方程中，为了保证较高的精度，还应该适当减小该值。
- AbsTol 是一个向量，其分量表示每一个状态变量允许的绝对误差，其默认值为 10^{-6}。可以自由设置其值，以改变求解精度。
- MaxStep 是求解方程最大允许的步长。
- Mass 为微分代数方程中的质量函数。
- Jacobian 为描述 Jacob 矩阵函数 $\dfrac{\partial f}{\partial x}$ 的函数名，如果已知该 Jacob 矩阵，则能加速仿真过程。

tspan = [t0 , tf]，其中，t0 和 tf 分别为用户指定的起始和终止计算时间。

函数 ode45()返回 t 和 x 两个变量。

方程函数的格式为

```
function xdot = 方程函数名（t , x , flag , 附加参数）
```

其中，t 为时间变量，x 为方程的状态变量，而 xdot 为状态变量的导数。注意，即使微分方程是非时变的，也需要在函数输入变量列表中写上 t。

例 2.9 设 Lorenz 模型的状态方程为

$$\begin{cases}\dot{x}_1(t) = -8x_1(t)/3 + x_2(t)x_3(t) \\ \dot{x}_2(t) = -10x_2(t) + 10x_3(t) \\ \dot{x}_3(t) = -x_1(t)x_2(t) + 28x_2(t) - x_3(t)\end{cases}$$

令初值 $x_1(0) = x_2(0) = x_3(0) = 0$，则可以编写一个 MATLAB 函数 lorenzeq.m 来描述系统的动态模型，具体如下：

```
% 函数文件 lorenzeq.m
function xdot=lorenzeq(t,x)
xdot=[-8/3*x(1)+x(2)*x(3);
        -10*x(2)+10*x(3);
        -x(1)*x(2)+28*x(2)-x(3)];
%Main
t_f=100;x0=[0;0;1e-10];
[t,x]=ode45('lorenzeq',[0,t_f],x0);
plot(t,x),
figure;
plot3(x(:,1),x(:,2),x(:,3));
axis([10 40 -20 20 -20 20]);
```

例 2.9 的输出如图 2.3.1 所示。

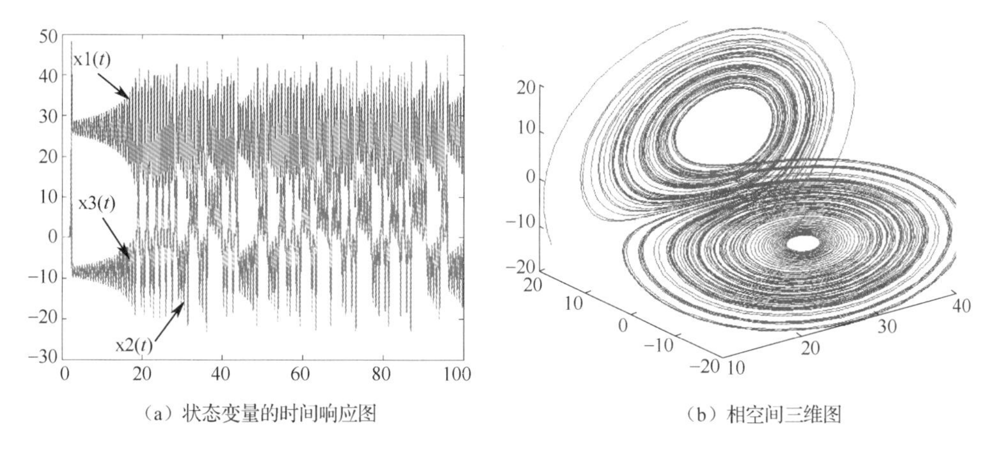

（a）状态变量的时间响应图　（b）相空间三维图

图 2.3.1　Lorenze 方程的仿真结果图示

在程序中，t_f 为设定的仿真终止时间，x0 为初始状态。图 2.3.1（a）为描述系统各个状态和时间关系的二维曲线图，图 2.3.1（b）为描述三个状态的相空间曲线。可见，利用 MATLAB，可轻易求得例 2.9 的三元一阶常微分方程组的数值解。

例 2.10　创建一个包含下列 MATLAB 命令，及名为 aver.m 的文件。

```
%函数文件 aver.m
function a=aver(x)
%求算术平均值
[m,n]=size(x);           %获得 x 的大小
sum_x=sum(sum(x));   %x 所有元素求和
a=sum_x/(m*n);          %求算术平均值
%main
clc;
clear;
close all;
x=[1,2,3;4,5,6;7,8,9];
a=aver(x)
输出：
a =
     5
```

2.4　编程实例

例 2.11　已知 $s^2+5s+6=0$，其中，$s=\dfrac{-b\pm\sqrt{b^2-4ac}}{2a}$，求 s 的值。

```
clc;
clear;
close all;
a=1;
b=5;
c=6;
x = -b/(2*a);
y = sqrt(b^2-4*a*c)/(2*a);
s1 = x+y
```

```
s2 = x-y
输出：
s1 =
      -2
s2 =
      -3
```

例 2.12 当时间小于 0 时，阶跃信号为 0，时间大于 0 时，跳变成 1。请自定义一个阶跃函数。

$$u(t)=\begin{cases}0, & t<0\\1, & t\geqslant 0\end{cases}$$

```
function u = step(t)                        % 单位阶跃函数
% u = step(t)
% u = 0 for t<0
% u = 1 for t>=0
u = zeros(size(t));
u(find(t>=0))=1;
```

例 2.13 已知 5×5 矩阵，请完成矩阵的左右翻转（左右镜像）。

```
clc;
clear;
close all;
a=[1 5 10 15 20;25 30 35 40 45;50 55 60 65 70;75 80 85 90 95;100 105 115 125 135];
%%%%%%%%%%%%%%%%%%%%%%方式 1
b=a;
b(:,1)=a(:,5);
b(:,2)=a(:,4);
b(:,4)=a(:,2);
b(:,5)=a(:,1);
%%%%%%%%%%%%%%%%%%%%%%方式 2
% [m,n]=size(a);
% b=zeros(m,n);
% for k=1:n
%       b(:,k)=a(:,n+1-k);
% end
%%%%%%%%%%%%%%%%%%%%%%方式 3
% [m,n]=size(a);
% for k=1:floor(n/2)
%       temp=a(:,k);
%       a(:,k)=a(:,n-k+1);
%       a(:,n-k+1)=temp;
% end
% %%%%%%%%%%%%%%%%%%%%方式 4
%[m,n]=size(a);
%b=a(:,n:-1:1);
% %%%%%%%%%%%%%%%%%%%%%方式 5
% B=fliplr(a);
```

上述 5 种编程方式都能完成将矩阵左右翻转的功能，显然在编程技巧上有区别。方式 1 逐条赋值，当矩阵很大时，显然难以操作；方式 2 和方式 3 通过循环来控制赋值，方式 2 循环了 *n* 次，方式 3 循环了 *n*/2 次，但是方式 2 未覆盖 *a*，方式 3 覆盖了原始的 *a*；方式 4 用 MATLAB 矩阵下标进行赋值控制；方式 5 调用 MATLAB 内置 fliplr()函数。显然，除方式 1，可根据具体的环境和需求进行程序设计。

例 2.14 对一幅图像做均值滤波，滤波窗口大小为 3×3 像素。

```
clc;clear;close all;
original=imread('cameraman.TIF');                %读图
[m,n]=size(original);                            %获得图像大小
original=double(original);                       %转换图像数据格式用于计算
noise=10*randn(m,n);                             %生成随机噪声
noiseimage=original+noise;                       %给图像添加噪声
figure(1);imshow(uint8(noiseimage));             %显示噪声图像
b=original;                                      %图像备份
%%%%%%%%%%%%%%%%%%%%%%%%%%%%%%%%%%%%%%%%
%方式一
for i=2:m-1
    for j=2:n-1
        b(i,j)=(noiseimage(i-1,j-1)+noiseimage(i-1,j)+...
            noiseimage(i-1,j+1)+noiseimage(i,j-1)+noiseimage(i,j)+...
            noiseimage(i,j+1)+noiseimage(i+1,j-1)+noiseimage(i+1,j)+...
            noiseimage(i+1,j+1))/9;
    end
end
%%%%%%%%%%%%%%%%%%%%%%%%%%%%%%%%%%%%%%%%%
%方式二
% for i=2:m-1
%     for j=2:n-1
%         temp=noiseimage(i-1:i+1,j-1:j+1);
%         b(i,j)=sum(sum(temp))/9;
%     end
% end
%%%%%%%%%%%%%%%%%%%%%%%%%%%%%%%%%%%%%%%%%
%方式三
temp=[1/9,1/9,1/9;1/9,1/9,1/9;1/9,1/9,1/9];
for i=2:m-1
    for j=2:n-1
        c=noiseimage(i-1:i+1,j-1:j+1);
        b(i,j)=sum(sum(temp.*c));
    end
end
figure(2);imshow(uint8(b));                     %输出去噪图像
```

输出结果如图 2.4.1 所示。

（a）带噪图像

（b）均值滤波去噪图像

图 2.4.1 均值滤波示例

比较上述三种方式。方式 1 需要逐项列写求取均值；方式 2 和方式 3 利用模板 temp 进行矩阵操作来求取均值。而方式 3 不仅能够求均值，而且可以通过改变系数，进行加权平均值滤波。然而，上述程序仍然存在以下不足：

（1）当滤波窗口大小改变时，需要对程序中所有相关的变量进行修改；

（2）当需要多次滤波，或在其他程序中使用滤波时，需要再次编写程序。

针对上述两个不足，可将均值滤波写成如下函数文件，以便随时调用。

```
function b=Mean_filter(a,k1,k2)
[m,n]=size(a);
b=a;
r1=(k1-1)/2;                        %滤波窗口的行半径
r2=(k2-1)/2;                        %滤波窗口的列半径
temp=ones(k1,k2);                   %用于提取滤波窗口覆盖下的图像区域
for i=r1+1:m-r1
    for j=r2+1:n-r2
        temp=a(i-r1:i+r1,j-r2:j+r2);
        b(i,j)=sum(sum(temp))/(k1*k2);
    end
end
```

主程序如下：

```
%main
clc;
clear all;
original=imread('cameraman.TIF');            %读图
[m,n]=size(original);                        %获得图像大小
original=double(original);                   %转换图像数据格式用于计算
noise=10*randn(m,n);                         %生成随机噪声
noiseimage=original+noise;                   %给图像添加噪声
```

```
figure(1);imshow(uint8(noiseimage));        %显示噪声图像
b=Mean_filter(noiseimage,5,5);              %调用函数 Mean_filter
figure(2);imshow(uint8(b));                 %输出去噪图像
```

本章小结

（1）介绍了 MATLAB 软件中的流控制语句、脚本和函数。

（2）通过编程实例分析编程技巧。

Chapter 2
MATLAB Programming

This chapter focuses on the basic knowledge of MATLAB programming, introducing the program design in the MATLAB environment, including the loop control, the generation and use of program script files, the writing and invoking of functions. Finally, the programming skills are analyzed through specific examples.

2.1 Flow Control

MATLAB has five flow control constructs.

1. *if* statements
2. *for* loops
3. *while* loops
4. *break* statements
5. *switch* statements

2.1.1 *if* Statements

The *if* statement evaluates a logical expression and executes a group of statements when the expression is true. The optional *else* or *elseif* keywords provide for the execution of alternate groups of statements. An *end* keyword, which matches the *if*, terminates the last group of statements. The groups of statements are delineated by the four keywords, no braces or brackets are involved.

```
if expression1
        statements1
elseif expression2
        statements2
...
else
      statements n
end
```

The above judgment flow is: if the condition expression 1 is satisfied, the statement 1 is executed; otherwise, it is judged whether the conditional expression 2 is satisfied; if it is satisfied, the statement segment 2 is executed; if it is not satisfied, the following condition is continuously judged. If all the conditions are not satisfied, statement n is executed. The conditional transfer structure ends with the keyword *end*.

2.1.2 *for* and *while* Statement

The *for* loop repeats a group of statements a fixed, predetermined number of times. A matching *end* delineates the statements.

The *while* loop repeats a group of statements an indefinite number of times under control of a logical condition. A matching *end* delineates the statements.

```
for loop variable=w                    while conditional expressions
    execute statement                      execute statement
end                                    end
```

```
Example 2.1                            Example 2.2
s=0;                                   s=0;
```

```
    for   i=1:100                         while (s<=100)
        s=s+1;                                s=s+1;
    end                                   end
s                                     s
Output:                               Output:
s =                                   s=
    100                                   101
```

In Example 2.1, use *for* to achieve the specified number of cycle control, a total of 100 cycles. Each time you enter a cycle *s* plus 1, the output *s* = 100. In Example 2.2, *while* using the specified conditions of the cycle control, a total of the cycle is 101 times. Every time you enter a cycle, add 1. When *s*=100, it still meets the condition. Therefore, it continues to enter the loop, executes the assignment statement, and obtains *s*=101. After that, the condition is judged, the condition is not satisfied, and the cycle ends. The final output is *s*=101.

```
Example 2.3                      Example 2.4                   Example 2.5
H=[1,2,3;4,5,6;7,8,9];           H=[1,2,3;4,5,6;7,8,9];        H=[1,2,3;4,5,6;7,8,9];
 for i=1:3                       for k=1:9                     H=H+1;
    for j=1:3                      H(k)=H(k)+1;                H
       H(i,j)=H(i+j)+1;          end
    end                          H
 end
H
Output:                          Output:                       Output:
H =                              H =                           H =
     2    3    4                     2    3    4                   2    3    4
     5    6    7                     5    6    7                   5    6    7
     8    9   10                     8    9   10                   8    9   10
```

Examples 2.3, 2.4, and 2.5 are a set of experiments that accomplish the same function. In Example 2.3, two-level loop nesting is used to control the matrix subscript rows and columns. In MATLAB, subscripts are marked in column order, so even a two-dimensional matrix, a single subscript can be used to control the operation of elements, as in Example 2.4. In Example 2.5, the program controlled by the loop can be implemented in MATLAB using a matrix assignment statement, which shows MATLAB's powerful matrix processing capabilities.

2.1.3 *break* Statements

In the *for* and *while* loop, adding a *break* statement and a *continue* statement can terminate the loop and proceed.

The role of *break* is to terminate the loop and jump out of the loop structure closest to it; while *continue* skips this loop and continues with the next decision and loop execution.

```
Example 2.6                      Output: s =
clc;                             1    3    6    10    15    21    28    36    45
clear;                            num=
close all;                       9
s=[];
num=0;
```

```
w=0;
for i=1:100
    w=w+i;
    if w>50
        break;
    end
    s=[s w];
  num=num+1;
end
s
num
```

In Example 2.6, the empty matrix *s* is defined to store the data *w* for each calculation, and the *w* is evaluated for each calculation. If *w*>50 is satisfied, the *break* statement terminates the loop and outputs *s* and *num*.

2.1.4 *switch* and *case*

The *switch* statement executes groups of statements based on the value of a variable or expression. The keywords *case* and *otherwise* delineate the groups. Only the first matching *case* is executed. There must always be an *end* to match the *switch.*

```
switch  switch expression
    case  expression1
      statements1
   case  { expression2, expression3}

      statements2
   ...
   otherwise
      statements
end
```

Example 2.7 Demand to output percentage score segments according to the level of test scores.

```
g=input('please input grade: ', 's');     % Entering the performance level by user interaction
switch(g)                                 % Judging user input
    case{'A', 'a'}
        disp('85~100');                   % When user input A, a, output 85~100
    case{'B', 'b'}
        disp('70~84');                    %When user input B, b, output 70~84
    case{'C', 'c'}
        disp('60~69');                    % When user inputs C, c, output 60~69
    case{'D', 'd'}
        disp('<60');                      % When the user inputs D, d, the output is <60
    otherwise
        disp('input error')               %When the user enters other information, it outputs
    end                                   %"input error"
```

2.2 Scripts

MATLAB is a powerful programming language as well as an interactive computational environment. Files that contain code in the MATLAB language are called M-files. You create M-files using a text editor, then use them as you would use any other MATLAB function or command.

There are two kinds of M-files:

(1) Scripts, which do not accept input arguments or return output arguments. They operate on data in the workspace.

(2) Functions, which can accept input arguments and return output arguments. Internal variables are local to the function.

When you invoke a script, MATLAB simply executes the commands found in the file. Scripts can operate on existing data in the workspace, or they can create new data on which to operate. Although scripts do not return output arguments, any variables that they create remain in the workspace, to be used in subsequent computations. In addition, scripts can produce graphical output using functions like *plot*.

In the script file, it is possible to modify the statements, debug breakpoints, and facilitate MATLAB programming. Because commands in the command window cannot be debugged with breakpoints, the executed statements cannot be modified, it is very important to master the use of script files.

Example 2.8 Creates a file named *sin.m* that contains the following MATLAB commands. The output is shown in Figure 2.2.1.

```
clc;
close all;
clear;
t=0:pi/100:2*pi;
y=sin(t);
y1=sin(t-0.25);
y2=sin(t-0.5);
plot(t,y,'-',t,y1,'--',t,y2,':')
```

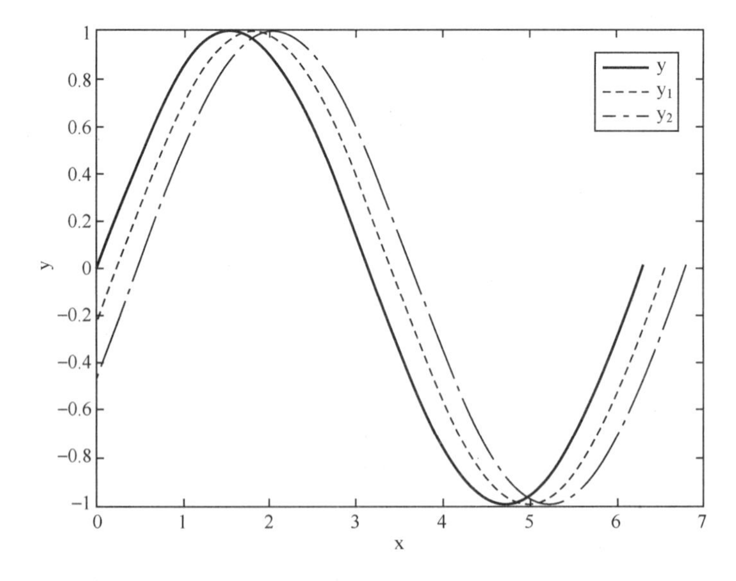

Figure 2.2.1 Output of Example 2.8

2.3 Functions

MATLAB offers a rich set of standard elementary math functions, including *abs* (absolute), *sqrt* (square root), *exp* (exponential), and *sin* (sine). There is no error in taking a square root or taking a logarithm of a negative number. In this case, MATLAB will automatically generate the corresponding complex result. MATLAB also provides many advanced math functions such as the *bessel* function and the *gamma* function. Most of these functions support complex parameters.

To get a list of elementary math functions, type *help elfun* and the output is

```
Elementary math functions.
  Trigonometric.
    sin        - Sine.
    sind       - Sine of argument in degrees.
    sinh       - Hyperbolic sine.
    asin       - Inverse sine.
    asind      - Inverse sine, result in degrees.
    asinh      - Inverse hyperbolic sine.
    cos             - Cosine.
    cosd            - Cosine of argument in degrees.
    cosh            - Hyperbolic cosine.
    acos       - Inverse cosine.
    acosd      - Inverse cosine, result in degrees.
    acosh      - Inverse hyperbolic cosine.
    tan        - Tangent.
    tand       - Tangent of argument in degrees.
    tanh       - Hyperbolic tangent.
    atan       - Inverse tangent.
    atand      - Inverse tangent, result in degrees.
    atan2      - Four quadrant inverse tangent.
    atanh      - Inverse hyperbolic tangent.
    sec        - Secant.
    secd       - Secant of argument in degrees.
    sech       - Hyperbolic secant.
    asec       - Inverse secant.
    asecd      - Inverse secant, result in degrees.
    asech      - Inverse hyperbolic secant.
    csc        - Cosecant.
    cscd       - Cosecant of argument in degrees.
    csch       - Hyperbolic cosecant.
    acsc       - Inverse cosecant.
    acscd      - Inverse cosecant, result in degrees.
    acsch      - Inverse hyperbolic cosecant.
    cot        - Cotangent.
    cotd       - Cotangent of argument in degrees.
    coth       - Hyperbolic cotangent.
    acot       - Inverse cotangent.
    acotd      - Inverse cotangent, result in degrees.
```

```
    acoth        - Inverse hyperbolic cotangent.
    hypot        - Square root of sum of squares.
  Exponential.
    exp          - Exponential.
    expm1        - Compute exp(x)-1 accurately.
    log          - Natural logarithm.
    log1p        - Compute log(1+x) accurately.
    log10        - Common (base 10) logarithm.
    log2         - Base 2 logarithm and dissect floating point number.
    pow2         - Base 2 power and scale floating point number.
    realpow      - Power that will error out on complex result.
    reallog      - Natural logarithm of real number.
    realsqrt     - Square root of number greater than or equal to zero.
    sqrt         - Square root.
    nthroot      - Real n-th root of real numbers.
    nextpow2     - Next higher power of 2.

  Complex.
    abs          - Absolute value.
    angle        - Phase angle.
    complex      - Construct complex data from real and imaginary parts.
    conj         - Complex conjugate.
    imag         - Complex imaginary part.
    real         - Complex real part.
    unwrap       - Unwrap phase angle.
    isreal       - True for real array.
    cplxpair     - Sort numbers into complex conjugate pairs.
Rounding and remainder.
    fix          - Round towards zero.
    floor        - Round towards minus infinity.
    ceil         - Round towards plus infinity.
    round        - Round towards nearest integer.
    mod          - Modulus (signed remainder after division).
    rem          - Remainder after division.
    sign         - Signum.
```

Similarly, to get a list of advanced math functions and matrix functions, type *help specfun* and *help elmat*.

MATLAB has built-in functions such as *sqrt* and *sin*. They are the core part of MATLAB and are very efficient to use, but the computational details are not easily accessible. Other functions, such as *gamma* and *sinh*, are executed in the M file. you can see the code, and you can even modify if necessary.

In addition to using MATLAB's built-in functions, it usually requires the user to write some functions to complete the required functions. The definition of the function has specific requirements.

Function definition line

```
function [return variable list] = function name (input variable list)
```

Function body: the statement segment of a function, with necessary commentary explanation (guided by %).

On the definition lines of a function, when there are multiple return variables, they are separated by commas. When there are multiple input variables, they are also separated by commas. It is generally required that the function name be consistent with the corresponding M file name.

In the process of writing a function, each time the function is modified, the function needs to be saved, and then the function is called and executed. Otherwise, the unmodified function is still executed. In addition, the function only executes variables in its own workspace, and the function's workspace is separate from the workspace accessed through the MATLAB command prompt.

Here is a brief introduction to the *ode45*() function used in this book after wards.

MATLAB provides a function *ode45*() for solving an ordinary differential equation. The function uses a fourth-order five-level RKF method and uses an adaptive variable step length solution method. The function call format is

```
[t, x] = ode45 (equation function name, tspan, x0, option, additional parameters)
```

The "options" can be set by two functions *odeget*() and *odeset*(). The specific common options are as follows:

RelTol is the relative error tolerance upper limit. The default value is 0.001 (that is, 0.1% relative error). In some special differential equations, in order to ensure high accuracy, the value should be properly reduced.

AbsTol is a vector whose component represents the absolute error allowed by each state variable. The default value is 10^{-6}. You can freely set its value to change the solution accuracy.

MaxStep is the maximum allowable step when solving the equation.

Mass is a mass function in a differential algebraic equation.

Jacobian is the function name describing the *Jacob* matrix function $\frac{\partial f}{\partial x}$. If the *Jacob* matrix is known, the simulation process can be accelerated.

tspan=[*t*0, *tf*], where *t*0 and *tf* are the start time and end time of calculation specified by the user.

The function *ode45*() returns *t* and *x* two variables.

The format of the equation function is

```
function x dot = equation function name (t, x, flag, additional parameters)
```

Where *t* is the time variable, *x* is the state variable of the equation, and *xdot* is the derivative of the state variable. Note that even if the differential equation is time-invariant, you need to write *t* in the list of function input variables.

Example 2.9 Let the state equation of the Lorenz model be

$$\begin{cases}\dot{x}_1(t) = -8x_1(t)/3 + x_2(t)x_3(t) \\ \dot{x}_2(t) = -10x_2(t) + 10x_3(t) \\ \dot{x}_3(t) = -x_1(t)x_2(t) + 28x_2(t) - x_3(t)\end{cases}$$

Make initial value $x_1(0)=x_2(0)=x_3(0)=0$, the user can write a MATLAB function *lorenzeq.m* to describe the dynamic model of the system, as follows,

```
% The function file of lorenzeq.m
function xdot=lorenzeq(t,x)
xdot=[-8/3*x(1)+x(2)*x(3);
        -10*x(2)+10*x(3);
        -x(1)*x(2)+28*x(2) -x(3)];
% Main
t_f=100;x0=[0;0;1e-10];
[t,x]=ode45('lorenzeq',[0,t_f],x0);
plot(t,x),
figure;
plot3(x(:,1),x(:,2),x(:,3));
axis([10 40 -20 20 -20 20]);
```

The output of Example 2.9 is shown in Figure 2.3.1.

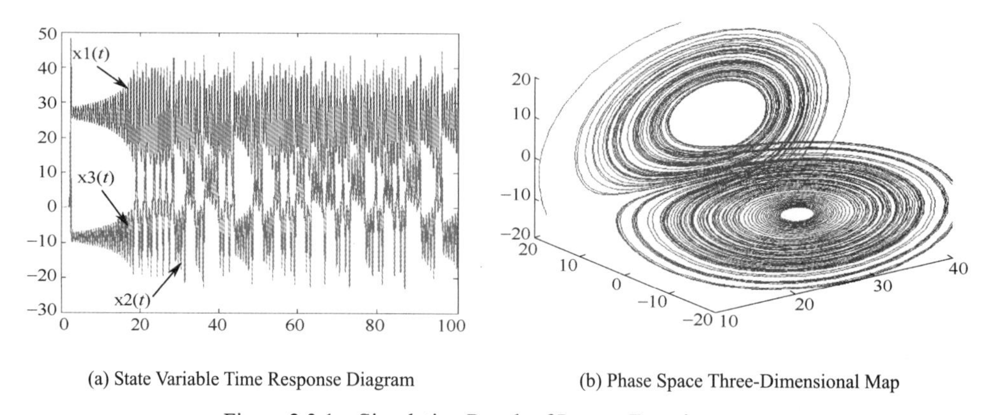

(a) State Variable Time Response Diagram (b) Phase Space Three-Dimensional Map

Figure 2.3.1 Simulation Result of Lorenz Equation

In the program, *t_f* is the set simulation termination time, and *x*0 is the initial state. Figure 2.3.1(a) depicts a two-dimensional graph of the various states and time relationships of the system. Figure 2.3.1(b) depicts the phase space curves of the three states. It can be seen that the numerical solution of the ternary first-order ordinary differential equations in Example 2.9 can be easily obtained by using MATLAB.

Example 2.10 Creates a file named *aver.m* that contains the following MATLAB commands.

```
% The function file of aver.m
function a=aver(x)
% Calculating the arithmetic mean
[m,n]=size(x);           % Get the size of x
sum_x=sum(sum(x));   % sum to All elements of x
a=sum_x/(m*n);          % Arithmetic average
%main
clc;
clear;
close all;
x=[1,2,3;4,5,6;7,8,9];
```

```
a=aver(x)
Output:
a =
       5
```

2.4 Examples

Example 2.11 Known $s^2 + 5s + 6 = 0$, where $s = \dfrac{-b \pm \sqrt{b^2 - 4ac}}{2a}$, find the value of s.

```
clc;
clear;
close all;
a=1;
b=5;
c=6;
x = -b/(2*a);
y = sqrt(b^2-4*a*c)/(2*a);
s1 = x+y
s2 = x-y
output:
s1 =
       -2
s2 =
       -3
```

Example 2.12 The step signal is 0 when the time is less than 0, and the jump becomes 1 when the time is greater than 0. Please customize a step signal function.

$$u(t) = \begin{cases} 0, & t < 0 \\ 1, & t \geqslant 0 \end{cases}$$

```
function u = step(t)                % STEP unit step function
% u = step(t)
% u = 0 for t<0
% u = 1 for t>=0
u = zeros(size(t));
u(find(t>=0))=1;
```

Example 2.13 Matrix 5×5 is known. Please complete the left and right flip of the matrix (Left and right mirror).

```
clc;
clear;
close all;
a=[1 5 10 15 20;25 30 35 40 45;50 55 60 65 70;75 80 85 90 95;100 105 115 125 135];
%%%%%%%%%%%%%%%%%%%%%%Method 1
b=a;
b(:,1)=a(:,5);
b(:,2)=a(:,4);
```

```
b(:,4)=a(:,2);
b(:,5)=a(:,1);
%%%%%%%%%%%%%%%%%%%%Method 2
% [m,n]=size(a);
% b=zeros(m,n);
% for k=1:n
%      b(:,k)=a(:,n+1-k);
% end
%%%%%%%%%%%%%%%%%%%%Method 3
% [m,n]=size(a);
% for k=1:floor(n/2)
%      temp=a(:,k);
%      a(:,k)=a(:,n-k+1);
%      a(:,n-k+1)=temp;
% end
% %%%%%%%%%%%%%%%%%%%%Method 4
%[m,n]=size(a);
%b=a(:,n:-1:1);
% %%%%%%%%%%%%%%%%%%%%Method 5
% B=fliplr(a);
```

The above five programming methods can complete the function of flipping the matrix left and right. Obviously there are differences in programming skills. Method 1 assigns values one by one, but when the matrix is large, it is obviously difficult to operate. Method 2 and Method 3 control assignments through loops. Mode 2 loops n times. Mode 3 loops $n/2$ times. However, Method 2 does not cover a, while Method 3 covers the original a. Method 4 uses MATLAB matrix subscripts for assignment control. Method 5 calls MATLAB's built-in *fliplr*() function. Obviously, in addition to Method 1, programs can be designed based on specific environments and requirements.

Example 2.14 Perform mean filtering on an image with a filter window sized 3×3 pixels.

```
clc; clear; close all;
original=imread('cameraman.TIF');          %Read image
[m,n]=size(original);                      % Get image size
original=double(original);                     % Convert image data format for
calculation noise=10*randn(m,n);               % Generate random noise
noiseimage=original+noise;                     % Add noise to the image
figure(1);imshow(uint8(noiseimage));           % Display noise image
b=original;                                    % Backup image
%%%%%%%%%%%%%%%%%%%%%%%%
%Method 1
for i=2:m-1
    for j=2:n-1
        b(i,j)=(noiseimage(i-1,j-1)+noiseimage(i-1,j)+...
            noiseimage(i-1,j+1)+noiseimage(i,j-1)+noiseimage(i,j)+...
            noiseimage(i,j+1)+noiseimage(i+1,j-1)+noiseimage(i+1,j)+...
            noiseimage(i+1,j+1))/9;
    end
```

```
end
%%%%%%%%%%%%%%%%%%%%%%%%%%%%%%%%%%
% Method 2
% for i=2:m-1
%       for j=2:n-1
%             temp=noiseimage(i-1:i+1,j-1:j+1);
%             b(i,j)=sum(sum(temp))/9;
%       end
% end
%%%%%%%%%%%%%%%%%%%%%%%%%%%%%%%%%%
% Method 3
temp=[1/9,1/9,1/9;1/9,1/9,1/9;1/9,1/9,1/9];
for i=2:m-1
    for j=2:n-1
        c=noiseimage(i-1:i+1,j-1:j+1);
        b(i,j)=sum(sum(temp.*c));
    end
end
figure(2);imshow(uint8(b));                    % Output denoising image
```

The output is shown in Figure 2.4.1.

(a) Noisy Image

(b) Mean Filter Denoising Image

Figure2.4.1 Example of Mean Filter

Comparing the above three methods. Method 1 finds the mean by item-by-item writing; Method 2 and 3 use the template *temp* to perform matrix to find the mean value. Method 3 can not only obtain the mean value, but also can change the coefficient of the template and achieve weighted average filtering. However, the above program displays the following deficiencies.

(1) When thc filter window size changes, all relevant variables in the program need to be modified;

(2) When more than one filter is needed, or when filtering is used in other programs, you need to write the program again.

For the above two deficiencies, we can write the average filter as the following function file so that we call it at any time.

```
function b=Mean_filter(a,k1,k2)
[m,n]=size(a);
b=a;
r1=(k1-1)/2;                    % Row radius of the filter window
r2=(k2-1)/2;                    % Column radius of the filter window
temp=ones(k1,k2);               % Used to place the image area covered by the filter window
for i=r1+1:m-r1
    for j=r2+1:n-r2
temp=a(i-r1:i+r1,j-r2:j+r2);
        b(i,j)=sum(sum(temp))/(k1*k2);
    end
end
```

The main program is as follows:

```
%main
clc;
clear;
close all;
original=imread('cameraman.TIF');           %Read image
[m,n]=size(original);                       %Get image size
original=double(original);                  % Convert image data format for calculation
noise=10*randn(m,n);                        % Generate random noise
noiseimage=original+noise;                  % Add noise to the image
figure(1);imshow(uint8(noiseimage));        % Display noise image
b=Mean_filter(noiseimage,5,5);              % Call function Mean_filter
figure(2);imshow(uint8(b));                 % Output denoising image
```

Summary

1. This chapter mainly introduces flow control statements, scripts, and functions in MATLAB software.

2. The programming skills are analyzed through specific examples.

生词注解

alternate	adj. 交替的，轮流的 n. 替换物
calculation	n. 计算，估算
delineate	vt. 勾画，描述
invoke	v. 调用
facilitate	v. 促进
segment	v. 分割
filtering	v. 过滤
flipping	v. 翻转

第 3 章 控制系统的传递函数模型

本章介绍如何利用 MATLAB 描述线性定常系统的数学模型，为后续线性定常系统的时域稳态分析、稳态误差分析、根轨迹分析以及频域分析奠定基础。

对于线性定常系统，其主要的数学模型为时域微分方程模型以及复域的传递函数模型。微分方程模型的 MATLAB 实现在第 2 章中已经介绍。本章主要介绍传递函数模型的 MATLAB 描述、模型之间的转换，以及系统性能的仿真分析。

3.1 线性定常系统简介

若系统输入输出的微分方程可以描述为如下线性形式：

$$a_0c^{(n)}(t)+a_1c^{(n-1)}(t)+\cdots+a_{n-1}c'(t)+a_nc(t)=b_0r^{(m)}(t)+b_1r^{(m-1)}(t)+\cdots+b_{m-1}r'(t)+b_mr(t),n\geqslant m \tag{3.1}$$

式中，$c(t)$为被控量，$r(t)$为系统输入量，系数$a_0,a_1,\cdots,a_n$，$b_0,b_1,\cdots,b_n$是常数，则称其为线性定常系统。

线性系统满足叠加原理。叠加原理说明，两个不同的外力函数同时作用于系统的响应，等于两个外力函数单独作用的响应之和。因此，对于线性系统可以通过一次处理一个输入，并将多个结果叠加，来计算对多个输入的响应。正是这一原理，使我们能够从简单的解建立线性微分方程的复杂解。在动态系统的实验研究中，输入量和输出量成正比，意味着满足叠加原理，因而可以把系统看成线性系统。

如果微分方程的系数是常数，或者仅仅是自变量的函数，则该微分方程是线性的。由线性定常参数元件构成的动态系统，可以用线性定常（常系数）微分方程描述，这类系统称为线性定常（或线性常系数）系统。如果系统的微分方程的系数是时间的函数，则称这类系统为线性时变系统。宇宙飞船控制系统就是时变控制系统的一个例子（宇宙飞船的质量随着燃料的消耗而变化）。

3.2 线性定常系统的传递函数

3.2.1 简介

线性定常系统的微分方程描述了系统输入输出随时间 t 的变化，是描述系统动态性能的数学模型。在第 2 章中可知，借助计算机能够快速地实现微分方程的求解，完成系统分析。

然而，在控制系统分析、改造以及设计的过程中，使用微分方程也存在不便之处。例如，当系统中某元件参数或系统结构发生改变，就要重写微分方程，并重新仿真或实测系统输入输出信号，以评估系统性能。此外，仅从输入输出信号观测和评估系统，难以分析系统参数、结构对性能的影响，不利于学习或发展控制理论。

在控制理论中，经常使用另一种数学模型来描述线性定常系统的输入输出关系，即传递函数模型。将时域的线性微分方程在零初始条件下进行拉普拉斯变换，便得到复域的传递函数模型，从而使输入输出之间的动态微分方程模型变换为代数模型，更有利于分析系统参数、结构对于性能的影响。

3.2.2 定义

线性定常系统的传递函数定义为：在全部初始条件为零的条件下，输出量的拉普拉斯变换与输入量的拉普拉斯变换之比。

考虑由下列微分方程描述的线性定常系统：

$$a_0c^{(n)}+a_1c^{(n-1)}+\cdots+a_{n-1}c'+a_nc=b_0r^{(m)}+b_1r^{(m-1)}+\cdots+b_{m-1}r'+b_mr \quad (n\geqslant m) \tag{3.2}$$

式中，r 是输入，c 是输出，在全部初始条件为零时，输出量与输入量的拉普拉斯变换之比，就是这个系统的传递函数 $G(s)$：

$$G(s)=\frac{C(s)}{R(s)}=\frac{b_0s^m+b_1s^{m-1}+\cdots+b_{m-1}s+b_m}{a_0s^n+a_1s^{n-1}+\cdots+a_{n-1}s+a_n} \tag{3.3}$$

利用传递函数的概念，可以用以 s 为变量的代数方程表示系统的动态特性。如果传递函数的分母中，s 的最高阶次为 n，则称该系统为 n 阶系统。

3.2.3　相关说明

传递函数概念的适用范围限于线性定常系统。当然，在这类系统的分析和设计中，传递函数的方法的应用是很广泛的。下面列出了传递函数的一些重要说明（注意，此处所指的系统仅限线性定常微分方程描述的系统）。

（1）系统的传递函数是一种数学模型，它是一种表示将输出与输入变量相关联的微分方程的运算方法；

（2）传递函数是系统本身的一种属性，与输入量或者驱动函数的大小和性质无关；

（3）传递函数包含联系输入量与输出量所必需的单元，但是它不提供与系统物理结构有关的任何信息（许多物理上完全不同的系统，可以具有相同的传递函数）；

（4）若系统的传递函数已知，则可以针对各种不同形式的输入量，研究系统的输出或响应，以便掌握系统的性质；

（5）若系统的传递函数未知，则可以通过引入已知输入量，并测量其相应输出量的实验方法，确定系统的传递函数。系统的传递函数一旦被确定，它就能够对系统的动态特性进行充分描述，不同于对系统的物理描述。

例 3.1　试求图 3.2.1 所示运算放大器电路的传递函数 $U_o(s)/U_i(s)$。

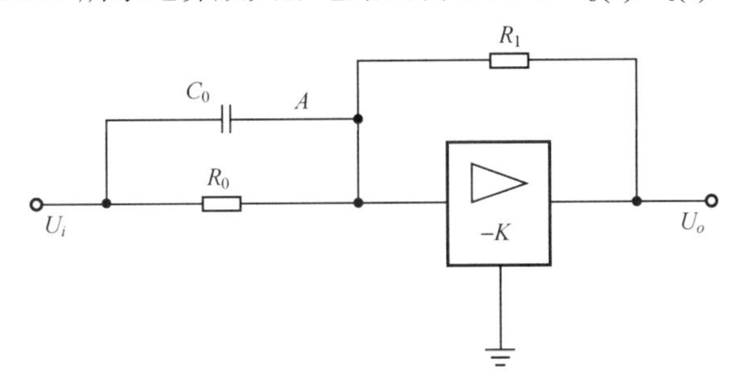

图 3.2.1　运算放大电路

解：利用复阻抗，A 点处有电压等效

$$\frac{U_i(s)}{\dfrac{R_0/C_0s}{R_0+1/C_0s}}=-\frac{U_o(s)}{R_1} \tag{3.4}$$

因此，传递函数是

$$\frac{U_o(s)}{U_i(s)}=-R_1\left(C_0s+\frac{1}{R_0}\right) \tag{3.5}$$

3.3 基于 MATLAB 的传递函数描述

对于线性定常系统，公式（3.3）中传递函数$G(s)$可进一步整理为“首 1 标准型”和“尾 1 标准型”：

$$G(s)=\frac{b_0(s-z_1)(s-z_2)\cdots(s-z_m)}{a_0(s-p_1)(s-p_2)\cdots(s-p_n)}=K^*\frac{\prod_{i=1}^{m}(s-z_i)}{\prod_{j=1}^{n}(s-p_j)} \quad \text{首 1 标准型} \tag{3.6}$$

$$G(s)=\frac{b_m(\tau_1 s+1)(\tau_2^2 s^2+2\zeta\tau_2 s+1)\cdots(\tau_i s+1)}{a_n(T_1 s+1)(T_2^2 s^2+2\zeta T_2 s+1)\cdots(T_j s+1)} \quad \text{尾 1 标准型} \tag{3.7}$$

式（3.6）所示的“首 1 标准型”也称为传递函数的零极点模型，其能直观地展现出系统的零极点个数和取值。对于系统的闭环传递函数，可直接利用该形式分析系统的稳定性；对于开环传递函数，可利用根轨迹分析法来分析系统性能。

式（3.7）所示的“尾 1 标准型”也称为传递函数的典型因子乘积的形式。当系统结构未知时，其开环传递函数可整理为该形式，能够直观地展现系统所包含的典型模式，对系统的整体性能做定性的了解，也可以利用频率法定量分析系统性能。

3.3.1 传递函数的降幂排列标准形式

在 MATLAB 中，传递函数由分子和分母系数构成的两个向量唯一地确定。分子系数向量由 num（numerator，分子）表示，分母系数向量由 den（denominator，分母）表示，利用 MATLAB 工具箱函数 tf() 能够表示传递函数 G。

num $=[b_1,b_2,\cdots,b_m,b_{m+1}]$；

den $=[a_1,a_2,\cdots,a_n,a_{n+1}]$；（注意：各系数都是按 s 的降幂进行排列的）

G=tf (num, den)。

函数 tf()的用法及说明见表 3.1 和表 3.2。

表 3.1 函数 tf()的用法及说明

函 数 用 法	说 明
Gtf = tf(num,den)	返回变量 G_{tf} 为连续系统传递函数模型
Gtfz = tf(num,den,Ts)	返回变量 G_{tfz} 为离散系统传递函数模型。T_s 为采样周期，当 T_s=−1 或 T_s=[]时，表示系统采样周期未定义
Gtf = tf('s')	定义 s 变换算子，以原形式输入传递函数
Gtfz = tf('z',TS)	定义 z 变换算子及采样时间 T_s，以原形式输入传递函数

求取系统传递函数相关的函数用法及其说明见表 3.2。

表 3.2 求取系统传递函数相关的函数用法及说明

函 数 用 法	说 明
printsys(num,den,'s')	将系统传递函数以分式的形式打印出来，'s' 表示传递函数变量
printsys(num,den,'z')	将系统传递函数以分式的形式打印出来，'z' 表示传递函数变量

续表

函数用法	说　明
get(sys)	可获得传递函数模型对象 sys 的所有信息
set(sys,'Property',Value,⋯)	为系统不同属性设定值
[num,den]= tfdata(sys,'v')	以行向量的形式返回传递函数的分子分母多项式
C = conv(A,B)	多项式 A 和 B 的系数以行向量表示，进行相乘，结果 C 仍以行向量表示

此外，MATLAB 还支持另一种传递函数描述格式，用 s = tf('s')先定义传递函数的算子，然后，将类似的数学表达式直接输入系统的传递函数模型。

例 3.2　请将传递函数模型 $G(s)=\dfrac{C(s)}{R(s)}=\dfrac{12s^3+24s^2+12s+20}{2s^4+4s^3+6s^2+2s+2}$ 输入到 MATLAB 中。

例 3.2 的解见图 3.3.1。

```
%例 3.2 的方法一：
num=[12 24 12 20];
den=[2 4 6 2 2];
G=tf(num,den)
```

```
%例 3.2 的方法一结果：
Transfer function:
   12 s^3 + 24 s^2 + 12 s + 20
-------------------------------
2 s^4 + 4 s^3 + 6 s^2 + 2 s + 2
```

```
%例 3.2 的方法二：
s=tf('s');
G=(12*s^3+24*s^2+12*s+20)/(2*s^4+4*s^3+6*s^2+2*s+2)
```

```
%例 3.2 的方法二结果：
Transfer function:
   12 s^3 + 24 s^2 + 12 s + 20
-------------------------------
2 s^4 + 4 s^3 + 6 s^2 + 2 s + 2
```

图 3.3.1　例 3.2 的程序及其结果

例 3.3　请将传递函数模型 $G(s)=\dfrac{3(s^2+3)}{(s+2)^3(s^2+2s+1)(s^2+5)}$ 输入到 MATLAB 中。

例 3.3 的解见图 3.3.2。

```
%例 3.3 的方法一：
num=conv(3,[1,0,3]);
den1=conv([1,2],conv([1,2],[1,2]));
den2=conv([1,2,1],[1,0,5]);
den=conv(den1,den2);
G=tf(num,den)
```

图 3.3.2　例 3.3 的程序及其结果

```
%例 3.3 的方法一结果：
Transfer function:
                    3 s^2 + 9
--------------------------------------------------------------
s^7 + 8 s^6 + 30 s^5 + 78 s^4 + 153 s^3 + 198 s^2 + 140 s + 40
```

```
%例 3.3 的方法二：
s=tf('s');
G=3*(s^2+3)/((s+2)^3*(s^2+2*s+1)*(s^2+5))
```

```
%例 3.3 的方法二结果：
Transfer function:
                    3 s^2 + 9
--------------------------------------------------------------
s^7 + 8 s^6 + 30 s^5 + 78 s^4 + 153 s^3 + 198 s^2 + 140 s + 40
```

图 3.3.2 例 3.3 的程序及其结果（续）

由例 3.3 可见，当传递函数不是以降幂排列的标准形式给出时，方法一需要将其整理为标准形式，此时，可借助 conv 函数完成多个多项式的相乘。方法二则无须转换为标准形式，在处理非标准形式的传递函数时更为方便。

3.3.2 传递函数的零极点模型

式（3.6）中的“首 1 标准型”传递数函 G，在 MATLAB 中可由采用零极点模型的矩阵 $[\boldsymbol{z},\boldsymbol{p},\boldsymbol{K}]$ 表示，即

$$\boldsymbol{z}=[z_1;z_2;\cdots z_m]$$

$$\boldsymbol{p}=[p_1;p_2;\cdots p_n]$$

$$\boldsymbol{K}=[k]$$

利用函数 zpk()能够表示传递函数 G。

例 3.4 请将函数 $G(s)=\dfrac{s(s+5)(s+6)}{(s+1)(s+2)(s+3-4\text{j})(s+3+4\text{j})}$ 输入到 MATLAB 中。

例 3.4 的解见图 3.3.3。

```
%例 3.4 的方法一：
z=[0;-5;-6];
p=[-1;-2;-3+4*j;-3-4*j];
k=[1];
G=zpk(z,p,k)
```

```
%例 3.4 的方法一结果：
Zero/pole/gain:
        s (s+5) (s+6)
---------------------------
(s+1) (s+2) (s^2 + 6s + 25)
```

图 3.3.3 例 3.4 的程序及其结果

```
%例 3.4 的方法二：
s=zpk('s');
G=s*(s+5)*(s+6)/((s+1)*(s+2)*(s^2 + 6*s + 25))
```

```
%例 3.4 的方法二结果：
Zero/pole/gain:
        s (s+5) (s+6)
---------------------------
(s+1) (s+2) (s^2 + 6s + 25)
```

图 3.3.3 例 3.4 的程序及其结果（续）

在例 3.4 中，方法一直接将零极点向量和增益赋值给函数 zpk()，方法二是先定义拉普拉斯算子，再输入零极点模型。

方法二需要注意的是，当系统传递函数中存在复数零极点时，不能直接输入一阶复数因子相乘，而需要输入其乘积结果，即二阶多项式来表示这两个因式。

通常，在已知传递函数的情况下，往往希望求取系统的零极点和增益，以便直观、快速地判断系统稳定与否。利用函数 tf2zp()以及函数 zpkdata()可求取传递函数零极点和增益，详见表 3.3 和表 3.4。

表 3.3 zpk 函数的用法及说明

函 数 用 法	说 明
Gzpk = zpk(z,p,K)	得到连续系统的零极点模型
Gzpkz = zpk(z,p,K,Ts)	得到连续系统的零极点模型，采样时间为 T_s
Gzpk = zpk('s')	定义拉普拉斯变换算子，按原格式输入系统
Gtfz = tf('z',Ts)	定义 z 变换算子，采样时间为 T_s，按原格式输入系统

与零极点模型相关的函数见表 3.4。

表 3.4 零极点模型的相关函数用法及说明

函 数 用 法	说 明
[z,p,k]=zpkdata(sys,'v')	得到系统的零极点和增益，参数 'v' 以向量形式表示
[p,z]=pzmap(sys)	返回系统零极点
pzmap(sys)	得到系统零极点分布图

例 3.5 求取传函数 $G(s)=\dfrac{s^3+11s^2+30s}{s^4+9s^3+45s^2+87s+50}$ 的零极点和增益，并求系统的零极点模型。

例 3.5 的解见图 3.3.4 和图 3.3.5。

```
%例 3.5 的方法一：
num=[1,11,30,0];
den=[1,9,45,87,50];
G=tf(num,den)
```

图 3.3.4 例 3.5 的程序 1 及其结果 1

```
% G=tf(num,den)结果：
Transfer function:
        s^3 + 11 s^2 + 30 s
--------------------------------
s^4 + 9 s^3 + 45 s^2 + 87 s + 50
```

```
%输入：
[z,p,k]=tf2zp(num,den)
```

```
%[z,p,k] = tf2zp(num,den)的结果：
z =
         0
   -6.0000
   -5.0000


p =
  -3.0000 + 4.0000i
  -3.0000 - 4.0000i
  -2.0000
  -1.0000


k =
     1
```

```
%输入：
G
```

```
%G 的结果：
Transfer function:
        s^3 + 11 s^2 + 30 s
--------------------------------
s^4 + 9 s^3 + 45 s^2 + 87 s + 50
```

```
%输入：
Gzpk = zpk(G) %将 G 变为零极点型，注意该句也可以写成 zpk(z,p,k)
```

```
%Gzpk = zpk(G)的结果：
Zero/pole/gain:
          s (s+6) (s+5)
---------------------------
(s+2) (s+1) (s^2 + 6s + 25)
```

图 3.3.4　例 3.5 的程序 1 及其结果 1（续）

该示例有助于理解函数的使用方法，程序使用了函数 tf2zpk()求取系统的零极点和增益。需要强调以下几点：

1．tf2zpk()分子分母的输入只能是向量形式，而不能是传递函数变量 G 的形式；

2．tf2zpk()只完成系统零极点和增益的求取，而不改变系统传递函数的形式；

3．zpk()的输入可以是降幂排列标准形式的传递函数变量 G，也可以是零极点变量 z,p,k。

```
%例 3.5 的方法二：
G=tf([1,11,30,0], [1,9,45,87,50])
```

```
%上句输出：
Transfer function:
          s^3 + 11 s^2 + 30 s
--------------------------------
s^4 + 9 s^3 + 45 s^2 + 87 s + 50
```

```
%输入：
[z,p,k]=zpkdata(G,'v')
```

```
% 上句输出：
z =
              0
     -6.0000
     -5.0000


p =
   -3.0000 + 4.0000i
   -3.0000 - 4.0000i
   -2.0000
   -1.0000


k =
     1
```

```
%输入：
G
```

```
% 上句输出：
Transfer function:
          s^3 + 11 s^2 + 30 s
--------------------------------
s^4 + 9 s^3 + 45 s^2 + 87 s + 50
```

```
%输入：
Gzpk=zpk(z,p,k)
```

```
%上句输出：
Zero/pole/gain:
           s (s+6) (s+5)
---------------------------
(s+2) (s+1) (s^2 + 6s + 25)
```

```
%输入：
pzmap(G)
```

图 3.3.5 例 3.5 的程序 2 及其结果 2

系统的零极点分布结果如图 3.3.6（a）所示。

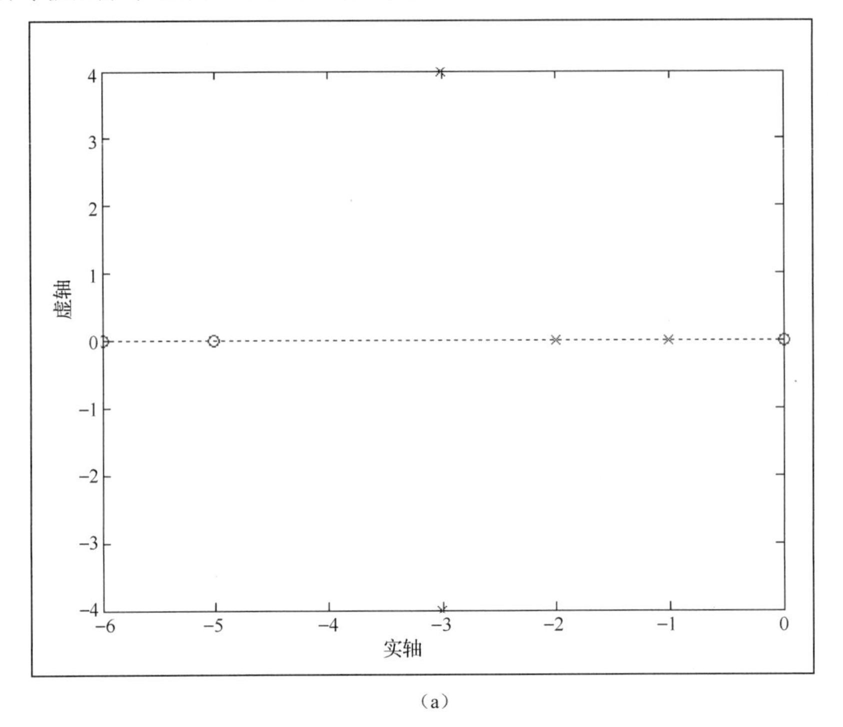

（a）

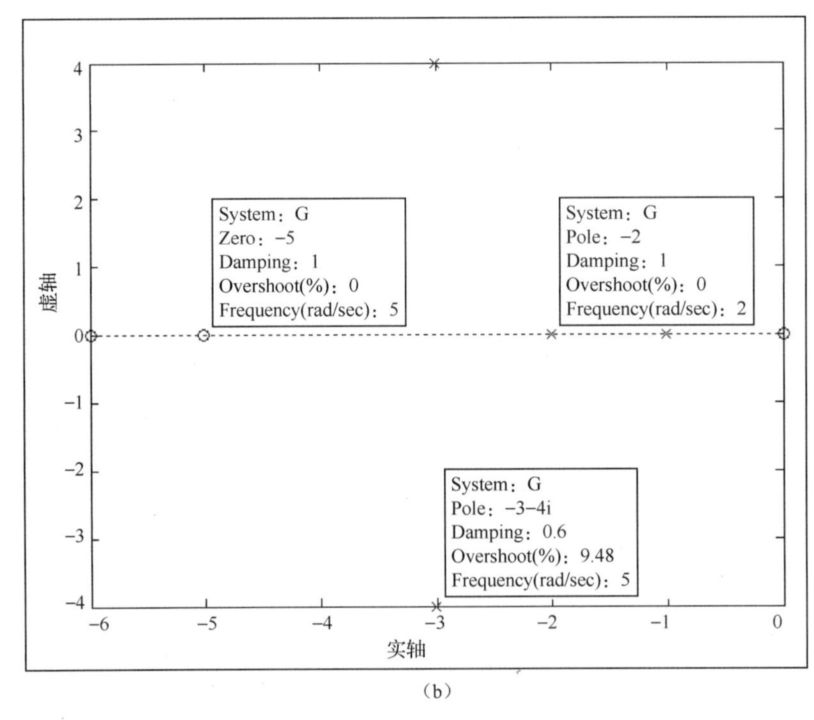

（b）

图 3.3.6 例 3.5 零极点分布结果

方法二使用了函数 zpkdata()来求取传递函数的零极点和增益。与方法一不同的是，函数 zpkdata()的输入只能是传递函数变量 G，而不能是零极点变量函数。函数 zpkdata()需指定参数“v”以确保得到向量形式的零极点和增益。与方法一相同的是，函数 zpkdata()也不改变的传递函数形式。

最后，函数 pzmap()完成了系统零极点分布图的绘制。如图 3.3.6（a）所示，在图上单击各零极点时，将显示其属性及属性值。

若将上述程序改写为如图 3.3.7 所示，即当 pzmap()具有返回值时，它不再绘制零极点分布图，而是返回零极点值。

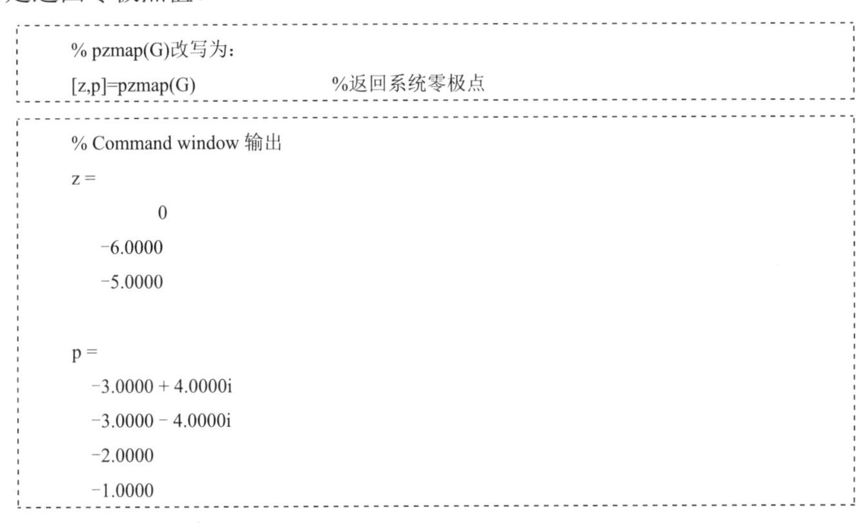

```
% pzmap(G)改写为:
[z,p]=pzmap(G)                    %返回系统零极点
```

```
% Command window 输出
z =
         0
   -6.0000
   -5.0000

p =
  -3.0000 + 4.0000i
  -3.0000 - 4.0000i
  -2.0000
  -1.0000
```

图 3.3.7　例 3.5 的程序 3 及其结果 3

3.3.3　传递函数的部分分式展开

若已知系统总的传递函数模型，想求取系统模型中包含哪些基本的因式，就要对系统传递函数进行部分分式展开，将其分解为基本因式的形式；相反，若已知系统模型中包含哪些基本因式，想求取系统总的传递函数模型，则可以通过将部分分式展开式转换为 s 降幂排列标准形式来实现。

利用函数 residue()能够实现多项式的部分分式展开。

[r,p,k]=residue(b,a)可对公式（3.3）中的传递函数进行部分分式展开。输入变量为向量 b 和 a，其分别为按 s 的降幂排列的分子分母多项式系数。输出变量 r 为余项构成的列向量，p 为极点构成的列向量，k 为余项构成的行向量。

[b,a]=residue(r,p,k)可以将部分分式展开转化为 s 降幂排列标准形式。

例 3.6　已知传递函数 $G(s)=\dfrac{2s^3+9s+1}{s^3+s^2+4s+4}$，求其部分分式展开形式。

例 3.6 的解见图 3.3.8。

```
%例 3.6 程序:
num=[2,0,9,1];
den=[1,1,4,4];
[r,p,k]=residue(num,den)
```

```
%% Command window 输出
r =
   -0.0000 - 0.2500i
   -0.0000 + 0.2500i
   -2.0000

p =
    0.0000 + 2.0000i
    0.0000 - 2.0000i
   -1.0000

k =
     2
```

图 3.3.8 MATLAB 程序和例 3.6 的结果

其结果可表示为 $G(s)=2+\frac{-0.25i}{s-2i}+\frac{0.25i}{s+2i}+\frac{-2}{s+1}$。

例 3.7 已知 $G(s)=\frac{-1.4}{s-1.5}-\frac{0.6}{s+1.1}+\frac{1.3}{s+0.4}-1.2$，求 $G(s)$的降幂排列标准形式。

例 3.7 的解见图 3.3.9。

```
%例 3.7 程序：
r =[-1.4,-0.6,1.3]';
p =[1.5,-1.1,-0.4]';
k =[-1.2];
[b,a] = residue(r,p,k)
```

```
% Command window 输出
b =
-1.2000    -0.7000    0.2120    -1.6090

a =

    1.0000    0.0000    -1.8100    -0.6600
```

图 3.3.9 MATLAB 程序和例 3.7 的结果

其结果可表示为 $G(s)=\frac{-1.2s^3-0.7s^2+0.212s-1.609}{s^3-1.81s-0.66}$。

3.4 结构图的连接与化简

实际控制系统通常包括被控对象、控制器和反馈等多个环节。每一个环节有其各自的数学模型，譬如传递函数模型。因此，整个控制系统由不同的单元连接而成。连接的主要形式有串联、并联和反馈等。

对于结构复杂的系统，可以通过结构图的化简，或者利用梅森公式来求得系统的闭环传递函数。表 3.5 给出了控制系统结构的基本连接形式和等效结构图。

表 3.5　结构图简化规则

原方框图	等效方框图	等效运算关系
R → $G_1(s)$ → $G_2(s)$ → C	R → $G_1(s)G_2(s)$ → C	（1）串联等效 $C(s)=G_1(s)G_2(s)R(s)$
R → $G_1(s)$、$G_1(s)$ 并联 ± → C	R → $G_1(s)\pm G_2(s)$ → C	（2）并联等效 $C(s)=\left[G_1(s)\pm G_2(s)\right]R(s)$
R ± → $G_1(s)$ → C，反馈 $G_1(s)$	R → $\frac{G_1(s)}{1\mp G_1(s)G_2}$ → C	（3）反馈等效 $C(s)=\dfrac{G_1(s)R(s)}{1\mp G_1(s)G_2(s)}$

3.4.1　相关函数

MATLAB 提供了系统模型连接化简的不同函数，其中主要函数及功能说明见表 3.6。

表 3.6　相关函数及功能说明

函　　数	功 能 说 明
sys = parallel(sys1, sys2) sys = parallel(sys1, sys2, inp1, inp2, out1, out2)	并联两个系统，等效于 sys=sys1+sys2，对 MIMO 系统，表示 sys1 的输入 inp1 与 sys2 的输入 inp2 相连，sys1 的输出 out1 与 sys2 的输出 out2 相连
sys = series(sys1, sys2) sys = series(sys1, sys2, inp1, inp2, out1, out2)	串联两个系统，等效于 sys=sys1*sys2，对 MIMO 系统，表示 sys1 的输出 outputs1 与 sys2 的输入 outputs2 相连
sys = feedback(sys1, sys2) sys = feedback(sys1, sys2, sign) sys = feedback(sys1, sys2, feedin, feedout, sign)	两系统负反馈连接，默认格式 sign=−1 表示负反馈，sign=1 表示正反馈，等效于 sys=sys1/(1±sys1*sys2) 对 MIMO 系统，部分反馈连接是指，sys1 的指定输出 feedout 连接到 sys2 的输入，而 sys2 的输出连接到 sys1 的指定输入 feedin，以此构成闭环系统
[numc, denc] = cloop(num, den, sign)	指由传递函数形式表示的开环系统构成闭环系统

3.4.2　连接与化简实例

例 3.8　已知系统的 $G_1(s)=\dfrac{1}{s^2+4s+4}$，$G_2(s)=\dfrac{1}{s+5}$，求 $G_1(s)$ 和 $G_2(s)$ 分别串联、并联、负反馈和闭环连接后的系统模型。

例 3.8 的解如图 3.4.1 所示。

```
%例 3.8 串联方式一：
clc;
clear all;
num1=[1];
den1=[2,4,4];
num2=[1];
den2=[1,5];
[num,den]=series(num1,den1,num2,den2);
G=tf(num,den)
```

图 3.4.1　MATLAB 程序 1 和例 3.8 的结果 1

```
%例 3.8 串联方式一输出结果：
Transfer function:
                1
--------------------------
2 s^3 + 14 s^2 + 24 s + 20
```

```
%例 3.8 串联方式二：
clc;
clear all;
num1=[1];
den1=[2,4,4];
G1=tf(num1,den1);
num2=[1];
den2=[1,5];
G2=tf(num2,den2);
G=G1*G2
```

```
%例 3.8 串联方式二输出结果：
Transfer function:
                1
--------------------------
2 s^3 + 14 s^2 + 24 s + 20
```

```
%例 3.8 串联方式三：
clc;
clear all;
num1=[1];
den1=[2,4,4];
G1=tf(num1,den1);
num2=[1];
den2=[1,5];
G2=tf(num2,den2);
G=series(G1,G2)
```

```
%例 3.8 串联方式三输出结果：
Transfer function:
                1
--------------------------
2 s^3 + 14 s^2 + 24 s + 20
```

图 3.4.1 MATLAB 程序 1 和例 3.8 的结果 1（续）

同理，并联也可通过上述三种方式完成。关键语句如图 3.4.2 所示。

```
%例 3.8 并联方式一：
[num,den]=parallel(num1,den1,num2,den2);
G=tf(num,den);
%并联方式二：
G=G1+G2;
%并联方式三
G=parallel(G1,G2);
```

图 3.4.2 MATLAB 程序 2 和例 3.8 的结果 2

```
%例 3.8 并联方式输出结果:
Transfer function:
        2 s^2 + 5 s + 9
--------------------------
2 s^3 + 14 s^2 + 24 s + 20
```

图 3.4.2　MATLAB 程序 2 和例 3.8 的结果 2（续）

负反馈方式表达如图 3.4.3 所示。

```
%例 3.8 负反馈方式一:
G=feedback(G1,G2) %注意：G1 在前向通道，G2 在反馈回路
%负反馈方式二:
G=G1/（1+G1*G2）
```

```
%例 3.8 负反馈方式输出结果:
Transfer function:
              s + 5
--------------------------
2 s^3 + 14 s^2 + 24 s + 21
```

```
%例 3.8 闭环连接方式:
clc;
clear all;
num1=[1];
den1=[2,4,4];
num2=[1];
den2=[1,5];
num=conv(num1,num2);
den=conv(den1,den2);
[num1,den1]=cloop(num,den);%注：由 G1 和 G2 串联构成开环传递函数，以此构造单位负反馈
                            %闭环系统
G=tf(num1,den1)
```

```
%例 3.8 闭环连接方式输出结果:
Transfer function:
                1
--------------------------
2 s^3 + 14 s^2 + 24 s + 21
```

图 3.4.3　MATLAB 程序 3 和例 3.8 的结果 3

3.5　建模与仿真实例

本节给出两个综合建模仿真实例——RLC 电路及机械位移系统。分别建立两个传递函数模型，并对其进行简单的时域分析。

3.5.1 RLC电路的传递函数建模

可以用几种方法来设计系统数学模型，选择最佳形式的数学模型可以使仿真执行更快、更准确。例如，考虑一个RLC电路，如图3.5.1所示。

$u_i(t)$为输入电压，

L为电感（H），

R为电阻（Ω），

C为电容（F），

$u_o(t)$为输出电压（V），

i为通过电阻的电流（A），

i_1为通过电感的电流（A），

i_2为通过电容的电流（A），

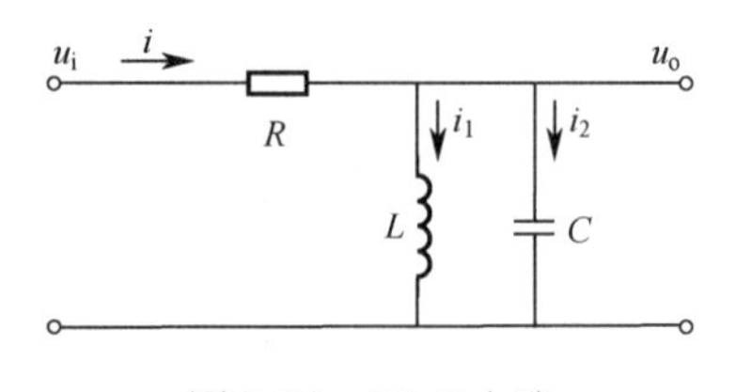

图3.5.1 RLC电路

使用基尔霍夫电路定律，得到

$$Ri(t)+u_o(t)=u_i(t) \tag{3.8}$$

$$i_1(t)+i_2(t)=i(t) \tag{3.9}$$

其中

$$u_o(t)=\frac{1}{C}\int i_2(t)\mathrm{d}t=L\frac{\mathrm{d}i_1(t)}{\mathrm{d}t} \tag{3.10}$$

利用复阻抗，在零初始条件下，对式（3.8）、式（3.9）和式（3.10）进行拉普拉斯变换，有

$$R\cdot I(s)+U_O(s)=U_I(s) \tag{3.11}$$

$$I_1(s)+I_2(s)=I(s) \tag{3.12}$$

$$U_O(s)=\frac{1}{Cs}I_2(s)=LsI_1(s) \tag{3.13}$$

则

$$I_1(s)=\frac{U_O(s)}{Ls}\quad ,\quad I_2(s)=Cs\cdot U_O(s) \tag{3.14}$$

将式（3.14）代入式（3.11），替换掉中间变量$I(s)$得到

$$R\left(\frac{U_O(s)}{Ls}+Cs\cdot U_O(s)\right)+U_O(s)=U_I(s) \tag{3.15}$$

从输入$U_I(s)$到输出$U_O(s)$的传递函数为

$$G(s)=\frac{U_O(s)}{U_I(s)}=\frac{s/(RC)}{s^2+s/(RC)+1/(LC)} \tag{3.16}$$

至此，RLC电路的传递函数模型建立完毕，下一节将用MATLAB进行系统时域仿真分析。

3.5.2 RLC电路的仿真分析

当输入是单位阶跃即输入电压$u_i(t)=1\text{V}$时，稳态输出电压$u_o(t)=0\text{V}$，RLC电路的阶跃响应如图3.5.2所示，程序如图3.5.3所示。

从图 3.5.1 的 RLC 电路可知，电感和电容是并联，电感与电容两端的电压相同，系统输出电压就是这个电压。电压开始从零上升，在大约 0.5s 附近上升到最高点 0.45V，然后开始下降。经过微弱的振荡后，电压在零伏趋于平稳。

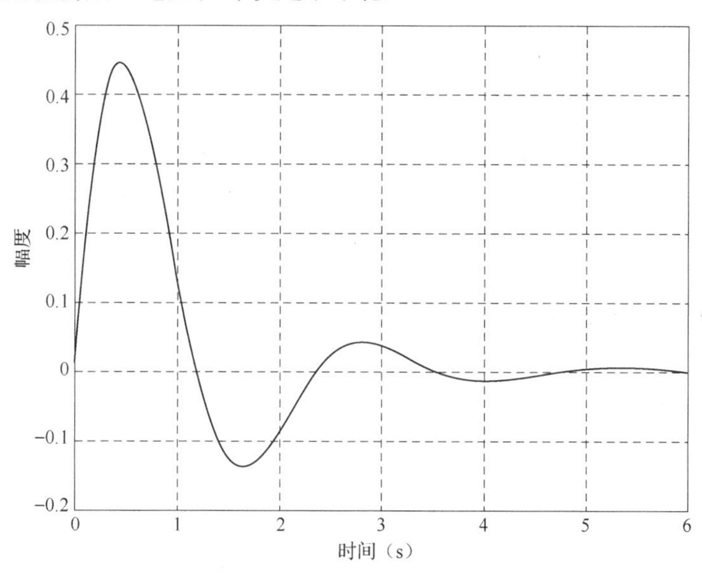

图 3.5.2　RLC 电路阶跃响应

```
%图 3.5.2 的程序
%模型参数
R=1;            %单位Ω
L=0.25;         %单位 H
C=0.5;          %单位 F
%传递函数模型
num=[1/(R*C),0];
den=[1,1/(R*C),1/(L*C)];
G=tf(num,den);
%仿真后的阶跃响应
step(G)
grid
```

图 3.5.3　MATLAB 程序

3.5.3　机械位移系统的传递函数建模

在图 3.5.4 所示的机械位移系统中，假设系统是线性的，外力 $u(t)$ 是系统的输入量，质量的位移 $y(t)$ 是系统的输出量。位移 $y(t)$ 从无外力作用时的平衡位置开始计算。该系统是一个单输入单输出系统。其中，k 为弹簧系数，m 为木块质量，b 为阻尼系数，$u(t)$ 为系统输入量，$y(t)$ 为系统输出量。

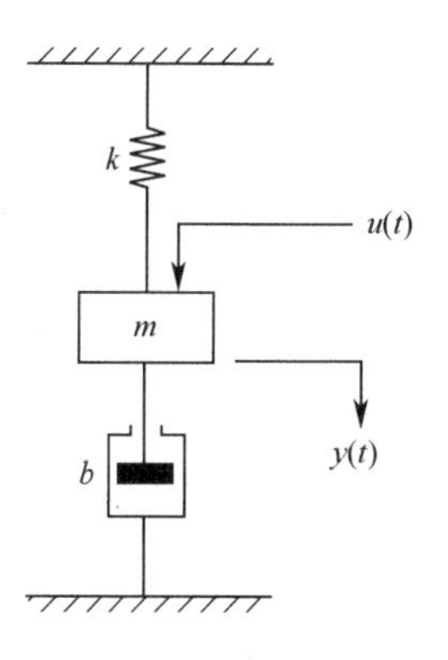

图 3.5.4　机械位移系统

阻尼器的阻力为

$$u_1(t) = b\frac{\mathrm{d}y(t)}{\mathrm{d}t}$$

弹簧弹力为

$$u_2(t) = ky(t)$$

根据牛顿第二定律

$$m\frac{\mathrm{d}^2 y(t)}{\mathrm{d}t^2} = u(t) - u_1(t) - u_2(t) \tag{3.17}$$

整理得

$$m\frac{\mathrm{d}^2 y(t)}{\mathrm{d}t^2} + b\frac{\mathrm{d}y(t)}{\mathrm{d}t} + ky(t) = u(t) \tag{3.18}$$

在零初始条件下，对式（3.18）做拉普拉斯变换，得到机械位移系统的传递函数为

$$G(s) = \frac{Y(s)}{U(s)} = \frac{1}{ms^2 + bs + k} = \frac{1/m}{s^2 + (b/m)s + k/m} \tag{3.19}$$

3.5.4 机械位移系统的仿真分析

用 MATLAB 程序仿真机械位移系统的阶跃响应如图 3.5.5 所示。程序见图 3.5.6。

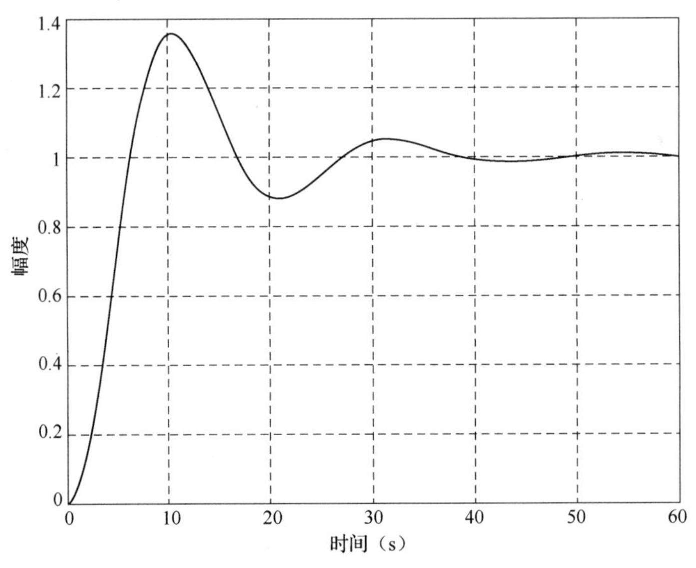

图 3.5.5 机械位移系统的阶跃响应

```
%图 3.5.5 的 MATLAB 程序：
%模型参数
m=10;
k=1;
b=2;
%传递函数模型
num=[1/m];
den=[1,b/m,k/m];
G=tf(num,den);
%仿真后的阶跃响应
step(G)
grid
```

图 3.5.6 MATLAB 程序

取 $m = 10\text{kg}$，$k = 1\text{N/m}$，$b = 2$。输入是单位阶跃，即输入外力 $u(t) = 1\text{N}$ 时，稳态输出位移是 $y(t) = 1\text{m}$。

从图 3.5.5 看出，当输入外力为单位阶跃时，系统的输出是质量块的位移。位移开始从零变化到峰值，再缓慢下降，最后逐渐在 1m 的稳定值趋于平稳。系统超调是小于 40%的，系统调节时间大约为 45s。

本章小结

（1）本章主要介绍了三种传递函数模型的 MATLAB 描述；

（2）运用实例说明了常用三种连接方式的 MATLAB 实现；

（3）通过两个实例，说明了简单系统的建模过程。

Chapter 3
Transfer Function Model of Control System

This chapter introduces how to use MATLAB to describe the mathematical model of linear, time-invariant systems. It lays a foundation for the subsequent time domain steady-state analysis, steady-state error analysis, root locus analysis and frequency domain analysis of the linear time-invariant system.

The main mathematical models of linear, time-invariant systems are differential equations in the time-domain and transfer function models in complex-domain. The MATLAB implementation of the differential equation model has been introduced in Chapter 2. The present chapter mainly focuses on the MATLAB description of transfer function models, the transformation between models, and the simulation analysis of system performance.

3.1 Introduction to Linear Time-Invariant Systems

If the input and output differential equation of system can be described in the following linear form

$$a_0c^{(n)}(t)+a_1c^{(n-1)}(t)+\cdots+a_{n-1}c'(t)+a_nc(t)=b_0r^{(m)}(t)+b_1r^{(m-1)}(t)+\cdots+b_{m-1}r'(t)+b_mr(t), n\geqslant m \tag{3.1}$$

In the formula, $c(t)$ is the controlled variable, $r(t)$ is the system's input, and the coefficient $a_0, a_1, \cdots, a_n, b_0, b_1, \cdots, b_n$ are constants. This is called linear time-invariant systems.

A linear system satisfies the principle of superposition. The principle of superposition states that the response produced by the simultaneous application of two different forcing functions is the sum of the two individual responses. Hence, for the linear system, the response to several inputs can be calculated by treating one input at a time and adding the results. It is this principle that allows us to build up complicated solutions to the linear differential equation from simple solutions. In an experimental investigation of a dynamic system, if cause and effect are proportional, thus implying that the principle of superposition holds, then the system can be considered linear.

A differential equation is linear if the coefficients are constants or functions only of the independent variable. Dynamic systems that are composed of linear time-invariant lumped-parameter components may be described by linear time-invariant differential equations—that is, constant-coefficient differential equations. Such systems are called linear time-invariant (or linear constant-coefficient) systems. Systems that are represented by differential equations whose coefficients are functions of time are called linear time-varying systems. An example of a time-varying control system is a spacecraft control system. (The mass of a spacecraft changes due to fuel consumption.)

3.2 Transfer Function of Linear Time-Invariant Systems

3.2.1 Introduction

The differential equation of the linear time-invariant system describes the change of the system's input and output due to time *t*, and is a mathematical model describing the dynamic performance of the system. In the second chapter, we can see that the computer can solve differential equations quickly and complete system analysis.

However, there are inconveniences of using differential equations in the process of analysis, modification and design of control systems. For example, when the component parameters or system structure changes, it is necessary to rewrite the differential equation, and re-simulate or measure the input and output signal of system to evaluate system performance. In addition, it is difficult to analyze the influence of system parameters and structure on performance only from the observation and evaluation of input and output signals, which is unfavorable for learning or developing control theory.

In control theory, another mathematical model is often used to describe the input and output relationship of the linear time-invariant system, which is called transfer function model. By applying the Laplace transform of the linear differential equations in the time-domain when the initial condition is *zero*, we get the transfer function model of the complex-domain, so that the input and output dynamic differential equation model is transformed into an algebraic model, which is more conducive to the analysis of system parameters and the impact of structure on performance.

3.2.2 Definition

The transfer function of linear time-invariant systems is defined as the ratio of the Laplace transform of the output to the Laplace transform of the input under the assumption that all initial conditions are *zero*.

Consider the linear time-invariant system defined by the following differential equation

$$a_0c^{(n)}+a_1c^{(n-1)}+\cdots+a_{n-1}c'+a_nc=b_0r^{(m)}+b_1r^{(m-1)}+\cdots+b_{m-1}r'+b_mr\ (n\geqslant m) \tag{3.2}$$

Where c is the output of the system and r is the input. The ratio of the Laplace transform of the output to the Laplace transform of the input when all initial conditions are zero is the transfer function $G(s)$ of the system

$$G(s)=\frac{C(s)}{R(s)}=\frac{b_0s^m+b_1s^{m-1}+\cdots+b_{m-1}s+b_m}{a_0s^n+a_1s^{n-1}+\cdots a_{n-1}s+a_n} \tag{3.3}$$

By using the concept of transfer function, it is possible to represent system dynamics by algebraic equations in variable s. If the highest power of s in the denominator of the transfer function is equal to n, the system is called an *nth-order* system.

3.2.3 Related Instructions

The applicability of the concept of the transfer function is limited to linear, time-invariant, systems. The transfer function approach, however, is extensively used in the analysis and design of such systems. In what follows, we shall list important instructions concerning the transfer function. (Note that a system referred to in the list is one described by a linear, time-invariant, differential equation.)

(1) The transfer function of a system is a mathematical model in that it is an operational method of expressing the differential equation that relates the output variable to the input variable.

(2) The transfer function is a property of a system itself, independent of the magnitude and nature of the input or driving function.

(3) The transfer function includes the units necessary to relate the input to the output. However, it does not provide any information concerning the physical structure of the system. (The transfer functions of many physically different systems can be identical.)

(4) If the transfer function of a system is known, the output or response can be studied for various forms of inputs with a view toward understanding the nature of the system.

(5) If the transfer function of a system is unknown, it may be established experimentally by

introducing known inputs and studying the output of the system. Once established, a transfer function gives a full description of the dynamic characteristics of the system, as distinct from its physical description.

Example 3.1 Try to find the transfer function $U_o(s)/U_i(s)$ of the operational amplifier circuit as shown in Figure 3.2.1.

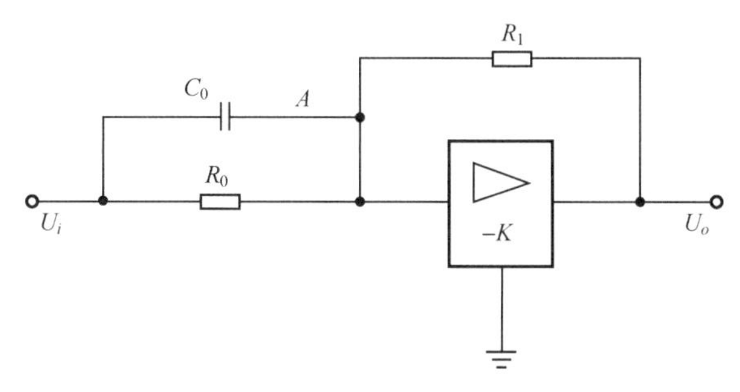

Figure 3.2.1 Operational Amplifier Circuit

Solution: Using complex impedance, the voltage equivalence at point A is

$$\frac{U_i(s)}{\dfrac{R_0 / C_0 s}{R_0 + 1/C_0 s}} = -\frac{U_o(s)}{R_1} \tag{3.4}$$

Hence the transfer function is

$$\frac{U_o(s)}{U_i(s)} = -R_1\left(C_0 s + \frac{1}{R_0}\right) \tag{3.5}$$

3.3 Transfer Function Description in MATLAB

For a linear, time-invariant system, the transfer function $G(s)$ given by Equation (3.3) can be organized into the following "First 1 Standard Type" and "End 1 Standard Type".

$$G(s) = \frac{b_0(s-z_1)(s-z_2)\cdots(s-z_m)}{a_0(s-p_1)(s-p_2)\cdots(s-p_n)} = K^* \frac{\prod_{i=1}^{m}(s-z_i)}{\prod_{j=1}^{n}(s-p_j)} \quad \text{First 1 Standard Type} \tag{3.6}$$

$$G(s) = \frac{b_m(\tau_1 s+1)(\tau_2^2 s^2 + 2\zeta\tau_2 s+1)\cdots(\tau_i s+1)}{a_n(T_1 s+1)(T_2^2 s^2 + 2\zeta T_2 s+1)\cdots(T_j s+1)} \quad \text{End 1 Standard Type} \tag{3.7}$$

The former shown in Equation (3.6) is also called the zero-pole model of the transfer function, which can intuitively show the number of poles and values of the system. For the closed-loop transfer function of the system, this can be directly used to analyze the stability of the system; for the open-loop transfer function, the root-locus method can be used to analyze the system performance.

The latter shown in Equation (3.7) also becomes a form of typical factor product of the transfer function. When the structure of the system is unknown, the open-loop transfer function can be sorted into this form, which can intuitively represent the typical model of the system, make a qualitative understanding of the overall performance of the system, and use the frequency analysis to quantitatively analyze the system's performance.

3.3.1 Standard Form in Descending Power of Transfer Functions

In MATLAB, the transfer function can be uniquely determined by two vectors consisting of molecular and denominator coefficients. The molecular coefficient vector is represented by *num*(numerator), the denominator coefficient vector is represented by *den*(denominator), and the MATLAB toolbox function tf() can be used to represent the transfer function G.

$num=[b_1,b_2,\cdots,b_m,b_{m+1}]$；

$den=[a_1,a_2,\cdots,a_n,a_{n+1}]$； (Note: The coefficients are all arranged in descending power order of s)

$G=tf(num,den)$.

The usage and instructions of the function *tf*() is shown in Table 3.1 and Table 3.2.

Table 3.1 Usage and Instructions of *tf*() Function

Usage	Instructions
Gtf = tf(num,den)	The return variable G_{tf} is a continuous system transfer function model.
Gtfz = tf(num,den,T_s)	The return variable G_{tfz} is a transfer function model of discrete system. T_s is the sampling period. When T_s=−1 or T_s=[], the system sampling period is not defined.
Gtf = tf('s')	Define the Laplace operator s and input the transfer function in its original form.
Gtfz = tf('z',T_s)	Define transform operator z and sampling time T_s, input transfer function in original form.

The usage and instructions of functions related to transfer functions of system is shown in Table 3.2.

Table 3.2 Usage and Instructions of Functions Related to Transfer Functions

Usage	Instructions
printsys(num,den,'s')	Print system transfer functions in fractional form, where 's' is the transfer function variables.
printsys(num,den,'z')	Print the system transfer function in fractional form, where 'z' is the transfer function variable.
get(sys)	Get all the information of the transfer function model object *sys*.
set(sys,'Property',Value,⋯)	Set values for different properties of system.
[num,den]= tfdata(sys,'v')	Return the transfer function molecular and denominator polynomial in the form of row vector.
C = conv(A,B)	The coefficients of Polynomials A and B are represented by row vectors and are multiplied. Result C is still represented by a row vector.

In addition, MATLAB also supports another description format of transfer function. The operator of the transfer function is defined by using s= tf('s'). Then, the transfer function model of the system is input into a similar mathematical expression directly.

Example 3.2 Please input the transfer function model $G(s)=\dfrac{C(s)}{R(s)}=\dfrac{12s^3+24s^2+12s+20}{2s^4+4s^3+6s^2+2s+2}$ into MATLAB.

The solution of Example 3.2 is shown in Figure 3.3.1.

```
%Method one of Example 3.2:
    num=[12 24 12 20];
    den=[2 4 6 2 2];
    G=tf(num,den)
```

```
%The results of Method one of Example 3.2:
Transfer function:
  12 s^3 + 24 s^2 + 12 s + 20
------------------------------
2 s^4 + 4 s^3 + 6 s^2 + 2 s + 2
```

```
%Method two of Example 3.2:
s=tf('s');
G=(12*s^3+24*s^2+12*s+20)/(2*s^4+4*s^3+6*s^2+2*s+2)
```

```
%The results of Method two of Example 3.2:
Transfer function:
  12 s^3 + 24 s^2 + 12 s + 20
------------------------------
2 s^4 + 4 s^3 + 6 s^2 + 2 s + 2
```

Figure 3.3.1 MATLAB Script and Results of Example 3.2

Example 3.3 Please input the transfer function model $G(s)=\dfrac{3(s^2+3)}{(s+2)^3(s^2+2s+1)(s^2+5)}$ into MATLAB.

The solution of Example 3.3 is shown in Figure 3.3.2.

```
%Method one of Example 3.3:
num=conv(3,[1,0,3]);
den1=conv([1,2],conv([1,2],[1,2]));
den2=conv([1,2,1],[1,0,5]);
den=conv(den1,den2);
G=tf(num,den)
```

```
%The results of method one of Example 3.3:
Transfer function:
                          3 s^2 + 9
-------------------------------------------------------------
s^7 + 8 s^6 + 30 s^5 + 78 s^4 + 153 s^3 + 198 s^2 + 140 s + 40
```

```
%Method two of Example 3.3:
s=tf('s');
G=3*(s^2+3)/((s+2)^3*(s^2+2*s+1)*(s^2+5))
```

```
%The results of method two of Example 3.3:
Transfer function:
                          3 s^2 + 9
-------------------------------------------------------------
s^7 + 8 s^6 + 30 s^5 + 78 s^4 + 153 s^3 + 198 s^2 + 140 s + 40
```

Figure 3.3.2 MATLAB Script and Results of Example 3.3

Example3.3 shows that when the transfer function is not given in the standard form of a descending power, Method One needs to be organized into a standard form. At this time, the multiplication of multiple polynomials can be accomplished by means of the *conv* function. Method Two does not need to be transformed into a standard form and is more convenient when dealing with non-standard transfer functions.

3.3.2 The Zero-Pole Model of Transfer Function

The transfer function G in First 1, Standard Type in Equation (3.6) can be represented by Matrix [$\boldsymbol{z}$, $\boldsymbol{p}$, $\boldsymbol{K}$] using a pole-zero gain model, as is shown as follows,

$\boldsymbol{z}=[z_1;z_2;\cdots z_m]$

$\boldsymbol{p}=[p_1;p_2;\cdots p_n]$

$\boldsymbol{K}=[k]$

The function zpk() can be used to represent the transfer function *G*.

Example 3.4 Please input $G(s)=\dfrac{s(s+5)(s+6)}{(s+1)(s+2)(s+3-4j)(s+3+4j)}$ into MATLAB.

The solution of Example 3.4 is shown in Figure 3.3.3.

```
%Method One of Example 3.4:
z=[0;-5;-6];
p=[-1;-2;-3+4*j;-3-4*j];
k=[1];
G=zpk(z,p,k)
```

```
%The results of method One of Example 3.4:
Zero/pole/gain:
        s (s+5) (s+6)
---------------------------
(s+1) (s+2) (s^2 + 6s + 25)
```

```
%Method Two of Example 3.4:
s=zpk('s');
G=s*(s+5)*(s+6)/((s+1)*(s+2)*(s^2 + 6*s + 25))
```

```
%The results of method Two of Example 3.4:
Zero/pole/gain:
        s (s+5) (s+6)
---------------------------
(s+1) (s+2) (s^2 + 6s + 25)
```

Figure 3.3.3 MATLAB Script and Results of Example 3.4

In Example 3.4, Method One assigns the zero-pole vector and gain to the function *zpk*() directly, and the second method defines the Laplace operator and then inputs zero-pole model.

Using the second method should be noted that when there are plural zero-poles in the system's transfer function, it is not possible to input the first-order plural factor multiplication directly.

Instead, it is necessary to input the result of multiplication, which is second-order polynomial, to represent the two factors.

Usually, in the case of known transfer functions, it is often desirable to determine the zeropole and gain of the system so that the system can be judged intuitively and quickly. Using the function *tf2zp*() and *zpkdata*(), the zeropole and gain of the transfer function can be determined. See Table 3.3 and Table 3.4.

Table 3.3 Usage and Instructions of *Zpk* Function

Usage	Instructions
Gzpk = zpk(z,p,K)	Obtain the zero-pole model for a continuous system.
Gzpkz = zpk(z,p,K,Ts)	Obtain a zero-pole model for a continuous system. The sampling time is T_s .
Gzpk = zpk('s')	Define Laplace operator and input into the system according to the original format.
Gtfz = tf('z',Ts)	Define z transform operator and the sampling time is T_s . Input it into system in the original format.

The function related to zero-pole model is shown in Table 3.4.

Table 3.4 Usage and Instructions of the Related Functions of Zero-Pole Model

Usage	Instructions
[z,p,k]=zpkdata(sys,'v')	Get the zero pole and gain of the system. The parameter 'v' is expressed in vector format.
[p,z] = pzmap(sys)	Return the system's zeropole.
pzmap(sys)	Get the system's zero-pole distribution.

Example 3.5 Find the zero pole and gain of the transfer function $G(s)=\dfrac{s^3+11s^2+30s}{s^4+9s^3+45s^2+87s+50}$, and Find the system's zero-pole model.

The solution of Example 3.5 is shown in Figure 3.3.4 and Figure 3.3.5.

```
%Method one of Example 3.5:
num=[1,11,30,0];
den=[1,9,45,87,50];
G=tf(num,den)
```

```
% G=tf(num,den)Result:
Transfer function:
        s^3 + 11 s^2 + 30 s
--------------------------------
s^4 + 9 s^3 + 45 s^2 + 87 s + 50
```

```
%Enter:
[z,p,k]=tf2zp(num,den)
```

Figure 3.3.4 MATLAB Script1 and Result 1 of Example 3.5

```
%Result of [z,p,k]=tf2zp(num,den):
z =
          0
    -6.0000
    -5.0000

p =
  -3.0000 + 4.0000i
  -3.0000 - 4.0000i
  -2.0000
  -1.0000

k =
     1
```

```
%Enter:
G
```

```
%Result of G:
Transfer function:
        s^3 + 11 s^2 + 30 s
--------------------------------
s^4 + 9 s^3 + 45 s^2 + 87 s + 50
```

```
% Enter:
Gzpk=zpk(G)
```

```
%Result of Gzpk=zpk(G):
Zero/pole/gain:
          s (s+6) (s+5)
---------------------------
(s+2) (s+1) (s^2 + 6s + 25)
```

Figure 3.3.4 MATLAB Script1 and Result 1 of Example 3.5 (continued)

The example helps understand how to use the function. The program uses the function *tf2zpk*() to find the zero pole and gain of the system. We need to emphasize the following points:

1. The input of the molecular and denominator of *tf2zpk*() can only be in vector form, rather than in the form of transfer function variable *G*;

2. *tf2zpk*() only gets the zero pole and gain of the system and does not change the form of the system transfer function;

3. The input of *zpk*() may be a variable *G* in the standard form of descending power, or a zero-pole variables z,p,k.

```
%Method two of Example 3.5:
G=tf([1,11,30,0], [1,9,45,87,50])
```

```
%The output of the previous sentence:
Transfer function:
        s^3 + 11 s^2 + 30 s
--------------------------------
s^4 + 9 s^3 + 45 s^2 + 87 s + 50
```

```
%Enter:
[z,p,k]=zpkdata(G,'v')
```

```
% The output of the previous sentence:
z =
        0
  -6.0000
  -5.0000

p =
 -3.0000 + 4.0000i
 -3.0000 - 4.0000i
 -2.0000
 -1.0000

k =
  1
```

```
% Enter:
G
```

```
% The output of the previous sentence:
Transfer function:
        s^3 + 11 s^2 + 30 s
--------------------------------
s^4 + 9 s^3 + 45 s^2 + 87 s + 50
```

```
% Enter:
Gzpk=zpk(z,p,k)
```

```
%The output of the previous sentence:
Zero/pole/gain:
        s (s+6) (s+5)
---------------------------
(s+2) (s+1) (s^2 + 6s + 25)
```

```
% Enter:
pzmap(G)
```

Figure 3.3.5 The MATLAB Script 2 and Result 2 of Example 3.5

The zero-pole distribution results of the system are shown in Figure 3.3.6(a).

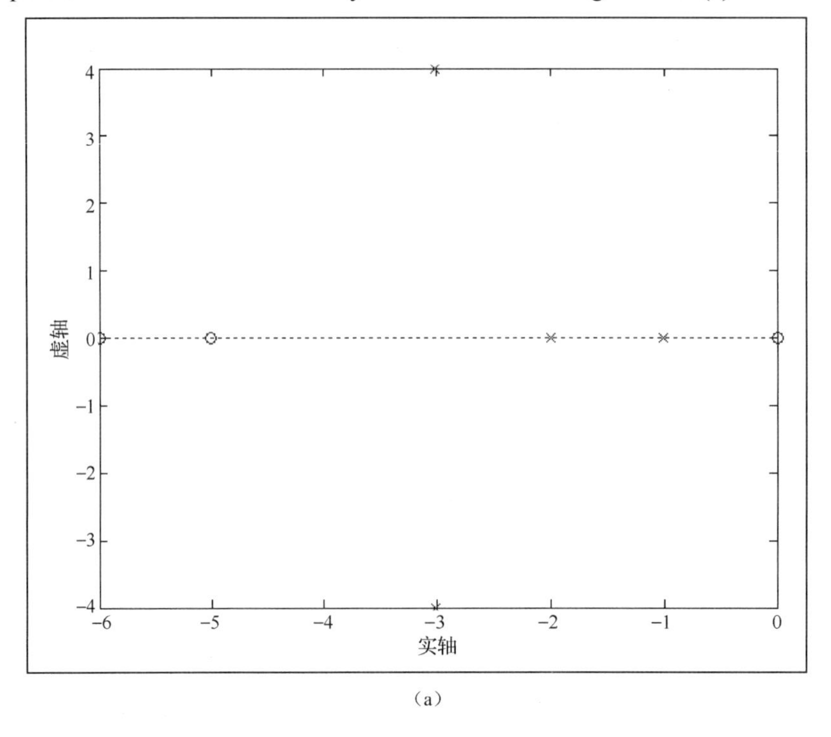

(a)

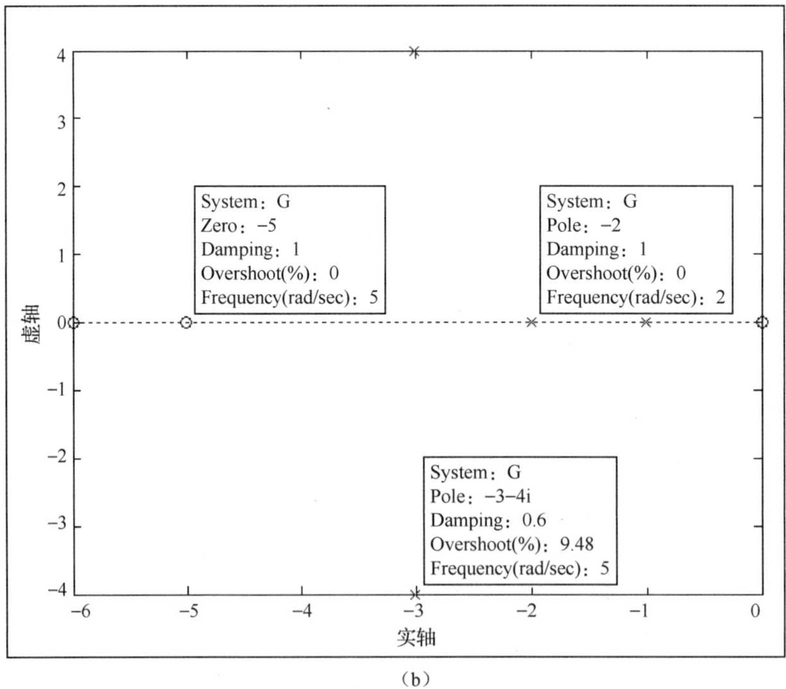

(b)

Figure 3.3.6 Zero-Pole Distribution of Example 3.5

This second method uses the function *zpkdata*() to obtain the zeropole and gain of the transfer function. Different from the first method, the input of *zpkdata*() can only be the transfer function variable *G*, not the zero-pole gain variable. *Zpkdata*() needs to specify the parameter “v” to ensure the zeropole and gain in vector form. Similar to the first method, *zpkdata*() does not change the form of transfer function.

Finally, the function *pzmap*() completes the drawing of the zero-pole distribution map of the system. As shown in Figure 3.3.6(a), when the zero-poles are clicked on the graph, their attributes and attribute values will be displayed.

The above program is rewritten as shown in Figure 3.3.7. When the function *pzmap*() has a return value, it no longer plots the pole-zero map, but only returns the pole-zero value.

```
% pzmap(G) Rewrite as:
[z,p]=pzmap(G)                % Return to system zero-pole
```

```
%results of Command window
 z =
         0
   -6.0000
   -5.0000

p =
  -3.0000 + 4.0000i
  -3.0000 - 4.0000i
  -2.0000
  -1.0000
```

Figure 3.3.7 MATLAB Script 3 and Result 3 of Example 3.5

3.3.3 Partial-Fraction Expansion of Transfer Function

If the total transfer function model of the system is known and we want to know which basic factors are included in the system model, then we must carry out partial fractional expansion of the system transfer function, and factorize it into the form of some basic factors. On the contrary, if the basic factors included in the system model are known and we want to get the overall transfer function model of the system we can implement this by converting the partial-fraction expansion to the standard form of s in descending power.

Using the function *residue*() can get the partial-fraction expansion.

[*r,p,k*]=*residue*(*b,a*) can get the partial-fraction expansion of transfer function in Eq.(3.3). The input variables are vectors b and a, which are the numerator and denominator polynomial coefficients arranged according to descending powers of s. The output variable r is a column vector consisting of residues, p is a column vector consisting of poles, and k is a row vector consisting of residues.

[*b,a*]=*residue*(*r,p,k*) can convert partial fractions to the standard form of s in descending power of standard form.

Example 3.6 The transfer function $G(s)=\dfrac{2s^3+9s+1}{s^3+s^2+4s+4}$ is known, Please find its partial-fraction expansion.

The solution of Example 3.6 are shown in Figure 3.3.8.

```
%Example 3.6 Program:
num=[2,0,9,1];
den=[1,1,4,4];
[r,p,k]=residue(num,den)
```

```
%results of Command window
r =
   -0.0000 - 0.2500i
   -0.0000 + 0.2500i
   -2.0000

p =
    0.0000 + 2.0000i
    0.0000 - 2.0000i
   -1.0000

k =
    2
```

Figure 3.3.8 MATLAB Script and Results of Example 3.6

The result can be expressed as $G(s)=2+\frac{-0.25i}{s-2i}+\frac{0.25i}{s+2i}+\frac{-2}{s+1}$

Example 3.7 The transfer function $G(s)=\frac{-1.4}{s-1.5}-\frac{0.6}{s+1.1}+\frac{1.3}{s+0.4}-1.2$ is known. Please find its standard form in descending power.

The solution of Example 3.7 are shown in Figure 3.3.9.

```
%Example 3.7 Script:
r =[-1.4,-0.6,1.3]';
p =[1.5,-1.1,-0.4]';
k =[-1.2];
[b,a] = residue(r,p,k)
```

```
%results of Command window
b =
-1.2000    -0.7000    0.2120    -1.6090


a =

    1.0000    0.0000    -1.8100    -0.6600
```

Figure 3.3.9 MATLAB Script and Results of Example 3.7

The result can be expressed as $G(s)=\frac{-1.2s^3-0.7s^2+0.212s-1.609}{s^3-1.81s-0.66}$

3.4 Connection and Simplification of Structure Diagram

The physical control system usually includes several units such as controlled objects, controllers and feedbacks, etc. Each unit has its own mathematical model, such as a transfer function model. Therefore, the control system is formed by connecting different units. The main forms of connection are series connection, parallel connection, and feedback, etc.

For a system with a complex structure, the closed-loop transfer function of the system can be obtained through the simplification of the structure diagram or using the Mason formula. Table 3.5 gives the basic connection and equivalent structure diagram.

Table 3.5 Simplification Rules of Structural Diagram

Original Block Diagram	Equivalent Block Diagram	Equivalent Computational Relationship
R → $G_1(s)$ → $G_2(s)$ → C	R → $G_1(s)G_2(s)$ → C	(1)Series Connection Equivalent $C(s)=G_1(s)G_2(s)R(s)$
R → $G_1(s)$, $G_1(s)$ → ± → C	R → $G_1(s)\pm G_2(s)$ → C	(2)Parallel Connection Equivalent $C(s)=[G_1(s)\pm G_2(s)]R(s)$
R → ± → $G_1(s)$ → C, feedback $G_1(s)$	R → $\dfrac{G_1(s)}{1\mp G_1(s)G_2}$ → C	(3)Feedback Equivalent $C(s)=\dfrac{G_1(s)R(s)}{1\mp G_1(s)G_2(s)}$

3.4.1 Related Functions

MATLAB provides different functions to simplify the connection of the system model. The main functions and their instructions are shown in Table 3.6.

Table 3.6 Related Functions and Their Instructions

Functions	Instructions
sys = parallel(sys1,sys2) sys = parallel(sys1,sys2,inp1,inp2,out1,out2)	Two systems are in parallel connection, which is equivalent to $\text{sys}=\text{sys1}+\text{sys2}$. For MIMO systems, the input $inp1$ of $sys1$ is connected to the input $inp2$ of $sys2$, and the output $out1$ of $sys1$ is connected to the output $out2$ of $sys2$.
sys = series(sys1,sys2) sys = series(sys1,sys2,inp1,inp2,out1,out2)	The two systems are in series connection, which is equivalent to $\text{sys}=\text{sys1}*\text{sys2}$. For MIMO systems, the output $outputs1$ of $sys1$ is connected to the input $outputs2$ of $sys2$.
sys = feedback(sys1,sys2) sys = feedback(sys1,sys2,sign) sys = feedback(sys1,sys2,feedin,feedout,sign)	Two systems are in negative feedback connection. The default format, $\text{sign}=-1$, represents negative feedback, while $\text{sign}=1$ represents positive feedback, equivalent to $\text{sys}={\text{sys1}}/{(1\pm\text{sys1}*\text{sys2})}$. For a MIMO system, partial feedback connection refers to that the specified output of $sys1$, feedout, is connected to the input of $sys2$, where as the output of $sys2$ is connected to the specified input $sys1$, feedin, thus forming a closed-loop system.

continued

Functions	Instructions
[numc,denc] = cloop(num,den,sign)	An open-loop system represented by a transfer function constitutes a closed-loop system.

3.4.2 Examples of Connection and Simplification

Example 3.8 $G_1(s)=\dfrac{1}{s^2+4s+4}$, $G_2(s)=\dfrac{1}{s+5}$ is known. Please find the system model in which $G_1(s)$ and $G_2(s)$ are connected in series connection, parallel connection, negative feedback, and closed-loop respectively.

The solution of Example 3.8 are shown in Figure 3.4.1.

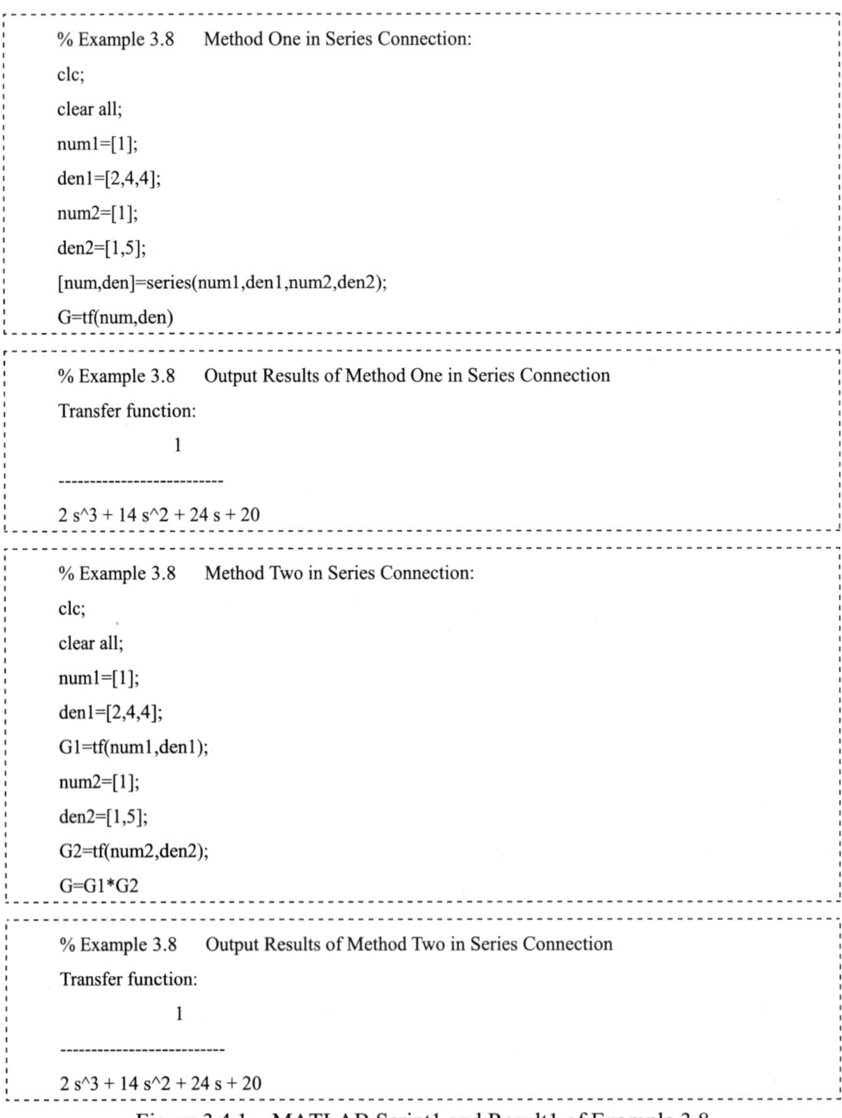

```
% Example 3.8    Method One in Series Connection:
clc;
clear all;
num1=[1];
den1=[2,4,4];
num2=[1];
den2=[1,5];
[num,den]=series(num1,den1,num2,den2);
G=tf(num,den)
```

```
% Example 3.8    Output Results of Method One in Series Connection
Transfer function:
              1
--------------------------
2 s^3 + 14 s^2 + 24 s + 20
```

```
% Example 3.8    Method Two in Series Connection:
clc;
clear all;
num1=[1];
den1=[2,4,4];
G1=tf(num1,den1);
num2=[1];
den2=[1,5];
G2=tf(num2,den2);
G=G1*G2
```

```
% Example 3.8    Output Results of Method Two in Series Connection
Transfer function:
              1
--------------------------
2 s^3 + 14 s^2 + 24 s + 20
```

Figure 3.4.1 MATLAB Script1 and Result1 of Example 3.8

```
% Example 3.8      Method Three in Series Connection:
clc;
clear all;
num1=[1];
den1=[2,4,4];
G1=tf(num1,den1);
num2=[1];
den2=[1,5];
G2=tf(num2,den2);
G=series(G1,G2)
```

```
% Example 3.8    Output Results of Method Three in Series Connection
Transfer function:
              1
--------------------------
2 s^3 + 14 s^2 + 24 s + 20
```

Figure 3.4.1 MATLAB Script1 and Result1 of Example 3.8 (continued)

Similarly, the parallel connection can also be completed in the above three ways. The key statements is shown in Figure 3.4.2.

```
% Example 3.8 Method One in Parallel Connection:
[num,den]=parallel(num1,den1,num2,den2);
G=tf(num,den);
% Method Two in Parallel Connection:
G=G1+G2;
% Method Three in Parallel Connection:
G=parallel(G1,G2);
```

```
% Example 3.8 Output Results of Method One in Parallel Connection
Transfer function:
     2 s^2 + 5 s + 9
--------------------------
2 s^3 + 14 s^2 + 24 s + 20
```

Figure 3.4.2 The MATLAB Script 2 and Result 2 of Example 3.8

Negative feedback can be expressed as follows,

```
%Example 3.8Method One in Negative Feedback:
G=feedback(G1,G2)   % Note: G1 is in the forward channel and G2 is in the feedback loop.
% Method Two in Negative Feedback
G=G1/（1+G1*G2）
```

```
% Example 3.8 Output Results of Negative Feedback
Transfer function:
           s + 5
--------------------------
2 s^3 + 14 s^2 + 24 s + 21
```

Figure 3.4.3 MATLAB Script 3 and Result 3 of Example 3.8

```
% Example 3.8Closed-loop:
clc;
clear all;
num1=[1];
den1=[2,4,4];
num2=[1];
den2=[1,5];
num=conv(num1,num2);
den=conv(den1,den2);
[num1,den1]=cloop(num,den);   % Note: The open loop transfer function is constructed by
                              % connecting G1 and G2 in series to construct a unit negative
                              % feedback closed loop system.
G=tf(num1,den1)
```

```
% Example 3.8 Output Results of Closed-loop
Transfer function:
            1
--------------------------
2 s^3 + 14 s^2 + 24 s + 21
```

Figure 3.4.3 MATLAB Script 3 and Result 3 of Example 3.8 (continued)

3.5 Modeling and Simulation Examples

This section gives two examples of modeling and simulation, the RLC circuit and the mechanical displacement system. Two transfer function models are established and a simple time-domain analysis is also made on this basis.

3.5.1 Transfer Function Modeling of an RLC Circuit

Different ways can be used to design mathematical models of system. Choosing the best form of mathematical model enables simulations to execute faster and more accurately. Consider the RLC circuit shown in Figure 3.5.1，where

$u_\mathrm{i}(t)$ is input voltage (volts) ,

L is inductance (H),

R is resistance (Ω),

C is capacitance (F),

$u_\mathrm{o}(t)$ is output voltage (Volts),

i is current across the resistance (Amps),

i_1 is current across the inductance (Amps),

i_2 is current across the capacitor (Amps).

Figure 3.5.1 RLC Circuit

Using Kirchhoff's Circuit Laws, we obtain

$$Ri(t)+u_o(t)=u_i(t) \tag{3.8}$$

$$i_1(t)+i_2(t)=i(t) \tag{3.9}$$

where

$$u_o(t) = \frac{1}{C}\int i_2(t)\mathrm{d}t = L\frac{\mathrm{d}i_1(t)}{\mathrm{d}t} \tag{3.10}$$

Take the Laplace transform of Eqs.(3.8), (3.9) and (3.10) under zero initial conditions by using a complex impedance and we have

$$R \cdot I(s) + U_O(s) = U_I(s) \tag{3.11}$$

$$I_1(s) + I_2(s) = I(s) \tag{3.12}$$

$$U_o(s) = \frac{1}{Cs}I_2(s) = LsI_1(s) \tag{3.13}$$

Then

$$I_1(s) = \frac{U_O(s)}{Ls} \quad , \quad I_2(s) = Cs \cdot U_O(s) \tag{3.14}$$

Substitute Eq. (3.14) into Eq. (3.11), replacing the intermediate variable and we have

$$R\left(\frac{U_O(s)}{Ls} + Cs \cdot U_O(s)\right) + U_O(s) = U_I(s) \tag{3.15}$$

The transfer function from Input $U_I(s)$ to Output $U_O(s)$ is

$$G(s) = \frac{U_O(s)}{U_I(s)} = \frac{s/(RC)}{s^2 + s/(RC) + 1/(LC)} \tag{3.16}$$

So far, the RLC circuit's transfer function model has been established. In the next section, we will use MATLAB to perform time-domain simulation analysis of the system.

3.5.2 Simulation Analysis of an RLC Circuit

When the input is a unit step, that is, $u_i(t) = 1\mathrm{V}$, the steady-state voltage is $u_o(t) = 0\mathrm{V}$.

The step response of the RLC circuit is shown in Figure 3.5.2.

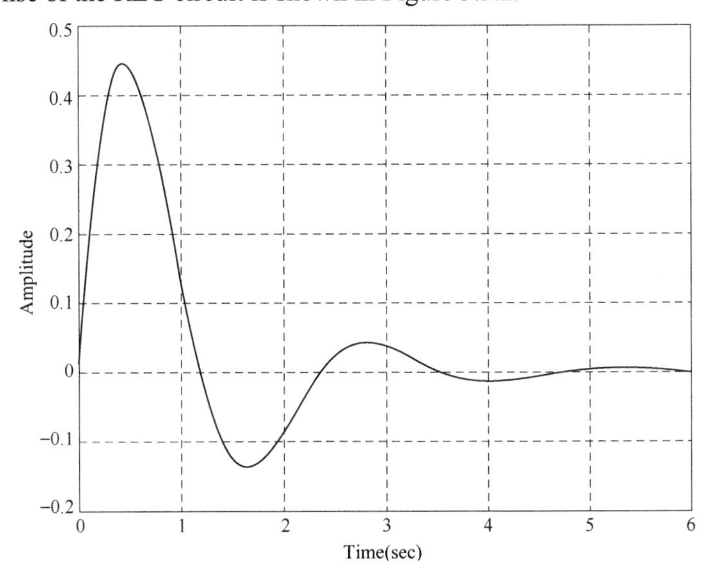

Figure 3.5.2 Step Response of RLC Circus

From Figure 3.5.1, the RLC circuit shows that the inductor and capacitor are connected in parallel, the voltage across the inductor and the capacitor is the same, and the system output voltage is this voltage. The voltage rises from zero at the beginning, reaches 0.45 volts at about 0.5 seconds to the highest point, before it starts to fall. After a weak oscillation, the voltage is stable at zero volts.

```
% The MATLAB script of Figure 3.5.2
%Model Parameters
R=1;      %R:Ohms(Ω )
L=0.25;  %L:Henry(H)
C=0.5;    %C:Farad(F)
%Transfer function Model
num=[1/(R*C),0];
den=[1,1/(R*C),1/(L*C)];
G=tf(num,den);
%Simulated step response
step(G)
grid
```

Figure 3.5.3 MATLAB Script

3.5.3 Transfer Function Modeling of a Mechanical Displacement System

The mechanical displacement system is shown in Figure 3.5.4. We assume that the system is linear. The external force $u(t)$ is the input of the system, and the displacement $y(t)$ of the mass is the output. The displacement $y(t)$ is measured from the equilibrium position in the absence of the external force. This system is a single input and single output system, where k is spring constant, m is mass of wood, b is damping coefficient, $u(t)$ is input of system, $y(t)$ is output of system.

Figure 3.5.4 Mechanical Displacement System

Damper resistance is

$$u_1(t) = b\frac{\mathrm{d}y(t)}{\mathrm{d}t}$$

Spring force is

$$u_2(t) = ky(t)$$

According to Newton's second law

$$m\frac{\mathrm{d}^2 y(t)}{\mathrm{d}t^2} = u(t) - u_1(t) - u_2(t) \tag{3.17}$$

Then we obtain

$$m\frac{\mathrm{d}^2 y(t)}{\mathrm{d}t^2} + b\frac{\mathrm{d}y(t)}{\mathrm{d}t} + ky(t) = u(t) \tag{3.18}$$

Under the zero initial condition, Take the Laplace transform of Equation (3.18) to obtain the transfer function of the mechanical displacement system

$$G(s)=\frac{Y(s)}{U(s)}=\frac{1}{ms^2+bs+k}=\frac{1/m}{s^2+(b/m)s+k/m} \tag{3.19}$$

3.5.4 Simulation Analysis of a Mechanical Displacement System

Simulation of the step response of the mechanical displacement system by MATLAB is shown in Figure 3.5.5. Its script is shown in Figure 3.5.6.

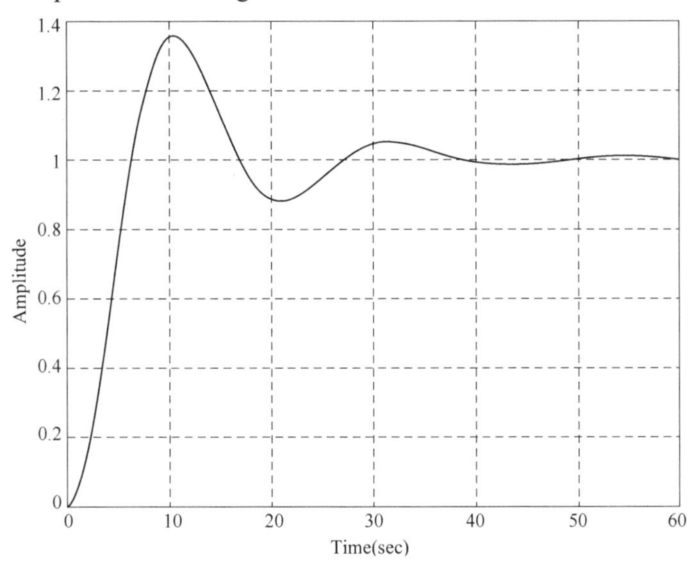

Figure 3.5.5 Step Response of Mechanical Displacement System

```
%The MATLAB script of Figure 3.5.5
%Model Parameters
m=10;
k=1;
b=2;
%Transfer function Model
num=[1/m];
den=[1,b/m,k/m];
G=tf(num,den);
%Simulated step response
step(G)
grid
```

Figure 3.5.6 MATLAB Script

Take $m=10\text{kg}$, $k=1\text{N/m}$, $b=2$. When the input is a unit step, that is, when $u(t)=1\text{N}$, the is steady-state output displacement is $y(t)=1\text{m}$.

As can be seen from Figure 3.5.5, when the input external force is a unit step, the output of the system is the displacement of the mass. The displacement began to vary from zero to peak, then slowly decreased, and finally stabilized at 1 meter. The system overshoot is less than 40% and the system setting time is approximately 45 seconds.

Summary

1. This chapter mainly introduces MATLAB description of three transfer function models.
2. Examples are given to illustrate the MATLAB implementation of three connection modes.
3. The modeling processes of two simple systems are clarified through two examples.

生词注解

subsequent	adj. 后来的，随后的
time-invariant	adj. 时不变的
superposition	n. 叠加，重合
algebraic	adj. 代数的
intuitively	adv. 直观的
molecular	adj. 分子的
coefficients	n. 系数
descending power	n. 降幂
simplification	n. 简单化
oscillation	n. 振荡，振动
approximately	adv. 大约，近似的
overshoot	n. 超调
setting time	n. 调节时间

第 4 章
状态空间模型

本章介绍如何利用 MATLAB 描述控制系统状态空间，为进行各种 MATLAB 分析与仿真奠定基础。

对于控制系统模型，其主要分为古典传递函数模型与状态空间模型。古典传递函数模型在第 3 章中已经介绍，本章主要介绍系统状态空间模型的 MATLAB 描述和模型之间的转换，两个简单的实际系统的状态空间建模及其仿真分析。

4.1 状态空间描述

本节介绍控制系统状态空间分析的基本知识。

4.1.1 现代控制理论

工程系统日益复杂化成为目前的发展趋势，主要原因是对任务复杂性和精度的要求越来越高。复杂系统可能是多输入和多输出，也可能是时变的。由于日益增长的对控制系统性能的严格要求，对系统的复杂程度的要求、易于大型计算机使用的要求，一种分析和设计复杂控制系统的新方法，即现代控制理论，从 20 世纪 60 年代开始逐渐发展起来。这种新方法基于状态这一概念。状态本身并不是一个新概念，它已经存在于经典力学和其他一些领域中有很长一段时间了。

4.1.2 现代控制理论与传统控制理论的比较

现代控制理论与传统控制理论形成鲜明的对照，前者适用于多输入多输出系统，系统可以是线性的或者非线性的，也可以是定常的或时变的；后者则仅仅适用于线性、定常和单输入单输出系统。此外，现代控制理论本质上既是一种时域方法也是频域方法，而传统控制理论则是一种复频域方法。前者不关注系统内部状态，只把系统当作黑匣子，只关注输入和输出的关系；后者则要定义系统内部状态，确定其状态变量。输入和输出之间的联系通过状态变量作为中间变量来实现的。

在介绍现代控制理论前，我们需要定义状态、状态变量、状态向量和状态空间。

4.1.3 状态

动态系统的状态是最小一组变量，称其为状态变量。只要知道了在 $t=t_0$ 时的一组变量和 $t \geqslant t_0$ 时的输入量，就能够完全确定在任一时间 $t \geqslant t_0$ 时的系统行为。

注意，状态这一概念绝不仅限于实际系统，它还适用于生物学系统、经济学系统、社会学系统以及其他系统。

4.1.4 状态变量

动态系统的状态变量是能够构成动态系统的最小一组变量。如果至少需要 n 个变量 $x_1, x_2, \cdots, x_n$ 才能描述动态系统的行为（一旦给出 $t \geqslant t_0$ 时的输入量，并指定 $t=t_0$ 时的初始状态，就可完全确定系统的未来状态），则这 n 个变量就是一组状态变量。

注意，状态变量未必是物理上可测量的或可观察的量。某些不代表物理量的变量，以及那些既不能测量，又不能观察的变量，也可以被选为状态变量。这种在选择状态变量方面的自由性，是状态空间法的一个优点。但是，从实用性而言，如果有可能，选择容易测量的量作为状态变量更方便，因为最佳控制规律需要反馈所有具有适当加权的状态变量。

4.1.5　状态向量

如果完全描述一个给定系统的行为需要 n 个状态变量，那么这 n 个状态变量可以看做是向量 $\boldsymbol{x}$ 的 n 个分量，该向量就称为状态向量。状态向量就是这样一种向量，一旦 $t=t_0$ 时的状态给定，并且给出 $t \geqslant t_0$ 时的输入量 $u(t)$，便可唯一确定任一时间 $t \geqslant t_0$ 时的系统状态 $x(t)$。

4.1.6　状态空间

设 $x_1, x_2, \cdots, x_n$ 为状态变量，那么由 x_1 轴，x_2 轴，…，x_n 轴所组成的 n 维空间称为状态空间。任何状态都可由状态空间中的一点表示。

4.1.7　状态空间方程

在状态空间分析中，涉及三种类型的变量，它们包含在动态系统的模型中，这三种变量是输入变量、输出变量和状态变量。对于一个给定的系统，其状态空间表达式不是唯一的。但是，对于相同系统任一不同状态空间表达式而言，其状态变量的数量是相同的。

动态系统必定包含着记忆元件，能够记忆在 $t \gg t_1$ 时输入量的值。因为在连续控制系统中，积分器作为记忆装置，所以这些积分器的输出量可以看做变量，这些变量确定了动态系统的内部状态。因此，积分器的输出量作为状态变量。能够完全确定系统动态特性的状态变量数目，则等于系统中包含的积分器数目。

假设多输入多输出系统中包含 n 个积分器，又设系统中有 r 个输入量 $u_1(t), u_2(t), \cdots, u_r(t)$ 和 m 个输出量 $y_1(t), y_2(t), \cdots, y_m(t)$，定义积分器的 n 个输出量为状态变量：$x_1(t), x_2(t), \cdots, x_n(t)$，则可以用下列方程描述系统

$$
\begin{aligned}
\dot{x}_1(t) &= f_1(x_1, x_2, \cdots, x_n; u_1, u_2, \cdots, u_r; t) \\
\dot{x}_2(t) &= f_2(x_1, x_2, \cdots, x_n; u_1, u_2, \cdots, u_r; t) \\
&\vdots \\
\dot{x}_n(t) &= f_n(x_1, x_2, \cdots, x_n; u_1, u_2, \cdots, u_r; t)
\end{aligned} \tag{4.1}
$$

系统的输出量 $y_1(t), y_2(t), \cdots, y_m(t)$ 可以表示为

$$
\begin{aligned}
y_1(t) &= g_1(x_1, x_2, \cdots, x_n; u_1, u_2, \cdots, u_r; t) \\
y_2(t) &= g_2(x_1, x_2, \cdots, x_n; u_1, u_2, \cdots, u_r; t) \\
&\vdots \\
y_m(t) &= g_m(x_1, x_2, \cdots, x_n; u_1, u_2, \cdots, u_r; t)
\end{aligned} \tag{4.2}
$$

如果定义

$$
x(t) = \begin{bmatrix} x_1(t) \\ x_2(t) \\ \vdots \\ x_n(t) \end{bmatrix}, f(x, u, t) = \begin{bmatrix} f_1(x_1, x_2, \cdots, x_n; u_1, u_2, \cdots, u_r; t) \\ f_2(x_1, x_2, \cdots, x_n; u_1, u_2, \cdots, u_r; t) \\ \vdots \\ f_n(x_1, x_2, \cdots, x_n; u_1, u_2, \cdots, u_r; t) \end{bmatrix},
$$

$$y(t)=\begin{bmatrix} y_1(t) \\ y_2(t) \\ \vdots \\ y_3(t) \end{bmatrix}, g(x,u,t)=\begin{bmatrix} g_1(x_1,x_2,\cdots,x_n;u_1,u_2,\cdots,u_r;t) \\ g_2(x_1,x_2,\cdots,x_n;u_1,u_2,\cdots,u_r;t) \\ \vdots \\ g_m(x_1,x_2,\cdots,x_n;u_1,u_2,\cdots,u_r;t) \end{bmatrix}, u(t)=\begin{bmatrix} u_1(t) \\ u_2(t) \\ \vdots \\ u_r(t) \end{bmatrix}$$

则方程（4.1）和方程（4.2）变成

$$\dot{x}(t)=f(x,u,t) \tag{4.3}$$

$$y(t)=g(x,u,t) \tag{4.4}$$

式中，方程（4.3）为状态方程，方程（4.4）则为输出方程。如果向量函数 $\boldsymbol{f}(\cdot)$ 和（或）$\boldsymbol{g}(\cdot)$ 中明显包含时间 t，则该系统称为时变系统。

如果方程（4.3）和方程（4.4）围绕着运行状态进行线性化，则有下列线性化状态方程和输出方程

$$\dot{x}(t)=\boldsymbol{A}(t)x(t)+\boldsymbol{B}(t)u(t) \tag{4.5}$$

$$y(t)=\boldsymbol{C}(t)x(t)+\boldsymbol{D}(t)u(t) \tag{4.6}$$

上式中，$\boldsymbol{A}(t)$ 为状态矩阵，$\boldsymbol{B}(t)$ 为输入矩阵，$\boldsymbol{C}(t)$ 为输出矩阵，$\boldsymbol{D}(t)$ 为直接传输矩阵。

4.2 MATLAB 中的状态空间模型

具有 n 个状态、r 个输入和 m 个输出的线性定常系统，用矩阵表示的状态空间模型是

$$\dot{x}(t)=\boldsymbol{A}(t)x(t)+\boldsymbol{B}(t)u(t)$$

$$y(t)=\boldsymbol{C}(t)x(t)+\boldsymbol{D}(t)u(t)$$

其中，状态向量 $\boldsymbol{x}(t)$ 是 n 维，输入向量 $\boldsymbol{u}(t)$ 是 r 维，输出向量 $\boldsymbol{y}(t)$ 是 m 维；状态矩阵 $\boldsymbol{A}$ 是 $n\times n$ 维，输入矩阵 $\boldsymbol{B}$ 是 $n\times r$ 维，输出矩阵 $\boldsymbol{C}$ 是 $m\times n$ 维，直接传输矩阵 $\boldsymbol{D}$ 是 $m\times r$ 维。对于一个线性定常系统，$\boldsymbol{A}$，$\boldsymbol{B}$，$\boldsymbol{C}$，$\boldsymbol{D}$ 都是常数矩阵。

4.2.1 状态空间模型相关的 MATLAB 函数

在 MATLAB 中，状态空间模型可以由方程的系数矩阵 $\boldsymbol{A},\boldsymbol{B},\boldsymbol{C},\boldsymbol{D}$ 来确定，利用 MATLAB 工具箱函数 ss()实现：

```
G = ss(A,B,C,D)
```

状态空间模型的几个相关函数说明见表 4.1。

表 4.1 状态空间模型的相关函数

函　数	说　明
sys=ss(A,B,C,D)	由矩阵 $\boldsymbol{A}$，$\boldsymbol{B}$，$\boldsymbol{C}$，$\boldsymbol{D}$ 直接返回变量 sys，即连续系统的状态空间模型
sys=ss(A,B,C,D,T_s)	由矩阵 $\boldsymbol{A}$，$\boldsymbol{B}$，$\boldsymbol{C}$，$\boldsymbol{D}$ 和采样时间 T_s（秒）直接返回变量 sys，即离散系统的状态空间模型。T_s 为采样周期，当 T_s=−1 或 T_s=[]时，表示系统采样周期未定义
[A,B,C,D]=ssdata(sys)	获取状态空间模型参数
[A,B,C,D, T_s]=ssdata(sys)	获取状态空间模型参数，且包括采样时间 T_s

4.2.2　运用函数建立状态空间模型

例 4.1　将下列系统的状态空间模型输入到 MATLAB 工作空间中。

$$\dot{x}=\begin{bmatrix}1 & 6 & 9 & 10\\ 3 & 12 & 6 & 8\\ 4 & 7 & 9 & 11\\ 5 & 12 & 13 & 14\end{bmatrix}x+\begin{bmatrix}4 & 6\\ 2 & 4\\ 2 & 2\\ 1 & 0\end{bmatrix}u$$

$$y=\begin{bmatrix}0 & 0 & 2 & 1\\ 8 & 0 & 2 & 2\end{bmatrix}x$$

由 $\boldsymbol{B}$ 和 $\boldsymbol{C}$ 矩阵维数可知，该系统是双输入双输出系统。程序的实现及其结果输出如图 4.2.1 所示。

```
%例 4.1 的 MATLAB 程序
A=[1 6 9 10; 3 12 6 8; 4 7 9 11; 5 12 13 14];%输入状态矩阵 A、B、C 和 D 的值
B=[4 6; 2 4; 2 2; 1 0];
C=[0 0 2 1; 8 0 2 2];
D=zeros(2,2);
G=ss(A,B,C,D)
```

```
%例 4.1 的 MATLAB 程序结果
G =

  a =
         x1   x2   x3   x4
    x1    1    6    9   10
    x2    3   12    6    8
    x3    4    7    9   11
    x4    5   12   13   14

  b =
         u1   u2
    x1    4    6
    x2    2    4
    x3    2    2
    x4    1    0

  c =
         x1   x2   x3   x4
    y1    0    0    2    1
    y2    8    0    2    2

  d =
         u1   u2
    y1    0    0
    y2    0    0

Continuous-time state-space model.
```

图 4.2.1　例 4.1 的 MATLAB 程序及其结果

在此例中，系统模型中并没有给出直接传输矩阵 $\boldsymbol{D}$。因此，必须构造矩阵 $\boldsymbol{D}$，$\boldsymbol{D}$ 矩阵元

素都为零值。需要注意，矩阵 ***D*** 的维数由系统的输出维数和输入维数决定。***D*** 矩阵的维数必须和其他的矩阵维数兼容，否则函数 *ss*()将给出明确的错误提示，并中断程序运行。

例 4.2 已知系统的状态空间模型，

$$\dot{x}=\begin{bmatrix}0 & 1\\ 1 & -2\end{bmatrix}x+\begin{bmatrix}0\\ 1\end{bmatrix}u$$

$$y=[1 \quad 3]x+u$$

求系统参数。

例 4.2 的程序和结果见图 4.2.2。

```
%例 4.2 的 MATLAB 程序 1
A=[0 1; -1 -2]; B=[0;1];          %输入状态矩阵 A、B、C 和 D 的值
C=[1,3];   D=[1];
G=ss(A,B,C,D)
```

```
%例 4.2 的 MATLAB 程序 1 结果
G =


  a =
         x1   x2
   x1     0    1
   x2    -1   -2

  b =
         u1
   x1     0
   x2     1

  c =
         x1   x2
   y1     1    3

  d =
         u1


   y1     1
Continuous-time state-space model.
```

```
%例 4.2 的 MATLAB 程序 2
[ad,bd,cd,dd]=ssdata(G)       %得到模型参数
```

```
%例 4.2 的 MATLAB 程序 2 结果
ad =


     0     1
    -1    -2
```

图 4.2.2 例 4.2 的 MATLAB 程序及其结果

```
bd =

     0
     1

cd =

     1     3

dd =

     1
```

```
%例 4.2 的 MATLAB 程序 3
get(G)                          %获取系统模型所有参数的信息
```

```
%例 4.2 的 MATLAB 程序 3 结果
[2x2 double]          a:
                      b: [2x1 double]
                      c: [1 3]
                      d: 1
                      e: []
                 Scaled: 0
              StateName: {2x1 cell}
              StateUnit: {2x1 cell}
          InternalDelay: [0x1 double]
             InputDelay: 0
            OutputDelay: 0
                     Ts: 0]
               TimeUnit: 'seconds'
              InputName: {''}
              InputUnit: {''}
             InputGroup: [1x1 struct]
             OutputName: {''}
             OutputUnit: {''}
            OutputGroup: [1x1 struct]
                   Name: ''
                  Notes: {}
               UserData: []
           SamplingGrid: [1x1 struct]
```

```
%例 4.2 的 MATLAB 程序 4
G.a                             %求取一个系统模型的参数
```

```
%例 4.2 的 MATLAB 程序 4 结果
ans =

     0     1
    -1    -2
```

图 4.2.2　例 4.2 的 MATLAB 程序及其结果（续）

由例 4.2 可知，获取系统状态空间模型的参数有不同方式实现。需要注意的是，这时候的参数均是矩阵格式。若不需要模型的所有参数信息，可采用 G.a 方式获得模型的单个参数信息。

4.3 模型间的转换

控制系统的模型主要有传递函数模型（tf）、零极点模型（zpk）和状态空间模型（ss）三种形式，它们之间存在着内在的联系，可以相互进行转换（见图 4.3.1）。MATLAB 提供了系统模型之间的相互转换的函数。

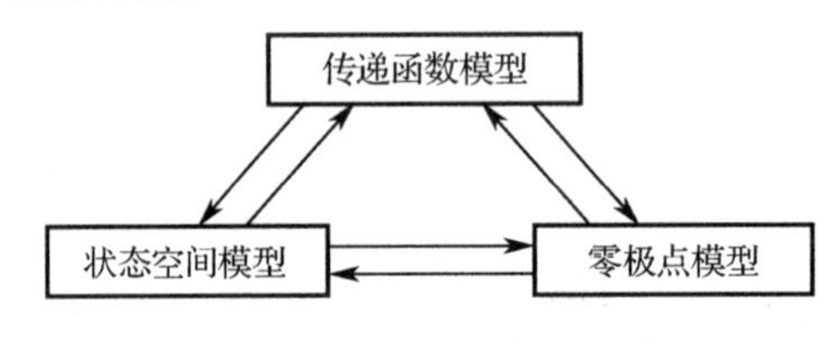

图 4.3.1 三个模型之间的转换

4.3.1 实现模型之间转换的 MATLAB 函数

系统模型的转换，主要分为两种。第一种是生成模型，把其他形式的模型转换为函数自身形式的模型，主要有图 4.3.1 所示的三种模型类型相对应的三个函数，其说明见表 4.2。第二种是把函数自身的参数表达转换成为其他函数类型的参数表达形式，主要有实现图 4.3.1 所示的三种模型之间的形式转换的有六种函数，其说明见表 4.3。

表 4.2 三种模型函数说明

函　数	说　明
G_{tf} =tf(sys)	将其他类型的模型转换成传递函数模型
G_{zpk} =zpk(sys)	将其他类型的模型转换成零极点模型
G_{ss} =ss(sys)	将其他类型的模型转换成状态空间模型

表 4.3 六种模型转换函数说明

函　数	说　明
[z,p,k]=tf2zp(num,den)	传递函数模型向零极点模型的转换
[num,den]=zp2tf(z,p,k)	零极点模型向传递函数模型的转换
[A,B,C,D]= zp2ss(z,p,k)	零极点模型向状态空间模型的转换
[z,p,k]=ss2zp(A,B,C,D,iu)	状态空间模型向零极点模型的转换，iu 表示是第 i 个输入
[A,B,C,D]= tf2ss(num,den)	传递函数模型向状态空间模型的转换
[num,den]=ss2tf(A,B,C,D,iu)	状态空间模型向传递函数模型的转换，iu 表示是第 i 个输入

4.3.2 模型之间的转换实例

例 4.3 已知系统状态空间模型为

$$\dot{x}=\begin{bmatrix}0 & 1\\1 & -2\end{bmatrix}x+\begin{bmatrix}0\\1\end{bmatrix}u$$
$$y=[1\quad 3]x+u$$

求其传递函数模型及零极点模型。

例 4.3 的程序和结果见图 4.3.2。

```
%例 4.3 的 MATLAB 程序 1
iu=1;
[z,p,k]=ss2zp(A,B,C,D,iu)
C=[1,3]; D=[1];
[num,den]=ss2tf(A,B,C,D,1)    %当只有一个输入时， iu=1，或者省略
```

```
%例 4.3 的 MATLAB 程序结果 1
num =

    1.0000    5.0000    2.0000

den =

     1     2     1
```

```
%例 4.3 的 MATLAB 程序 2
[z,p,k]=ss2zp(A,B,C,D)   %当只有一个输入时， iu=1，或者省略
```

```
%例 4.3 的 MATLAB 程序结果 2
z =

   -0.4384
   -4.5616

p =

    -1
    -1

k =

     1
```

图 4.3.2　例 4.3 的 MATLAB 程序及其结果

在例 4.3 中，需要注意的是，当从状态空间模型向传递函数模型和零极点模型转换时，必须确定函数中参数 iu 的值，一组 iu 值对应于模型中的一个输入输出端口。因为状态空间模型是可以表示多输入多输出系统的，而传递函数和零极点模型都表示的是单输入单输出系统。

例 4.4　已知一个单输入三输出系统的传递函数模型为

$$G_{11}(s)=\frac{y_1(s)}{u(s)}=\frac{-2}{s^3+6s^2+11s+6}$$

$$G_{21}(s)=\frac{y_2(s)}{u(s)}=\frac{-s-5}{s^3+6s^2+11s+6}$$

$$G_{31}(s)=\frac{y_3(s)}{u(s)}=\frac{s^2+2s}{s^3+6s^2+11s+6}$$

求系统的状态空间模型。

例 4.4 的程序和结果见图 4.3.3。

```
%例 4.4 的 MATLAB 程序
num=[0 0 -2;0 -1 -5;1 2 0];      %分子是 3×3 矩阵，第一行对应 G11(s) 中的分子 s 的系数
den=[1 6 11 6];                  %第二行对应 G21(s)  中的分子 s 的系数，以此类推
[A,B,C,D]=tf2ss(num,den)
```

```
%例 4.4 的 MATLAB 程序结果
A =

    -6   -11    -6
     1     0     0
     0     1     0


B =

     1
     0
     0


C =

     0     0    -2
     0    -1    -5
     1     2     0


D =

     0
     0
     0
```

图 4.3.3 例 4.4 的 MATLAB 程序及其结果

需要注意的是，例 4.4 中的系统是单输入三输出系统，因此传递函数是三行一列，分子多项式 num 不再是一个数，而是一个 3×3 矩阵。num 矩阵的行数由输出个数确定，列数由三个传递函数中分子 s 的最高次幂决定。此例中，第三个传递函数 $G_{31}(s)$ 中的分子 s 的次幂

是最高的，等于 2，因此分子的列数就是 3。num 矩阵中，3 行 3 列数值是 s 的系数，且按照 s 的降幂排列的，s 是三个传递函数的分子多项式中的参数。

例 4.5　已知系统的零极点模型为

$$G(s)=\frac{6(s+3)}{(s+1)(s+2)(s+5)}$$

求系统的传递函数模型及状态空间模型。

例 4.5 的程序和结果见图 4.3.4。

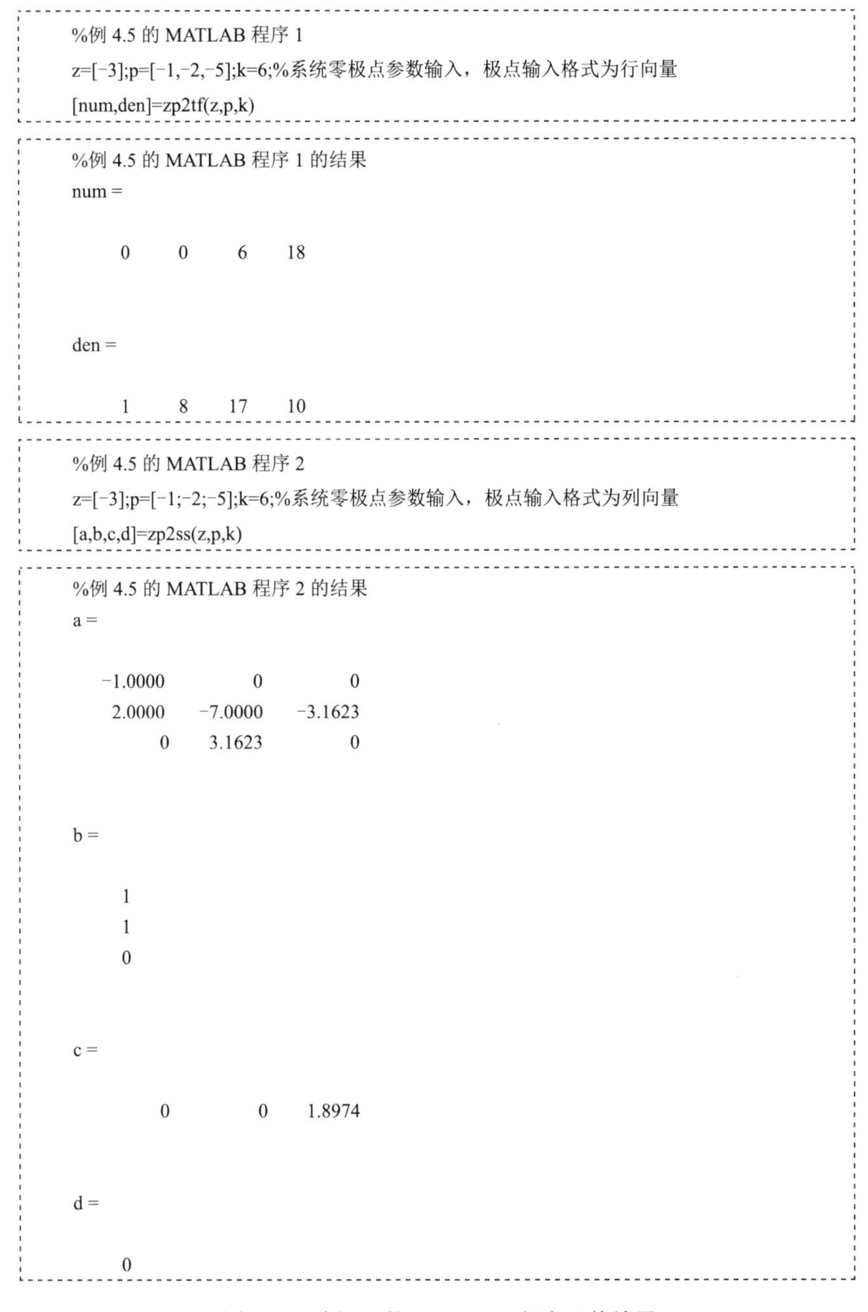

```
%例 4.5 的 MATLAB 程序 1
z=[-3];p=[-1,-2,-5];k=6;%系统零极点参数输入，极点输入格式为行向量
[num,den]=zp2tf(z,p,k)
```

```
%例 4.5 的 MATLAB 程序 1 的结果
num =

     0     0     6    18


den =

     1     8    17    10
```

```
%例 4.5 的 MATLAB 程序 2
z=[-3];p=[-1;-2;-5];k=6;%系统零极点参数输入，极点输入格式为列向量
[a,b,c,d]=zp2ss(z,p,k)
```

```
%例 4.5 的 MATLAB 程序 2 的结果
a =

   -1.0000         0         0
    2.0000   -7.0000   -3.1623
         0    3.1623         0


b =

     1
     1
     0


c =

         0         0    1.8974


d =

     0
```

图 4.3.4　例 4.5 的 MATLAB 程序及其结果

对于例 4.5，在程序 1 中，对极点矩阵 $\boldsymbol{p}$ 的输入采用了行向量；在程序 2 中，对极点矩阵 $\boldsymbol{p}$ 的输入采用了列向量。零点和极点的矩阵输入可以是行向量，也可以是列向量。

4.3.3 状态空间模型的化简

状态空间模型可以表示多输入多输出系统，对于模型串联、并联与反馈的三种连接，需要给出具体的连接端子标号，才能确保正确的连接。MATLAB 中实现模型的串联、并联与反馈的连接函数，都在第 3 章中介绍了，在这里仅以实例来说明在状态空间模型中的这些函数的具体使用方法。

例 4.6 已知系统 1 为

$$\dot{x}_1 = \begin{bmatrix} 0 & 1 \\ 1 & -2 \end{bmatrix} x_1 + \begin{bmatrix} 0 \\ 1 \end{bmatrix} u_1$$

$$y_1 = [1 \quad 3] x_1 + u_1$$

系统 2 为

$$\dot{x}_2 = \begin{bmatrix} 0 & 1 \\ -1 & -3 \end{bmatrix} x_2 + \begin{bmatrix} 0 \\ 1 \end{bmatrix} u_2$$

$$y_2 = [1 \quad 4] x_2$$

求按串联、并联、正反馈和负反馈连接时的系统状态空间模型及系统 1 按单位负反馈连接时的状态空间模型。

例 4.6 的程序和结果见图 4.3.5。

```
%例 4.6 的 MATLAB 程序
a1=[0 1;-1 -2];b1=[0;1];
c1=[1 3];d1=[1];
a2=[0 1; -1 -3];b2=[0;1];
c2=[1 4];d2=[0];
disp('串联连接')
[a,b,c,d]=series(a1,b1,c1,d1,a2,b2,c2,d2)          %串联连接
disp('并联连接')
[a,b,c,d]=parallel(a1,b1,c1,d1,a2,b2,c2,d2)        %并联连接
disp('正反馈连接')
[a,b,c,d]=feedback(a1,b1,c1,d1,a2,b2,c2,d2,+1)     %正反馈连接
disp('负反馈连接')
[a,b,c,d]=feedback(a1,b1,c1,d1,a2,b2,c2,d2)        %负反馈连接
disp('单位负反馈连接')
[a,b,c,d]=cloop(a1,b1,c1,d1)                       %单位负反馈连接
```

```
%例 4.6 的 MATLAB 程序实现结果
串联连接
a =
      0     1     0     0
     -1    -3     1     3
```

图 4.3.5 例 4.6 的 MATLAB 程序及其结果

```
      0      0      0      1
      0      0     -1     -2

 b =

      0
      1
      0
      1

 c =

      1      4      0      0

 d =

      0

 并联连接

 a =

      0      1      0      0
     -1     -2      0      0
      0      0      0      1
      0      0     -1     -3

 b =

      0
      1
      0
      1

 c =

      1      3      1      4

 d =

      1

 正反馈连接

 a =
```

图 4.3.5　例 4.6 的 MATLAB 程序及其结果（续）

```
      0     1     0     0
     -1    -2     1     4
      0     0     0     1
      1     3     0     1
b =

      0
      1
      0
      1

c =

      1     3     1     4

d =

      1


负反馈连接

a =

      0     1     0     0
     -1    -2    -1    -4
      0     0     0     1
      1     3    -2    -7
b =

      0
      1
      0
      1


c =

      1     3    -1    -4

d =

      1
```

图 4.3.5 例 4.6 的 MATLAB 程序及其结果（续）

```
单位负反馈连接

a =

         0    1.0000
   -1.5000   -3.5000


b =

         0
    0.5000


c =

    0.5000    1.5000


d =

    0.5000
```

图 4.3.5　例 4.6 的 MATLAB 程序及其结果（续）

例 4.6 是两个单输入单输出的状态空间模型，两个状态空间模型的串联、并联和反馈连接。此处的函数使用都与在传递函数中的函数连接使用相同，只是这里给函数赋值的是状态空间模型的参数。

例 4.7　已知系统 1 和系统 2 状态空间模型如下：

系统 1

$$\begin{bmatrix}\dot{x}_{11}\\\dot{x}_{12}\\\dot{x}_{13}\end{bmatrix}=\begin{bmatrix}1&4&4\\2&2&1\\3&6&2\end{bmatrix}\begin{bmatrix}x_{11}\\x_{12}\\x_{13}\end{bmatrix}+\begin{bmatrix}0&1&0\\1&0&0\\0&0&1\end{bmatrix}\begin{bmatrix}u_{11}\\u_{12}\\u_{13}\end{bmatrix}$$

$$\begin{bmatrix}y_{11}\\y_{12}\end{bmatrix}=\begin{bmatrix}0&0&1\\0&1&1\end{bmatrix}\begin{bmatrix}x_{11}\\x_{12}\\x_{13}\end{bmatrix}+\begin{bmatrix}0&1&0\\1&0&1\end{bmatrix}\begin{bmatrix}u_{11}\\u_{12}\\u_{13}\end{bmatrix}$$

系统 2

$$\begin{bmatrix}\dot{x}_{21}\\\dot{x}_{22}\\\dot{x}_{23}\end{bmatrix}=\begin{bmatrix}1&-1&0\\3&-2&1\\1&6&-1\end{bmatrix}\begin{bmatrix}x_{21}\\x_{22}\\x_{23}\end{bmatrix}+\begin{bmatrix}1&0&0\\0&1&0\\0&0&1\end{bmatrix}\begin{bmatrix}u_{21}\\u_{22}\\u_{23}\end{bmatrix}$$

$$\begin{bmatrix}y_{21}\\y_{22}\end{bmatrix}=\begin{bmatrix}0&1&0\\1&0&1\end{bmatrix}\begin{bmatrix}x_{21}\\x_{22}\\x_{23}\end{bmatrix}+\begin{bmatrix}1&1&0\\1&0&1\end{bmatrix}\begin{bmatrix}u_{21}\\u_{22}\\u_{23}\end{bmatrix}$$

求部分并联后的状态空间模型，要求 $\boldsymbol{u}_{11}$ 与 $\boldsymbol{u}_{22}$ 连接，$\boldsymbol{u}_{13}$ 与 $\boldsymbol{u}_{23}$ 连接，$\boldsymbol{y}_{11}$ 与 $\boldsymbol{y}_{21}$ 连接。

例 4.7 的程序和结果见图 4.3.6。

```
%例 4.7 的 MATLAB 程序
a1=[1 4 4;2 2 1;3 6 2];b1=[0 1 0;1 0 0;0 0 1];          %系统 1 参数输入
c1=[0 0 1;0 1 1];d1=[0 1 0;1 0 1];
a2=[1 -1 0;3 -2 1;1 6 -1];b2=[1 0 0;0 1 0;0 0 1];       %系统 2 参数输入
c2=[0 1 0;1 0 1];d2=[1 1 0;1 0 1];
% 部分并联后的状态空间模型，要求 u11 与 u22 连接，u13 与 u23 连接，y11 与 y21 连接
disp('部分并联后的状态空间模型')
[a,b,c,d]=parallel(a1,b1,c1,d1,a2,b2,c2,d2,[1 3],[2 3],1,1)
%input1=[1 3]   input2=[2 3]
%output1=1output2=1
```

```
%例 4.7 的 MATLAB 程序结果
部分并联连接后的状态方程
a =

     1     4     4     0     0     0
     2     2     1     0     0     0
     3     6     2     0     0     0
     0     0     0     1    -1     0
     0     0     0     3    -2     1
     0     0     0     1     6    -1

b =

     0     1     0     0
     1     0     0     0
     0     0     1     0
     0     0     0     1
     1     0     0     0
     0     0     1     0

c =

     0     0     1     0     1     0
     0     1     1     0     0     0
     0     0     0     1     0     1


d =

     1     1     0     1
     1     0     1     0
     0     0     1     1
```

图 4.3.6　例 4.7 的 MATLAB 程序及其结果

例 4.6 中的两个系统都是三输入两输出系统，要求两个系统实现并联。在使用并联函数 parallel 时，必须确定此时需要连接的端子。首先，两个系统的输入端，input1= [1 3]，input2= [2 3]，系统 1 的 1 号输入端子和系统 2 的 2 号输入端子是相连接的，系统 1 的 3 号输入端子与系统 2 的 3 号输入端子是相连接的，如此完成了系统 1 和系统 2 的输入端子的并联。

其次，输出端子的确定。两个系统的输出端子均有两个输出，但各自只需一个端子相连接

即可，output1 = 1，output2 = 1，系统 1 和系统 2 都是选择各自的 1 号端子进行连接。需要注意的是，当需要相连的端子个数大于等于两个时，此时定义端子的参数是数组，不是单独的数据，需要用方括号括起来，数组中的数据需与相应连接的端子编号对应，并注意数据的维数匹配。

4.4　RLC 电路及机械位移系统

RLC 电路和机械位移系统两个例子都已经在第 3 章中介绍了它们各自的传递函数模型的建立。本节继续以其为例，详细介绍它们的状态空间模型的建立过程。

4.4.1　RLC 电路的状态空间建模

下面分析 RLC 电路的响应过程。如图 4.4.1 所示，采用以下 RLC 电路。

V_{in} 为输入电压（V），V_{out} 为输出电压（V），L 为电感（H），R 为电阻（Ω），C 为电容（F），i 为通过电阻的电流（A），i_1 为通过电感的电流（A），i_2 为通过电容的电流（A）。

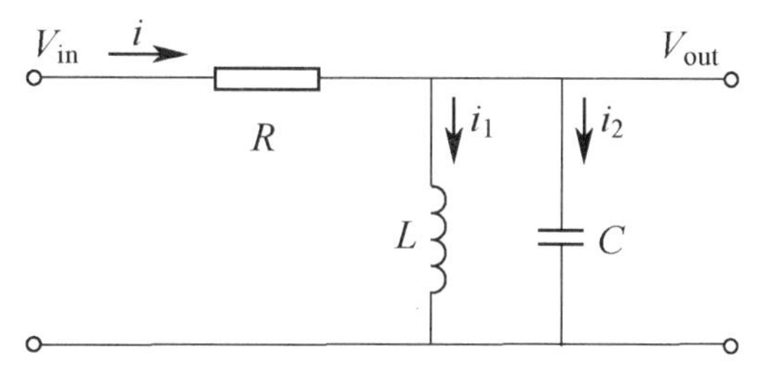

图 4.4.1　RLC 电路

使用基尔霍夫电路定律，可以得

$$V_{in} = Ri + V_{out} \tag{4.7}$$

$$i = i_1 + i_2 \tag{4.8}$$

其中

$$V_{out} = \frac{1}{C}\int i_2 \mathrm{d}t = L\frac{\mathrm{d}i_1}{\mathrm{d}t} \tag{4.9}$$

状态变量定义为

$$x_1 = V_{out}$$

$$x_2 = i_1$$

然后，利用式（4.7）、式（4.8）和式（4.9）对 x_1 求导，得

$$\dot{x} = -\frac{1}{RC}x_1 - \frac{1}{C}x_2 + \frac{1}{RC}V_{in} \tag{4.10}$$

再用式（4.9）对 x_2 求导，得

$$\dot{x}_2 = \frac{1}{L}x_1 \tag{4.11}$$

我们可以把式（4.10）和式（4.11）写成矩阵形式

$$\dot{x} = \boldsymbol{A}x + \boldsymbol{B}u$$

其中

$$x = \begin{pmatrix} x_1 \\ x_2 \end{pmatrix} = \begin{pmatrix} V_{out} \\ i_1 \end{pmatrix},\quad u = V_{in}$$

和

$$A=\begin{bmatrix}-\dfrac{1}{RC} & -\dfrac{1}{C}\\ \dfrac{1}{L} & 0\end{bmatrix},\qquad B=\begin{bmatrix}\dfrac{1}{RC}\\ 0\end{bmatrix}$$

取 R=1Ω，L=0.25H，C=0.5F，得

$$\dot{x}=\begin{bmatrix}-2 & -4\\ 4 & 0\end{bmatrix}x+\begin{bmatrix}2\\ 0\end{bmatrix}u$$

测量 V_{out} 的值，得

$$y=\boldsymbol{C}x+\boldsymbol{D}u$$

其中

$$\boldsymbol{C}=[1\quad 0]\text{和}\boldsymbol{D}=[0]$$

4.4.2 RLC 电路的仿真分析

用 MATLAB 脚本程序仿真 RLC 电路的阶跃响应如图 4.4.2 所示，纵坐标的幅度是输出电压 V_{out} 的大小。当输入是单位阶跃，即输入电压 $V_{\text{in}}(t)=1\text{V}$ 时，电容上的稳态输出电压 $V_{\text{out}}(t)=0\text{V}$ 。

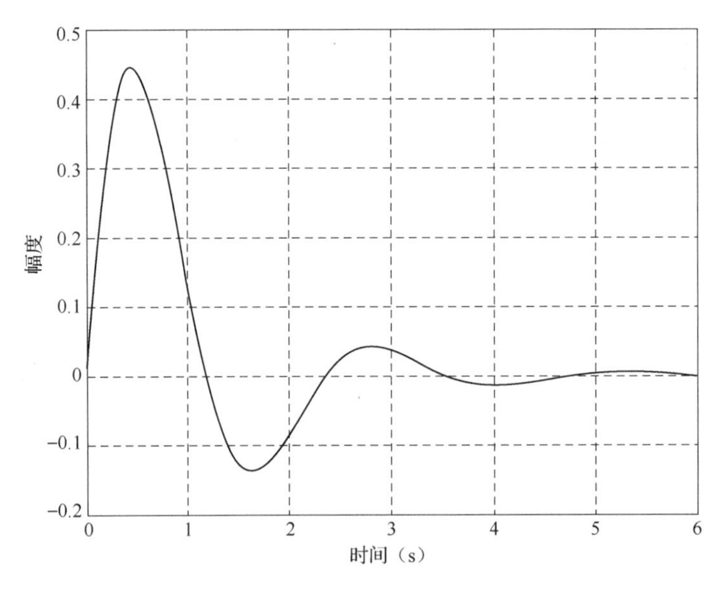

图 4.4.2 RLC 阶跃响应

从图 4.4.2 中可以看出，电容的电压跟随输入电压的变化，有个瞬态的小幅振荡后，在 5s 附近进入稳态的零值。

```
% 模型参数
R=1;
L=0.25;
C=0.5;
```

图 4.4.3 RLC 阶跃响应的程序

```
%状态空间模型
A=[-1/(R*C) -1/C;1/L 0];
B=[1/(R*C) 0]';
C=[1 0]; • 1
D=[0];
%模拟阶跃响应
step(A,B,C,D)
grid
```

图 4.4.3　RLC 阶跃响应的程序（续）

本例中，也可选择测量电感电流 i_1，仅需要把输出方程中的 **C** 矩阵值，改为 **C**=[0 1]，其他均不变，依旧采用图 4.4.3 的程序。重新设置 **C** 矩阵值，即可得到图 4.4.4 所示的系统阶跃响应。

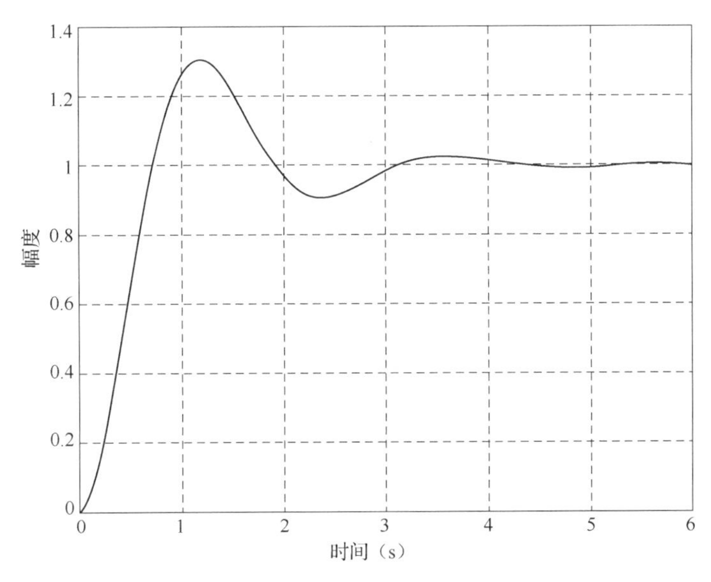

图 4.4.4　改动后的 RLC 阶跃响应

从图 4.4.4 中可以看出，输出改为电感电流时，电感的电流跟随输入电压的变化，输出也逐渐增加，一个小幅振荡后，在 4.5s 附近进入稳态的电流值 1A。在输入电压 $V_{\text{in}}(t)=1\text{V}$ 时，输出电流 $i(t)=1\text{A}$ 。

4.4.3　机械位移系统的状态空间建模

如图 4.4.5 所示的机械位移系统，假设系统是线性的。外力 $u(t)$是系统的输入量，质量块的位移 $y(t)$是系统的输出量。位移 $y(t)$从无外力作用时的平衡位置开始计算。该系统是一个单输入单输出系统。其中，k 为弹簧系数，m 为木块质量，b 为阻尼系数，$u(t)$为系统输入量，$y(t)$为系统输出量。

由图 4.4.5 可以得到系统方程

$$m\ddot{y}+b\dot{y}+ky=u \tag{4.12}$$

这是一个二阶系统，意味着该系统包括两个积分器。我们定义状态变量 $x_1(t)$和 $x_2(t)$为

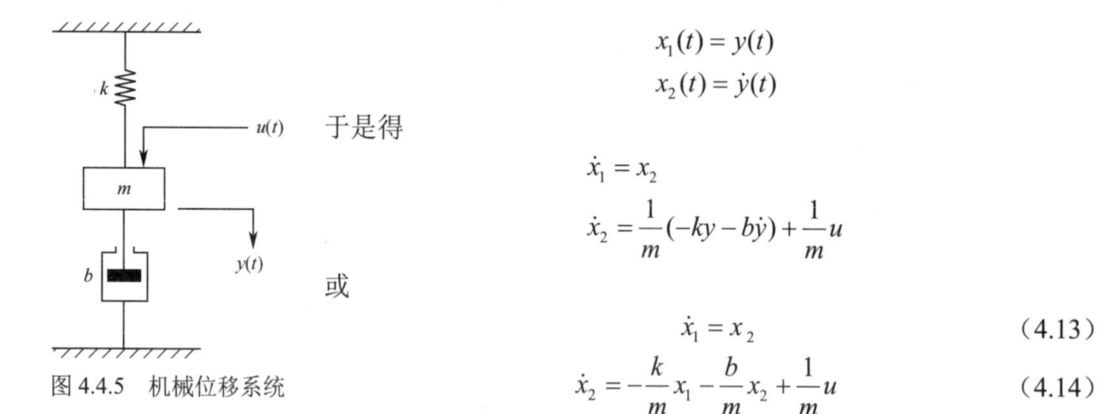

图 4.4.5 机械位移系统

$$x_1(t)=y(t)$$
$$x_2(t)=\dot{y}(t)$$

于是得

$$\dot{x}_1=x_2$$
$$\dot{x}_2=\frac{1}{m}(-ky-b\dot{y})+\frac{1}{m}u$$

或

$$\dot{x}_1=x_2 \tag{4.13}$$
$$\dot{x}_2=-\frac{k}{m}x_1-\frac{b}{m}x_2+\frac{1}{m}u \tag{4.14}$$

输出方程为

$$y=x_1 \tag{4.15}$$

若采用矩阵表示，则方程（4.13）和方程（4.14）可以写为

$$\begin{bmatrix}\dot{x}_1\\ \dot{x}_2\end{bmatrix}=\begin{bmatrix}0 & 1\\ -\dfrac{k}{m} & -\dfrac{b}{m}\end{bmatrix}\begin{bmatrix}x_1\\ x_2\end{bmatrix}+\begin{bmatrix}0\\ \dfrac{1}{m}\end{bmatrix}u \tag{4.16}$$

输出方程（4.15）可以写为

$$y=[1\quad 0]\begin{bmatrix}x_1\\ x_2\end{bmatrix} \tag{4.17}$$

方程（4.16）是上述系统的状态方程，方程（4.17）则是上述系统的输出方程。方程（4.16）和方程（4.17）可以写成标准形式

$$\begin{aligned}\dot{x}&=\boldsymbol{A}x+\boldsymbol{B}u\\ y&=\boldsymbol{C}x+\boldsymbol{D}u\end{aligned} \tag{4.18}$$

其中

$$\boldsymbol{A}=\begin{bmatrix}0 & 1\\ -\dfrac{k}{m} & -\dfrac{b}{m}\end{bmatrix},\quad \boldsymbol{B}=\begin{bmatrix}0\\ \dfrac{1}{m}\end{bmatrix},\quad \boldsymbol{C}=[1\quad 0],\quad \boldsymbol{D}=0$$

4.4.4 机械位移系统的仿真分析

用 MATLAB 仿真机械位移系统的脚本程序如图 4.4.6 所示。机械位移系统的阶跃输出结果如图 4.4.7 所示。取 $m=10\text{kg}$，$k=1\text{N/m}$，$b=2$，当输入是单位阶跃即输入外力是 $u(t)=1\text{N}$ 时，输出的稳态位移 $y(t)=1\text{m}$。

```
% 机械位移系统阶跃响应的 MATLAB 程序
m=10;
k=1;
b=2;
A=[0 1;-k/m -b/m];
B=[0 1/m]';
C=[1 0];
```

图 4.4.6 机械位移系统阶跃响应的程序

```
D=[0];
step(A,B,C,D)
grid
```

图 4.4.6　机械位移系统阶跃响应的程序（续）

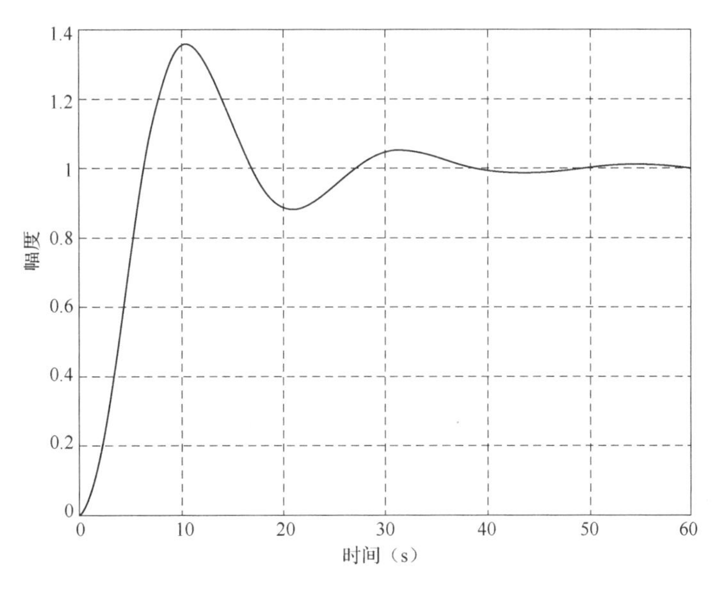

图 4.4.7　机械位移系统阶跃响应

从图 4.4.7 中可以看出，纵坐标的幅度是输出位移的大小。输出为质量块 m 的位移，当输入外力是单位阶跃时，质量块 m 在这个外力作用下，产生位移。位移逐渐增加，在一个小幅振荡后，在 40s 附近进入稳态值。在输入作用力 $u(t)=1\text{N}$ 时，输出的稳态位移 $y(t)=1\text{m}$ 。

本章小结

（1）本章详细介绍了状态空间方程的定义及其表达格式，以及相关函数的使用。

（2）本章阐明了三种模型形式及其相互之间的转换，并以实际例子说明了六个 MATLAB 转换函数的具体使用方式。

（3）本章以 RLC 电路和机械位移系统为例，给出了系统微分方程的推导过程，及其转化到状态空间模型的详细过程。

Chapter 4 State-Space Modeling

This chapter introduces how to use MATLAB to describe the state-space model of control systems, and lays foundation for its analysis and simulation of MATLAB.

The mathematical model of the control system is mainly divided into the classical transfer function model and the state-space model. The classical transfer function model has been introduced in Chapter 3. This chapter mainly focuses on the MATLAB description of the state-space model of the system, the conversion between models and the state-space modeling and simulation analysis of two simple physical systems.

4.1 State-Space Description

This section introduces the basic knowledge of state-space analysis of control systems.

4.1.1 Modern Control Theory

The modern trend in engineering systems is toward greater complexity, mainly due to the requirements of complex tasks and good accuracy. Complex systems may have multiple inputs and multiple outputs and may be time varying. Because of the necessity of meeting increasingly stringent requirements on the performance of control systems, the increase in system complexity, and easy access to large scale computers, modern control theory, a new approach to the analysis and design of complex control systems, has been developed since around 1960. This new approach is based on the concept of state. The concept of state by itself is not new, since it has been in existence for a long time in the field of classical mechanics and other fields.

4.1.2 Modern Control Theory Versus Conventional Control Theory

Modern control theory is contrasted with conventional control theory in that the former is applicable to multiple inputs and multiple outputs systems, which may be linear or nonlinear, time-invariant or time-varying, while the latter is applicable only to linear time-invariant single input, single output systems. Also, modern control theory is essentially time-domain approach and frequency-domain approach, while conventional control theory is a complex frequency-domain approach. The former is not concerned with the internal state of the system but regards the system as a black box and only pays attention to the relationship between input and output. The latter defines the internal state of the system and determines its state variables. The connection between input and output is achieved by using state variables as intermediate variables.

Before we proceed further, we must define state, state variables, state vector, and state space.

4.1.3 State

The state of a dynamic system is the smallest set of variables, called state variables. With the knowledge of the variables at $t=t_0$, together with the knowledge of the input for $t \geqslant t_0$ we can completely determine the behavior of the system for any time $t \geqslant t_0$.

Note that the concept of state is by no means limited to physical systems. It is applicable to biological systems, economic systems, sociological systems etc.

4.1.4 State Variables

The state variables of a dynamic system are the smallest set of variables that determine the state of the dynamic system. If at least *n* variables $x_1, x_2, \cdots, x_n$ are needed to completely describe the behavior of a dynamic system (so that once the input is given for $t \geqslant t_0$ and the initial state at

$t = t_0$ is specified, the future state of the system is completely determined), then such n variables are a set of state variables.

Note that state variables need not be physically measurable or observable quantities. Variables that do not represent physical quantities and those that are neither measurable nor observable can be chosen as state variables. Such freedom in choosing state variables is an advantage of the state-space methods. Practically, however, it is convenient to choose easily measurable quantities for the state variables, if this is possible at all, because optimal control laws will require the feedback of all state variables with proper weighting.

4.1.5 State Vector

If n state variables are needed to completely describe the behavior of a given system, then these n state variables can be considered the n components of a vector $\boldsymbol{x}$. Such a vector is called a state vector. A state vector is thus a vector that determines uniquely the system state $x(t)$ for any time $t \geqslant t_0$, once the state at $t = t_0$ is given and the input $u(t)$ for $t \geqslant t_0$ is specified.

4.1.6 State Space

The n-dimensional space whose coordinate axes consist of the x_1 axis, x_2 axis, $\cdots$, x_n axis, where $x_1, x_2, \cdots, x_n$ are state variables, is called a state space. Any state can be represented by a point in the state space.

4.1.7 State-Space Equations

In state-space analysis we are concerned with three types of variables that are involved in the modeling of dynamic systems: input variables, output variables, and state variables. The state-space representation for a given system is not unique, except that the number of state variables is the same for any of the different state-space representations of the same system.

The dynamic system must involve elements that memorize the values of the input for $t \gg t_1$. Since integrators in a continuous control system serve as memory devices, the outputs of such integrators can be considered as the variables that define the internal state of the dynamic system. Thus the outputs of integrators serve as state variables. The number of state variables to completely define the dynamics of the system is equal to the number of integrators involved in the system.

Assume that a multiple inputs and multiple outputs system involves n integrators. Assume also that there are r inputs $u_1(t), u_2(t), \cdots, u_r(t)$ and m outputs $y_1(t), y_2(t), \cdots, y_m(t)$. Define n outputs of the integrators as state variables: $x_1(t), x_2(t), \cdots, x_n(t)$. Then the system may be described by

$$
\begin{aligned}
\dot{x}_1(t) &= f_1(x_1, x_2, \cdots, x_n; u_1, u_2, \cdots, u_r; t) \\
\dot{x}_2(t) &= f_2(x_1, x_2, \cdots, x_n; u_1, u_2, \cdots, u_r; t) \\
&\vdots \\
\dot{x}_n(t) &= f_n(x_1, x_2, \cdots, x_n; u_1, u_2, \cdots, u_r; t)
\end{aligned}
\tag{4.1}
$$

The outputs $y_1(t), y_2(t), \cdots, y_m(t)$ of the system may be given by

$$\begin{aligned} y_1(t) &= g_1(x_1, x_2, \cdots, x_n; u_1, u_2, \cdots, u_r; t) \\ y_2(t) &= g_2(x_1, x_2, \cdots, x_n; u_1, u_2, \cdots, u_r; t) \\ &\vdots \\ y_m(t) &= g_m(x_1, x_2, \cdots, x_n; u_1, u_2, \cdots, u_r; t) \end{aligned} \tag{4.2}$$

If we define

$$x(t) = \begin{bmatrix} x_1(t) \\ x_2(t) \\ \vdots \\ x_n(t) \end{bmatrix}, f(x,u,t) = \begin{bmatrix} f_1(x_1, x_2, \cdots, x_n; u_1, u_2, \cdots, u_r; t) \\ f_2(x_1, x_2, \cdots, x_n; u_1, u_2, \cdots, u_r; t) \\ \vdots \\ f_n(x_1, x_2, \cdots, x_n; u_1, u_2, \cdots, u_r; t) \end{bmatrix},$$

$$y(t) = \begin{bmatrix} y_1(t) \\ y_2(t) \\ \vdots \\ y_3(t) \end{bmatrix}, g(x,u,t) = \begin{bmatrix} g_1(x_1, x_2, \cdots, x_n; u_1, u_2, \cdots, u_r; t) \\ g_2(x_1, x_2, \cdots, x_n; u_1, u_2, \cdots, u_r; t) \\ \vdots \\ g_m(x_1, x_2, \cdots, x_n; u_1, u_2, \cdots, u_r; t) \end{bmatrix}, u(t) = \begin{bmatrix} u_1(t) \\ u_2(t) \\ \vdots \\ u_r(t) \end{bmatrix}$$

then Equations (4.1) and (4.2) become

$$\dot{x}(t) = \boldsymbol{f}(x,u,t) \tag{4.3}$$

$$y(t) = \boldsymbol{g}(x,u,t) \tag{4.4}$$

where Equation (4.3) is the state equation and Equation (4.4) is the output equation. If vector functions $f(\cdot)$ and/or $g(\cdot)$ involve time t explicitly, then the system is called a time-varying system.

If Equations (4.3) and (4.4) are linearized about the operating state, then we have the following linearized state equation and output equation

$$\dot{x}(t) = \boldsymbol{A}(t)x(t) + \boldsymbol{B}(t)u(t) \tag{4.5}$$

$$y(t) = \boldsymbol{C}(t)x(t) + \boldsymbol{D}(t)u(t) \tag{4.6}$$

where $\boldsymbol{A(t)}$ is called the state matrix, $\boldsymbol{B(t)}$ the input matrix, $\boldsymbol{C(t)}$ the output matrix, and $\boldsymbol{D(t)}$ the direct transmission matrix.

4.2 State-Space Model in MATLAB

If there are n states, r inputs and m outputs in a linear time-invariant system, then the state-space model of the system represented by matrix is

$$\dot{x}(t) = \boldsymbol{A}(t)x(t) + \boldsymbol{B}(t)u(t)$$

$$y(t) = \boldsymbol{C}(t)x(t) + \boldsymbol{D}(t)u(t)$$

where the state vector $x(t)$ is n-dimensional, the input vector $u(t)$ r-dimensional, and the output vector $y(t)$ m-dimensional; the state matrix $\boldsymbol{A}$ is $n \times n$-dimensional, the input matrix $\boldsymbol{B}$ $n \times r$-dimensional, the output matrix $\boldsymbol{C}$ $m \times n$-dimensional, and the direct transmission matrix $\boldsymbol{D}$ $m \times r$-dimensional. In a time-invariant system $\boldsymbol{A}$, $\boldsymbol{B}$, $\boldsymbol{C}$ and $\boldsymbol{D}$ are constant matrices.

4.2.1 MATLAB Functions Related to State-Space Model

In MATLAB, the state-space model can be determined by the coefficient matrix $\boldsymbol{A}$, $\boldsymbol{B}$, $\boldsymbol{C}$, $\boldsymbol{D}$ of the equation, and created by the function *ss*() in MATLAB toolbox.

```
G = ss(A,B,C,D)
```

The description of several related functions of the state-space model are shown in Table 4.1.

Table 4.1 Related Functions of State-Space Models

Function	Description
sys=ss(A,B,C,D)	Return the variable sys directly from the matrix ***A***, ***B***, ***C***, ***D***. This is the space-state model of continuous system.
sys=ss(A,B,C,D,Ts)	Return the variable sys directly from the matrix ***A***, ***B***, ***C***, ***D*** and the sampling time T_s (seconds). This is the space-state mode of discrete system. T_s is the sampling period. When $T_s = -1$ or $T_s = []$, the system sampling period is not defined.
[A,B,C,D]=ssdata(sys)	Access space-state model data.
[A,B,C,D, T_s]=ssdata(sys)	Access space-state model data. Also include sample time T_s.

4.2.2 Create State-Space Model by Using Functions

Example 4.1 Enter the following state-space model of system into the MATLAB workspace.

$$\dot{x} = \begin{bmatrix} 1 & 6 & 9 & 10 \\ 3 & 12 & 6 & 8 \\ 4 & 7 & 9 & 11 \\ 5 & 12 & 13 & 14 \end{bmatrix} x + \begin{bmatrix} 4 & 6 \\ 2 & 4 \\ 2 & 2 \\ 1 & 0 \end{bmatrix} u$$

$$y = \begin{bmatrix} 0 & 0 & 2 & 1 \\ 8 & 0 & 2 & 2 \end{bmatrix} x$$

We can know the system is a double inputs and double outputs system from the dimensions of the ***B*** and ***C*** matrices. The MATLAB script and its results are shown in Figure 4.2.1.

```
% The MATLAB script of Example 4.1
A=[1 6 9 10; 3 12 6 8; 4 7 9 11; 5 12 13 14]; %Enter the values of state matrix A, B, C, and D
B=[4 6; 2 4; 2 2; 1 0];
C=[0 0 2 1; 8 0 2 2];
D=zeros(2,2);
G=ss(A,B,C,D)
```

```
% The results of Example 4.1
G =
  a =
        x1  x2  x3  x4
   x1    1   6   9  10
   x2    3  12   6   8
   x3    4   7   9  11
   x4    5  12  13  14
```

Figure 4.2.1 MATLAB Script and Results of Example 4.1

```
b =
      u1  u2
  x1   4   6
  x2   2   4
  x3   2   2
  x4   1   0
c =
      x1  x2  x3  x4
  y1   0   0   2   1
  y2   8   0   2   2
d =
      u1  u2
  y1   0   0
  y2   0   0
Continuous-time state-space model.
```

Figure 4.2.1 The MATLAB Script and Results of Example 4.1 (continued)

In this example, the direct transmission matrix ***D*** is not given in the system model. The matrix ***D*** must be created, where the matrix elements should be zero. Note that the dimension of the matrix ***D*** is determined by the system's output and input dimensions. The dimension of the matrix ***D*** must be compatible with the dimensions, of other matrix otherwise the function *ss()* will give a clear error message and interrupt the program operation.

Example 4.2 Given the state-space model of system

$$\dot{x} = \begin{bmatrix} 0 & 1 \\ 1 & -2 \end{bmatrix} x + \begin{bmatrix} 0 \\ 1 \end{bmatrix} u$$

$$y = [1 \quad 3]x + u$$

Find the system parameters.

The script and results of Example 4.2 are shown in Figure 4.2.2.

```
% The MATLAB script 1 of Example 4.2
A=[0 1; -1 -2];  B=[0;1];        % Enter the values of state matrix A, B, C, and D
C=[1,3];  D=[1];
G=ss(A,B,C,D)
```

```
% The results of MATLAB script 1 of Example 4.2
G =
  a =
       x1  x2
   x1   0   1
   x2  -1  -2

  b =
       u1
```

Figure 4.2.2 MATLAB Script and Results of Example 4.2

```
    x1   0
    x2   1

  c =
        x1  x2
    y1   1   3

  d =
        u1

  y1    1

Continuous-time state-space model.
```

```
% The MATLAB script 2 of Example 4.2
[ad,bd,cd,dd]=ssdata(G)        % Get model parameters
```

```
% The results of MATLAB script 2 of Example 4.2
ad =

     0     1
    -1    -2

bd =

     0
     1

cd =

     1     3

dd =

     1
```

```
% The MATLAB script 3 of Example 4.2
get(G)                          % Get information about all parameters of the system model
```

```
% The results of MATLAB script 3of Example 4.2
                 a: [2x2 double]
                 b: [2x1 double]
                 c: [1 3]
                 d: 1
                 e: []
            Scaled: 0
```

Figure 4.2.2 MATLAB Script and Results of Example 4.2（continued）

```
StateName: {2x1 cell}
        StateUnit: {2x1 cell}
    InternalDelay: [0x1 double]
       InputDelay: 0
      OutputDelay: 0
               Ts: 0
         TimeUnit: 'seconds'
        InputName: {''}
        InputUnit: {''}
       InputGroup: [1x1 struct]
       OutputName: {''}
       OutputUnit: {''}
      OutputGroup: [1x1 struct]
             Name: ''
            Notes: {}
         UserData: []
     SamplingGrid: [1x1 struct]
```

```
% The MATLAB script 4 of Example 4.2
G.a                          % Find the parameters of a system model
```

```
% The results of MATLAB script 4 of Example 4.2
ans =


     0     1
    -1    -2
```

Figure 4.2.2 MATLAB Script and Results of Example 4.2 (continued)

Example 4.2 illustrates different ways to obtain the parameters of the state-space model. It should be noted that the parameters are in matrix format. If all the parameter information of the model is not needed, the single parameter information of the model can be obtained by G.a .

4.3 Conversions Between Models

The model of control systems mainly includes three forms, i.e. transfer function model (tf), zero-pole model (zpk) and state-space model (ss). With intrinsic links each other, they can be converted (shown in Figure 4.3.1). MATLAB provides functions of conversions between models.

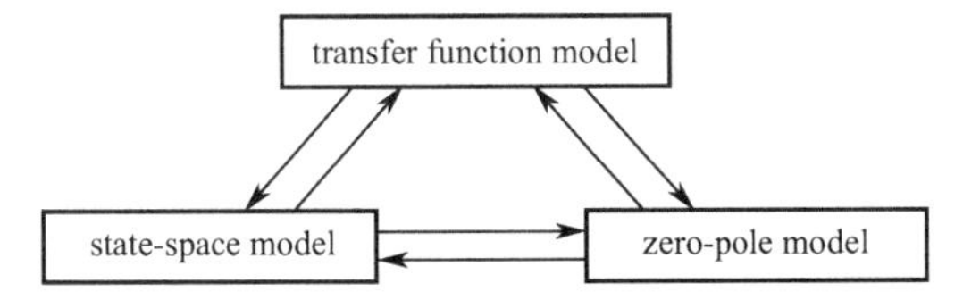

Figure 4.3.1 Conversions Between Three Models

4.3.1 Functions Related to Conversions Between Models

The conversion of the system model is mainly divided into two types. The first type is to create the model, converting other forms of the model into a model of the function itself. There are mainly three functions corresponding to three model types in the Figure 4.3.1, as shown in Table 4.2.The second type is to convert the parameter expression of the function itself into that of other functions. There are mainly six functions to realize the format conversion between three models in Figure 4.3.1, as showed in Table 4.3.

Table 4.2 Description of Three Model Functions

Function	Description
G_{tf} =tf(sys)	Convert other types of models to transfer function models
G_{zpk} =zpk(sys)	Convert other types of models to zero-pole models
G_{ss} =ss(sys)	Convert other types of models to state-space models

Table 4.3 Description of Six Model Conversion Functions

Function	Description
[z,p,k]=tf2zp(num,den)	Convert transfer function models to zero-pole models
[num,den]=zp2tf(z,p,k)	Convert zero-pole models to transfer function models
[A,B,C,D]= zp2ss(z,p,k)	Convert zero-pole models to state-space models
[z,p,k]=ss2zp(A,B,C,D,iu)	Convert state-space models to zero-pole-models. ***iu*** is the ***i*** input
[A,B,C,D]= tf2ss(num,den)	Convert transfer-function models to state-space models
[num,den]=ss2tf(A,B,C,D,iu)	Convert state-space models to transfer function models. ***iu*** is the ***i*** input

4.3.2 Examples of conversion Between Models

Example 4.3 Given the state-space model of system

$$\dot{x}=\begin{bmatrix}0 & 1\\ 1 & -2\end{bmatrix}x+\begin{bmatrix}0\\ 1\end{bmatrix}u$$

$$y=[1 \quad 3]x+u$$

Find its transfer function model and zero-pole model.

The script and results of Example 4.3 are shown in Figure 4.3.2.

```
% The MATLAB script 1 of Example 4.3
[z,p,k]=ss2zp(A,B,C,D,iu)
C=[1,3]; D=[1];
[num,den]=ss2tf(A,B,C,D,1)    % iu=1, or omitted, when there is only one input
```

Figure 4.3.2 MATLAB Script and Results of Example 4.3

```
% The first results of MATLAB script 1 of Example 4.3
num =

    1.0000    5.0000    2.0000

den =

     1     2     1
```

```
% The MATLAB script 2 of Example 4.3
[z,p,k]=ss2zp(A,B,C,D)   % iu=1, or omitted, when there is only one input
```

```
% The second results of MATLAB script 2 of Example 4.3
z =

   -0.4384
   -4.5616


p =

    -1
    -1


k =

     1
```

Figure 4.3.2 MATLAB Script and Results of Example 4.3（continued）

In Example 4.3, it should be noted that when state-space models are converted to transfer function models and zero-pole models, the value of ***iu***, the parameter in the function, must be defined, with one set of ***iu*** values corresponding to an input-output port. This is because the state-space model can represent multiple inputs and multiple outputs systems, while the transfer function model and the pole-zero model both represent single input and single output systems.

Example 4.4 Given a transfer function model of a single input three outputs system

$$G_{11}(s)=\frac{y_1(s)}{u(s)}=\frac{-2}{s^3+6s^2+11s+6}$$

$$G_{21}(s)=\frac{y_2(s)}{u(s)}=\frac{-s-5}{s^3+6s^2+11s+6}$$

$$G_{31}(s)=\frac{y_3(s)}{u(s)}=\frac{s^2+2s}{s^3+6s^2+11s+6}$$

Find the state-space model of system.

The script and results of Example 4.4 are shown in Figure 4.3.3.

```
%The MATLAB script of Example 4.4
num=[0 0 -2;0 -1 -5;1 2 0];     % The numerator is a 3×3 matrix, the first row corresponds to
                                % the coefficient of the molecule s in G11(s)
den=[1 6 11 6];                 % Corresponds to the coefficient of the molecular s in G21(s) ,
                                % and so on
[A,B,C,D]=tf2ss(num,den)
```

```
%The results of MATLAB script of Example 4.4
A =

    -6   -11    -6
     1     0     0
     0     1     0


B =

     1
     0
     0


C =

     0     0    -2
     0    -1    -5
     1     2     0


D =

     0
     0
     0
```

Figure 4.3.3 MATLAB Script and Results of Example 4.4

It should be noted that the system in Example 4.4 is a single input three outputs system, thus the transfer function being three rows and one column, and the molecular polynomials *num* being a 3×3 matrix rather than a number. The number of rows of *num* matrix is determined by the number of outputs, and the number of columns is determined by the highest power of the molecular s in the three transfer functions. In this example, the power of the molecular s in the third transfer function $G_{31}(s)$ is the highest (i.e. equal to 2) , so the number of columns of the molecular is 3. Three rows and three columns values in the num matrix correspond to the coefficients of s in descending powers. S is the parameters of molecular polynomials in three transfer functions.

Example 4.5 Given the zero-pole model of the system

$$G(s) = \frac{6(s+3)}{(s+1)(s+2)(s+5)}$$

Find the transfer function model and state-space model of the system.

The script and results of Example 4.5 are shown in Figure 4.3.4.

```
%The MATLAB script 1 of Example 4.5
z=[-3]; p=[-1, -2, -5]; k=6;    % System zero-pole parameter input, the pole input format is the
                                % row vector
[num,den]=zp2tf(z,p,k)
```

```
%The results of MATLAB script 1 for Example 4.5
num =

     0     0     6    18


den =

     1     8    17    10
```

```
%The MATLAB script 2 of Example 4.5
z=[-3]; p=[-1; -2; -5]; k=6;    % The input of pole-zero parameter input and the pole format of
                                % system is columns vector
[a,b,c,d]=zp2ss(z,p,k)
```

```
%The results of MATLAB script 2 for Example 4.5
a =

   -1.0000         0         0
    2.0000   -7.0000   -3.1623
         0    3.1623         0

b =
     1
     1
     0

c =
          0         0    1.8974

d =

     0
```

Figure 4.3.4 MATLAB Script and Results of Example 4.5

In Example 4.5, the pole matrix $\boldsymbol{p}$ was created by using the row vector in Script 1, while in Script 2, the pole matrix $\boldsymbol{p}$ was created by using the column vector. The input of zero matrixes and pole matrixes can be both in row vectors and in column vectors.

4.3.3 Simplification of State-Space Models

The state-space model can represent multiple inputs and multiple outputs systems. For the serial connection, parallel connection, and feedback of the model, a connection terminal label must be specified to ensure its correctness. In Chapter 3, the three link functions of serial connection, parallel connection and feedback in MATLAB are introduced. This section is to illustrate the specific use of these functions in the state-space model.

Example 4.6 Given System 1

$$\dot{x}_1 = \begin{bmatrix} 0 & 1 \\ 1 & -2 \end{bmatrix} x_1 + \begin{bmatrix} 0 \\ 1 \end{bmatrix} u_1$$

$$y_1 = [1 \quad 3] x_1 + u_1$$

And System 2

$$\dot{x}_2 = \begin{bmatrix} 0 & 1 \\ -1 & -3 \end{bmatrix} x_2 + \begin{bmatrix} 0 \\ 1 \end{bmatrix} u_2$$

$$y_2 = [1 \quad 4] x_2$$

Find the state-space model of system when connected in series, parallel, and positive/negative feedbacks.

The script and results of Example 4.6 are shown in Figure 4.3.5.

```
%The MTLAB script of Example 4.6
a1=[0 1; -1 -2];b1=[0;1];
c1=[1 3];d1=[1];
a2=[0 1; -1 -3];b2=[0;1];
c2=[1 4];d2=[0];
disp('Serial connection')
[a,b,c,d]=series(a1,b1,c1,d1,a2,b2,c2,d2)            % Serial connection
disp('Parallel connection')
[a,b,c,d]=parallel(a1,b1,c1,d1,a2,b2,c2,d2)          % Parallel connection
disp('Positive feedback connection')
[a,b,c,d]=feedback(a1,b1,c1,d1,a2,b2,c2,d2,+1)       % Positive feedback connection
disp('Negative feedback connection')
[a,b,c,d]=feedback(a1,b1,c1,d1,a2,b2,c2,d2)          % Negative feedback connection
disp('Unit negative feedback connection')
[a,b,c,d]=cloop(a1,b1,c1,d1)                         %Unit negative feedback connection
```

```
%The results of MTLAB script for Example 4.6
Serial connection
a =
     0     1     0     0
    -1    -3     1     3
     0     0     0     1
     0     0    -1    -2
b =

     0
     1
```

Figure 4.3.5 MATLAB Script and Results of Example 4.6

```
      0
      1

c =

      1    4    0    0

d =

      0

Parallel connection
a =

      0    1    0    0
     -1   -2    0    0
      0    0    0    1
      0    0   -1   -3

b =

      0
      1
      0
      1

c =
      1    3    1    4

d =

      1

Positive feedback connection
a =

      0    1    0    0
     -1   -2    1    4
      0    0    0    1
      1    3    0    1

b =
```

Figure 4.3.5 MATLAB Script and Results of Example 4.6（continued）

```
     0
     1
     0
     1

c =
     1     3     1     4

d =

     1

Negative feedback connection
a =

     0     1     0     0
    -1    -2    -1    -4
     0     0     0     1
     1     3    -2    -7

b =

     0
     1
     0
     1

c =
     1     3    -1    -4

d =

     1

Unit negative feedback connection
a =

         0    1.0000
   -1.5000   -3.5000
b =
         0
    0.5000

c =

    0.5000    1.5000
```

Figure 4.3.5 MATLAB Script and Results of Example 4.6（continued）

```
d =

    0.5000
```

Figure 4.3.5 MATLAB Script and Results of Example 4.6 (continued)

Example 4.6 illustrates two single input and single output state-space models in serial, parallel and feedback connections. The functions are applied in the same way as in the transfer function, except that they are valued by the parameters of the state-space model.

Example 4.7 Given the state-space model of System 1 and System 2

System 1

$$\begin{bmatrix} \dot{x}_{11} \\ \dot{x}_{12} \\ \dot{x}_{13} \end{bmatrix} = \begin{bmatrix} 1 & 4 & 4 \\ 2 & 2 & 1 \\ 3 & 6 & 2 \end{bmatrix} \begin{bmatrix} x_{11} \\ x_{12} \\ x_{13} \end{bmatrix} + \begin{bmatrix} 0 & 1 & 0 \\ 1 & 0 & 0 \\ 0 & 0 & 1 \end{bmatrix} \begin{bmatrix} u_{11} \\ u_{12} \\ u_{13} \end{bmatrix}$$

$$\begin{bmatrix} y_{11} \\ y_{12} \end{bmatrix} = \begin{bmatrix} 0 & 0 & 1 \\ 0 & 1 & 1 \end{bmatrix} \begin{bmatrix} x_{11} \\ x_{12} \\ x_{13} \end{bmatrix} + \begin{bmatrix} 0 & 1 & 0 \\ 1 & 0 & 1 \end{bmatrix} \begin{bmatrix} u_{11} \\ u_{12} \\ u_{13} \end{bmatrix}$$

System 2

$$\begin{bmatrix} \dot{x}_{21} \\ \dot{x}_{22} \\ \dot{x}_{23} \end{bmatrix} = \begin{bmatrix} 1 & -1 & 0 \\ 3 & -2 & 1 \\ 1 & 6 & -1 \end{bmatrix} \begin{bmatrix} x_{21} \\ x_{22} \\ x_{23} \end{bmatrix} + \begin{bmatrix} 1 & 0 & 0 \\ 0 & 1 & 0 \\ 0 & 0 & 1 \end{bmatrix} \begin{bmatrix} u_{21} \\ u_{22} \\ u_{23} \end{bmatrix}$$

$$\begin{bmatrix} y_{21} \\ y_{22} \end{bmatrix} = \begin{bmatrix} 0 & 1 & 0 \\ 1 & 0 & 1 \end{bmatrix} \begin{bmatrix} x_{21} \\ x_{22} \\ x_{23} \end{bmatrix} + \begin{bmatrix} 1 & 1 & 0 \\ 1 & 0 & 1 \end{bmatrix} \begin{bmatrix} u_{21} \\ u_{22} \\ u_{23} \end{bmatrix}$$

Find a partial parallel state-space model which requires connection between $\boldsymbol{u}_{11}$-and-$\boldsymbol{u}_{22}$, $\boldsymbol{u}_{13}$-and-$\boldsymbol{u}_{23}$, and $\boldsymbol{y}_{11}$-and-$\boldsymbol{y}_{21}$.

The script and results of Example 4.7 are shown in Figure 4.3.6.

```
%The MTLAB script of Example 4.7
a1=[1 4 4;2 2 1;3 6 2];   b1=[0 1 0;1 0 0;0 0 1];        % Parameter input of the first system
c1=[0 0 1;0 1 1]; d1=[0 1 0;1 0 1];
a2=[1 -1 0;3 -2 1;1 6 -1];    b2=[1 0 0;0 1 0;0 0 1];     % Parameter input of the second system
c2=[0 1 0;1 0 1]; d2=[1 1 0;1 0 1];
% Partial parallel state-space requires u11 to connect to u22, u13 connect to u23, and y11 connect
% to y21
disp('Partial parallel state-space model')
[a,b,c,d]=parallel(a1,b1,c1,d1,a2,b2,c2,d2,[1 3],[2 3],1,1)
%input1=[1 3]   input2=[2 3]
%output1=1      output2=1
```

Figure 4.3.6 MATLAB Script and Results of Example 4.7

```
%The results of MTLAB script of Example 4.7
Partial parallel state-space equations

a =

     1     4     4     0     0     0
     2     2     1     0     0     0
     3     6     2     0     0     0
     0     0     0     1    -1     0
     0     0     0     3    -2     1
     0     0     0     1     6    -1


b =

     0     1     0     0
     1     0     0     0
     0     0     1     0
     0     0     0     1
     1     0     0     0
     0     0     1     0


c =

     0     0     1     0     1     0
     0     1     1     0     0     0
     0     0     0     1     0     1


d =

     1     1     0     1
     1     0     1     0
     0     0     1     1
```

Figure 4.3.6 MATLAB Script and Results of Example 4.7（continued）

The two systems in Example 4.6 are both three inputs and two outputs systems requiring the parallel connection. We must determine the connection terminals when using the parallel function. First, at the input terminals of the two systems, input1=[1 3] and input 2=[2 3], Input Terminal 1 in System 1 and Input Terminal 2 in System 2 are connected while Input Terminal 3 in System 1 and Input Terminal 3 in System 2 are connected, thus connecting the input terminals in parallel of the two systems.

Next, determine the output terminals. There are two outputs in both systems, but only one terminal is required respectively for connection. Here with output1=1 and output2=1, both systems select their output 1 for connection. It should be noted that when the number of terminals to be connected is larger than or equal to 2, the parameters to define terminals are arrays rather than separated numbers, which needs to be enclosed in square brackets. The data in the array must agree with the label of its corresponding terminal. Pay attention to match all these to the dimensions of the data.

4.4 RLC Circuits and Mechanical Displacement Systems

Both the RLC circuit and the mechanical displacement systems have been introduced in Chapter 3 about their respective transfer function modeling. This section goes further to illustrate the process of building their state-space model.

4.4.1 State-Space Modeling of RLC Circuit

Let's analyze the Response of an RLC Circuit.

Consider the RLC Circuit shown in Figure 4.4.1, where

V_{in} is input voltage (volts) ,

V_{out} is output voltage (volts),

L is inductance (H),

R is resistance (Ω),

C is capacitance (F),

i is current across the resistance (amps),

i_1 is current across the inductance (amps),

i_2 is current across the capacitor (amps).

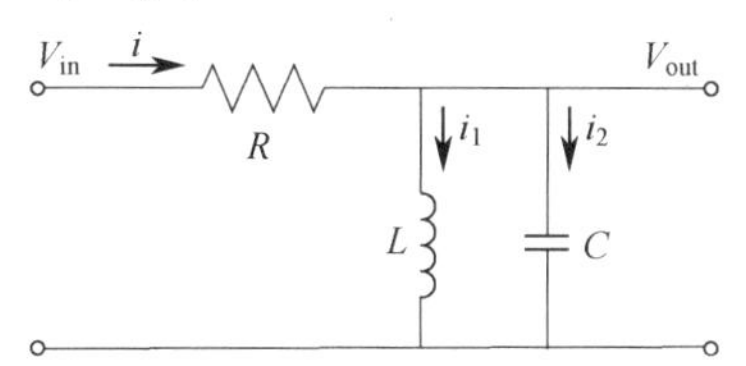

Figure 4.4.1 The RLC Circuit

Using Kirchhoff's Circuit Laws, we obtain

$$V_{in} = Ri + V_{out} \tag{4.7}$$

$$i = i_1 + i_2 \tag{4.8}$$

where

$$V_{out} = \frac{1}{C}\int i_2 \mathrm{d}t = L\frac{\mathrm{d}i_1}{\mathrm{d}t} \tag{4.9}$$

We define the state variable as

$$x_1 = V_{out}$$

$$x_2 = i_1$$

Then taking the derivative of x_1 by means of Eq.(4.7), (4.8) and (4.9) yields

$$\dot{x}_1 = -\frac{1}{RC}x_1 - \frac{1}{C}x_2 + \frac{1}{RC}V_{in} \tag{4.10}$$

Also, taking the time derivative of x_2 by means of Eq.(4.9) yields

$$\dot{x}_2 = \frac{1}{L}x_1 \tag{4.11}$$

We can write Eqs.(4.10) and (4.11) in the form of matrix as

$$\dot{x} = \boldsymbol{A}x + \boldsymbol{B}u$$

where

$$x = \begin{pmatrix} x_1 \\ x_2 \end{pmatrix} = \begin{pmatrix} V_{\text{out}} \\ i \end{pmatrix},\ u = V_{\text{in}},$$

And

$$\boldsymbol{A} = \begin{bmatrix} -\dfrac{1}{RC} & -\dfrac{1}{C} \\ \dfrac{1}{L} & 0 \end{bmatrix}, \qquad \boldsymbol{B} = \begin{bmatrix} \dfrac{1}{RC} \\ 0 \end{bmatrix}$$

With R=1 Ω, L=0.25H and C=0.5F, we have

$$\dot{x} = \begin{bmatrix} -2 & -4 \\ 4 & 0 \end{bmatrix} x + \begin{bmatrix} 2 \\ 0 \end{bmatrix} u$$

If we measure V_{out}, then we have

$$y = \boldsymbol{C}x + \boldsymbol{D}u$$

where

$$\boldsymbol{C} = [1 \quad 0], \text{ and } \boldsymbol{D} = [0].$$

4.4.2 Simulation Analysis of an RLC Circuit

The MATLAB script used to simulate the step response of the RLC circuit is shown in Figure 4.4.2. The magnitude of the ordinate is the magnitude of the output voltage V_{out}. When the input is a unit step, that is, when $V_{\text{in}}(t) = 1\text{V}$, the steady-state voltage is $V_{\text{out}}(t) = 0\text{V}$.

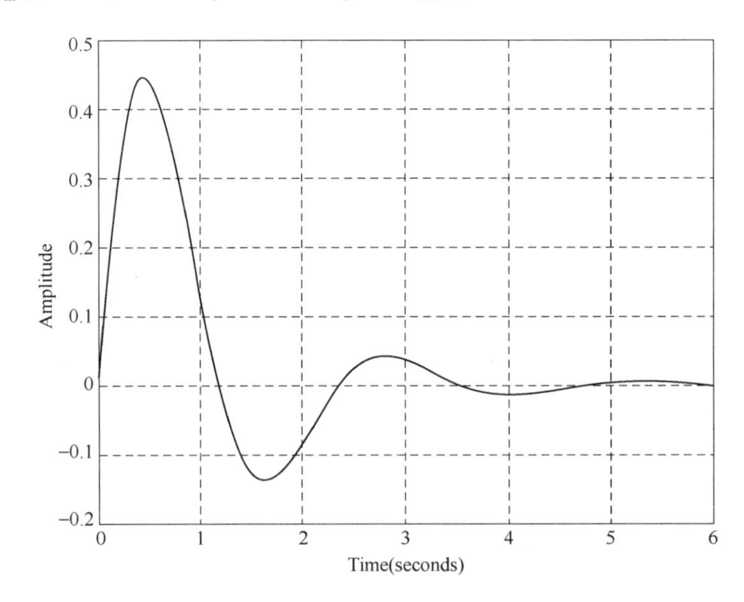

Figure 4.4.2 RLC Step Respond

From Fig. 4.4.2, it can be seen that the voltage of the capacitor follows the variation of the input voltage, and has a small oscillation of the transient before it enters a steady state zero value in the vicinity of 5 seconds.

```
%Model Parameters
R=1;
L=0.25;
C=0.5;
%State-Space Model
A=[-1/(R*C) -1/C;1/L 0];
B=[1/(R*C) 0]';
C=[1 0];
D=[0];
%Simulated step response
step(A,B,C,D)
grid
```

Figure 4.4.3 Script of RLC Step Respond

In this example, we can also choose to output the inductor current i_1. Change the ***C*** matrix value in the output equation to ***C***=[0 1], with the others being unchanged and still by means of the script in Figure 4.4.2. Reset the ***C*** matrix value, the system step response is shown in Figure 4.4.4.

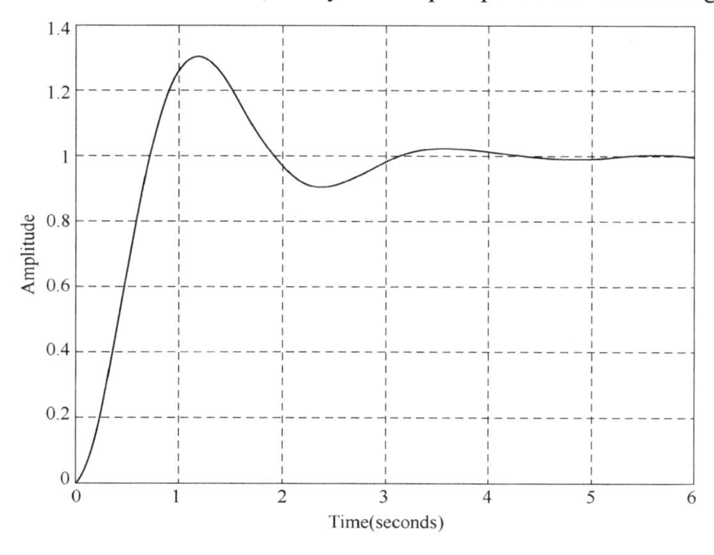

Figure 4.4.4 Changed RLC Step Response

It can be seen from Figure 4.4.3 that when the output is changed to the inductor current, the inductor current follows the variation of input voltage, and the output gradually increases. After a small oscillation, it enters the steady-state current value 1A near 4.5 seconds. At the input voltage $V_{\text{in}}(t)=1\text{V}$, the output current $i(t)=1\text{A}$.

4.4.3 State-Space Modeling of Mechanical Displacement Systems

The mechanical displacement system is shown in Figure 4.4.5. We assume that the system is linear. The external force $u(t)$ is the input to the system, and the displacement $y(t)$ of the mass is the output. The displacement $y(t)$ is measured from the equilibrium point in the absence of the external force. This system is a single input and single output system, where k is spring constant, m is mass of wood, b is damping coefficient, $u(t)$ is input of system, $y(t)$ is output of system.

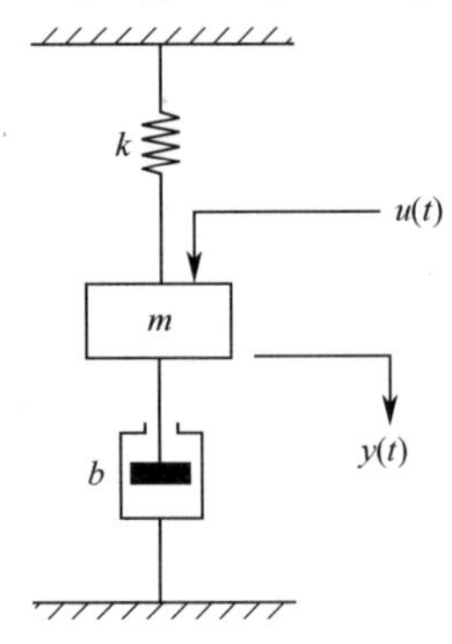

Figure 4.4.5 Mechanical Displacement System

From Figure 4.4.5, we get the system equation

$$m\ddot{y}+b\dot{y}+ky=u \tag{4.12}$$

This is a second order system. It means that the system involves two integrators. Let us define state variables $x_1(t)$ and $x_2(t)$ as

$$x_1(t)=y(t)$$
$$x_2(t)=\dot{y}(t)$$

Then we obtain

$$\dot{x}_1=x_2$$
$$\dot{x}_2=\frac{1}{m}(-ky-b\dot{y})+\frac{1}{m}u$$

or

$$\dot{x}_1=x_2 \tag{4.13}$$
$$\dot{x}_2=-\frac{k}{m}x_1-\frac{b}{m}x_2+\frac{1}{m}u \tag{4.14}$$

The output equation is

$$y=x_1 \tag{4.15}$$

In the form of matrix, Eqs. (4.13) and (4.14) can be written as

$$\begin{bmatrix}\dot{x}_1\\ \dot{x}_2\end{bmatrix}=\begin{bmatrix}0 & 1\\ -\frac{k}{m} & -\frac{b}{m}\end{bmatrix}\begin{bmatrix}x_1\\ x_2\end{bmatrix}+\begin{bmatrix}0\\ \frac{1}{m}\end{bmatrix}u \tag{4.16}$$

The output Eq. (4.15) can be written as

$$y=[1\quad 0]\begin{bmatrix}x_1\\ x_2\end{bmatrix} \tag{4.17}$$

Eq. (4.16) is a state equation and Eq. (4.17) is an output equation for the system. They can be written in the standard form

$$\begin{aligned}\dot{x}&=\boldsymbol{A}x+\boldsymbol{B}u\\ y&=\boldsymbol{C}x+\boldsymbol{D}u\end{aligned} \tag{4.18}$$

where

$$A=\begin{bmatrix} 0 & 1 \\ -\frac{k}{m} & -\frac{b}{m} \end{bmatrix}, B=\begin{bmatrix} 0 \\ \frac{1}{m} \end{bmatrix}, C=[1 \quad 0], D=0$$

4.4.4 Simulation Analysis A Mechanical Displacement System

The simulation script of the mechanical displacement system by MATLAB is shown in Figure 4.4.6. The step output of the mechanical displacement system is shown in Figure 4.4.7. With $m=10\text{kg}$, $k=1\text{N/m}$, $b=2$, when the input is a unit step, that is, when the external force of the input is $u(t)=1\text{N}$, the steady-state displacement of the output $y(t)=1\text{m}$.

```
%MATLAB script of mechanical displacement system step response
m=10;
k=1;
b=2;
A=[0 1;-k/m -b/m];
B=[0 1/m]';
C=[1 0];
D=[0];
step(A,B,C,D)
grid
```

Figure 4.4.6 MATLAB Script of Mechanical Displacement System Step Response

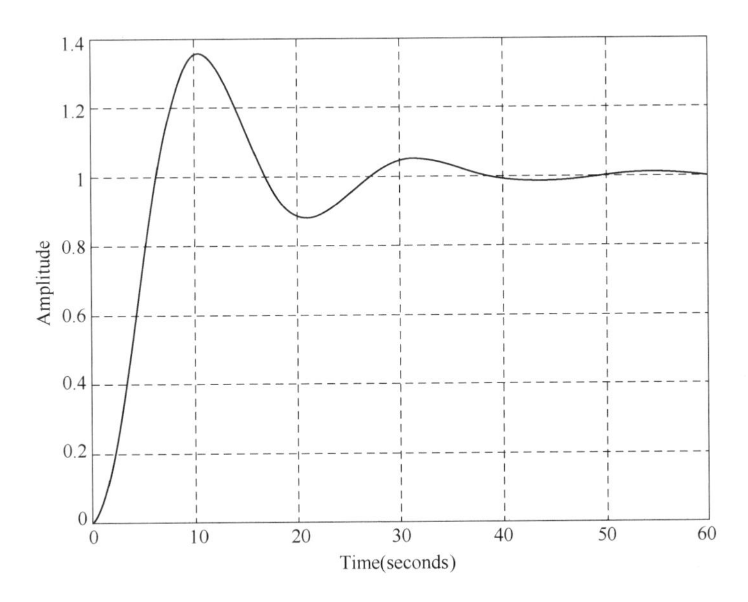

Figure 4.4.7 Mechanical Displacement System Step Response

As can be seen from Figure 4.4.7, the magnitude of the ordinate is the magnitude of the output displacement, the output is the displacement of the mass ***m***. When the external force of the input is a unit step, the mass ***m*** is displaced under the external force. The displacement gradually increased and entered a steady state value around 40 seconds after a small oscillation.When the external force of the input $u(t)=1\text{N}$, the steady-state displacement of the output $y(t)=1\text{m}$.

Summary

1. This chapter specifies the definition and expression of the state-space equation as well as the use of related functions.

2. Three model types and their transformations are clarified and the use of six MATLAB Transformation functions are also illustrated in this chapter.

3. The derivation process of the system differential equations is given by taking RLC circuits and mechanical displacement systems as examples. Besides, the detailed process of transforming differential equations into state-space equations is also given in this chapter.

生词注解

practically	adv. 实际上，几乎，实际地
components	n. 部件；组件；成分
coordinate	n. 坐标
	adj. 并列的，等同的
	v. 调整，整合
axes	n. 轴心，轴线
integrators	n. 积分器，积分电路
dimensional	adj. 空间的，尺寸的
intrinsic	adj. 本质的，固有的
molecular	adj. 分子的
polynomials	n. 多项式
derivative	n. 衍生物，派生物
oscillation	n. 振荡
equilibrium	n. 均衡，平衡

第 5 章 双容水箱液位系统建模

第 3 章和第 4 章重点介绍了如何使用 MATLAB 软件对给定模型的控制系统进行仿真分析。在此基础上，本章重点介绍建立实际物理系统的数学模型过程。以双容水箱液位系统为例，建立其非线性动态微分方程模型，将其线性化得到传递函数模型，最后对非线性和线性模型进行了简单分析。

5.1 建模方法简介

在控制系统设计过程中，关键步骤之一是对被控对象进行数学建模。本章讨论双容水箱液位系统数学模型的建立。

控制系统的建模步骤如下：

（1）定义系统及其组成；

（2）进行必要的假设，列出物理等效的数学表达式；

（3）推导系统微分方程模型；

（4）确定输入和输出变量；

（5）求出输出变量的解；

（6）验证上述结果和假设。

本章的核心内容是建立物理系统的数学模型，以确定控制系统的组成作为出发点。由于控制系统由控制器、执行器和被控对象等多个环节构成，因此确定系统的组成需要考虑更多的因素。本章仅考虑获得被控对象模型的过程，控制器和执行器的选择及建模相关问题均不讨论。以双容水箱液位系统为例，推导其物理过程，得到非线性模型，通过线性化得到线性模型。最后，采用 MATLAB 仿真加以分析。

5.2 双容水箱液位系统

液位的高低在生产中是一个重要的参数，对其进行控制有着广泛的需求。例如，需测量油罐等容器内的液面高度以计算产品产量和原料消耗；化工反应塔内需保持一定的液位并取得较高的生产率。双容水箱液位系统是工业生产过程中常见的被控对象，由两个具有自平衡能力的单容水箱上下串联而成，通常要求对其下水箱液位进行控制，液位为被控量，选取上水箱的进水流量为给定量。双容水箱一般表现出二阶特性。

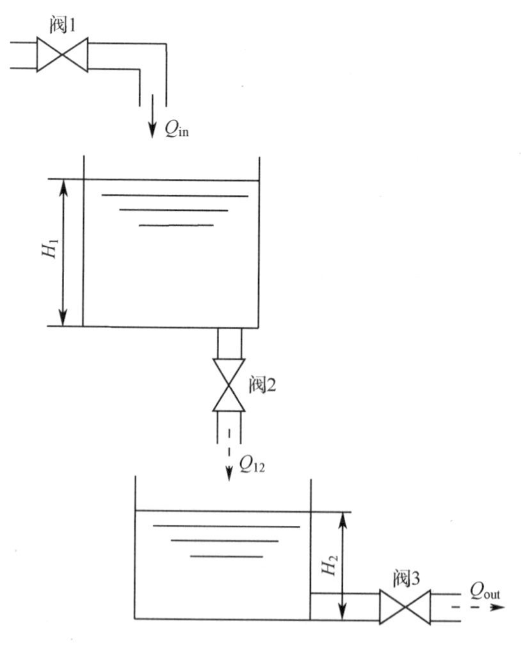

图 5.2.1 双容水箱液位系统

双容水箱液位系统如图 5.2.1 所示。上水箱和下水箱串联构成液位系统，水通过由阀 1 控制的管道供给上水箱，上水箱里面的水由阀 2 控制的管道排入下水箱，最终由阀 3 控制的管道流出。

图 5.2.1 中 Q_{in}、Q_{12} 和 Q_{out} 分别为各管道的瞬时流量，本章以 Q_{in} 为系统输入，研究由输入引起的水位高度 H_1、H_2 和水流量 Q_{12}、Q_{out} 的变化。所涉及的变量有流体速度 V（m/s），水位高度 H（m）和压强 P（N/m^2）。

5.3 假设

通常，在对物理系统建模时，需要根据其物理特性进行适当的假设，从而降低数学模型的复杂性，提高建模的可行性。本章针对流体进行建模，需要做如下假设。

5.3.1 压缩性

假设水箱中水是无压缩的，则具有恒定的密度ρ（kg/m^3）。现实中，在一定程度内，所有的液体都是可压缩的，压缩系数k是液体可压缩性的量度，k值越小说明可压缩性越小。空气压缩系数$k_{\text{air}}=0.98\text{m}^2/\text{N}$，水的压缩系数是$k_{\text{water}}=4.9\times10^{-10}\,\text{m}^2/\text{N}=50\times10^{-6}\,\text{atm}^{-1}$。在本例中可假设水是无压缩的。

5.3.2 黏度

假设流体是流动的，液体相邻层的初始流速是不同的。那么在层间就形成内摩擦和动量的交换，将其称为黏度。固体比流体更黏滞，流体比气体更黏滞。衡量黏度程度用黏度系数μ（Nsec/m^2）来表示。较大的黏度系数意味着较高的黏度。空气的黏度系数为（在标准条件 20℃下）

$$\mu_{\text{air}}=0.178\times10^{-4}\,\text{N}\,\text{s}/\,\text{m}^2$$

水的黏度系数为

$$\mu_{\text{H}_2\text{O}}=1.054\times10^{-3}\,\text{N}\,\text{s}/\,\text{m}^2$$

水黏度约为空气黏度的 60 倍。黏度取决于温度而不是压力。对低黏度的液体，如水，只有边界层摩擦力才显著，由于水箱内层和输出管道上仅存在很薄的边界层，因此可以忽略黏度，称流体是无黏度的。

5.3.3 无旋流

在液体中，如果液体每个点上都没有净角速度，此时的液体被视为无旋流。这里可以假设在水箱中的水是无旋流的。

5.3.4 稳定流动

如果在液体中每一个点的速度都是恒定值，不随时间改变，则液体是稳定流动的。稳定状态可以在低流速下实现。在本例中，水箱出水口面积较小，形成水箱内水流速缓慢，因此，可假设水箱内的水是稳定流动的。

5.4 运动微分方程

在任意给定时间里上水箱中水的质量为

$$m_1=\rho A_1 H_1 \tag{5.1}$$

式中，A_1是上水箱的横截面积，ρ是水的密度，H_1是上水箱中水的高度。

在下面的式子中，下标 1 表示上水箱的量值，下标 2 表示下水箱的量值，下标 in 表示阀 1 管道即输入管道的量值，下标 12 表示阀 2 管道的量值，下标 out 表示阀 3 即输出管道的量值。

对式（5.1）做时间微分，得到上水箱水质量的瞬时变化率

$$\dot{m}_1 = Q_{in} - Q_{12} = \rho A_1 \dot{H}_1 \tag{5.2}$$

即容器内存储的水的质量变化量等于输入量减去输出量。式中，Q_{12} 是阀 2 控制的管道的流量，其表示为面积 A_{12} 和流速 V_{12} 的函数

$$Q_{12} = \rho A_{12} V_{12} \tag{5.3}$$

把式（5.3）代入式（5.2），得

$$\rho A_1 \dot{H}_1 = Q_{\text{in}} - \rho A_{12} V_{12} \tag{5.4}$$

依据伯努利方程，对上水箱可得

$$\frac{1}{2}\rho V_{\text{in}}^2 + P_1 + \rho g H_1 = \frac{1}{2}\rho V_{12}^2 + P_{12} \tag{5.5}$$

式中，V_{in} 是水箱内的水从上层向下流动的速度，P_1 和 P_{12} 是上水箱输入口和输出口各自的压强，则有 $P_1 = P_{12}$，其同为大气压强。在容器液面高度变化缓慢时，在每个微小时间间隔内，可以应用动量守恒公式

$$A_1 V_{\text{in}} = A_{12} V_{12}$$

因为 $A_1 \gg A_{12}$，故 $V_{\text{in}} \ll V_{12}$，可忽略式（5.5）中的 V_{in}，得

$$V_{12} = \sqrt{2gH_1} \tag{5.6}$$

双容水箱液位系统的常数列于表 5.1 中。

表 5.1　水箱物理常数

ρ (kg/m^3)	g (m/s^2)	A_1 (m^2)	A_{12} (m_2)	A_2 (m_2)	A_{out} (m_2)
1000	9.8	4π	0.04π	π	0.04π
H^* (m)	Q_{in}^* (kg/s)				
1	556.32				

把式（5.6）代入式（5.4），求得 $\dot{H}_1$

$$\dot{H}_1 = -\frac{A_{12}\sqrt{2g}}{A_1}\sqrt{H_1} + \frac{1}{\rho A_1}Q_{\text{in}} = f_1(H_1, Q_{\text{in}}) \tag{5.7}$$

把式（5.6）代入式（5.3），得到上水箱出口流量

$$Q_{12} = (\rho A_{12}\sqrt{2g})\sqrt{H_1} = y_1(H_1) \tag{5.8}$$

式（5.7）和式（5.8）即为上水箱动态微分方程。可见，上述方程为一阶非线性常微分方程，非线性出自 $\sqrt{H_1}$。将式（5.7）和式（5.8）定义为分别具有函数形式 $f_1(H_1, Q_{\text{in}})$ 和 $y_1(H_1)$。

同理，分析下水箱，有

$$m_2 = \rho A_2 H_2 \tag{5.9}$$

对式（5.9）求导，可得

$$\dot{m}_2 = Q_{12} - Q_{\text{out}} = \rho A_2 \dot{H}_2 \tag{5.10}$$

因为有

$$Q_{out} = \rho A_{out} V_{out} \tag{5.11}$$

把式（5.8）和式（5.11）代入式（5.10），得

$$\rho A_2 \dot{H}_2 = \rho A_{12}\sqrt{2gH_1} - \rho A_{out} V_{out} \tag{5.12}$$

根据伯努利方程，对下水箱可得

$$\frac{1}{2}\rho V_{12}^2 + P_{12} + \rho g H_2 = \frac{1}{2}\rho V_{out}^2 + P_{out} \tag{5.13}$$

与上水箱同理，因为$P_{12} = P_{out}$，$A_2 \gg A_{out}$，故$V_{12} \ll V_{out}$，可忽略式（5.13）中的V_{12}，得

$$V_{out} = \sqrt{2gH_2} \tag{5.14}$$

把式（5.14）代入式（5.12），得

$$\dot{H}_2 = \frac{A_{12}\sqrt{2g}}{A_2}\sqrt{H_1} - \frac{A_{out}\sqrt{2g}}{A_2}\sqrt{H_2} = f_2(H_1, H_2) \tag{5.15}$$

把式（5.14）代入式（5.11），得

$$Q_{out} = (\rho A_{out}\sqrt{2g})\sqrt{H_2} = y_2(H_2) \tag{5.16}$$

式（5.15）和式（5.16）即为下水箱动态微分方程。上述方程仍为一阶非线性常微分方程，非线性出自$\sqrt{H_1}$和$\sqrt{H_2}$。将式（5.15）和式（5.16）定义为分别具有函数形式$f_2(H_1, H_2)$和$y_2(H_2)$。

至此，得到双容水箱液位系统的非线性微分方程模型

$$\begin{cases} \dot{H}_1 = -\dfrac{A_{12}\sqrt{2g}}{A_1}\sqrt{H_1} + \dfrac{1}{\rho A_1}Q_{in} \\ Q_{12} = (\rho A_{12}\sqrt{2g})\sqrt{H_1} \\ \dot{H}_2 = \dfrac{A_{12}\sqrt{2g}}{A_2}\sqrt{H_1} - \dfrac{A_{out}\sqrt{2g}}{A_2}\sqrt{H_2} \\ Q_{out} = (\rho A_{out}\sqrt{2g})\sqrt{H_2} \end{cases}$$

5.5　线性化和传递函数

5.5.1　线性化

上述推导得到了上水箱的动态微分方程——式（5.7）和式（5.8），以及下水箱的动态微分方程——式（5.15）和式（5.16）。但其均为非线性微分方程，为了简化模型和分析，需要对其进行线性化处理。

当双容水箱液位系统处于平衡状态时，即足够的水进入水箱中，以弥补通过出口处离开的水量，则有

$$\begin{cases} \dot{H}_1 = f_1(H_1^*, Q_{in}^*) = 0 \\ \dot{H}_2 = f_2(H_1^*, H_2^*) = 0 \end{cases} \tag{5.17}$$

式中，上水箱的平衡点为(H_1^*, Q_{in}^*)，下水箱的平衡点为(H_1^*, H_2^*)。根据式（5.7），在平衡点，有

$$\sqrt{H_1^*} = \frac{Q_{\text{in}}^*}{\rho A_{12}\sqrt{2g}} \tag{5.18}$$

把式（5.18）代入式（5.15），在平衡点，得

$$\sqrt{H_2^*} = \frac{Q_{\text{in}}^*}{\rho A_{\text{out}}\sqrt{2g}} \tag{5.19}$$

可以用各变量的平衡点值和相对于平衡值的微小偏差 Δ 来表示其瞬时值

$$\begin{cases} Q_{\text{in}} = Q_{\text{in}}^* + \Delta Q_{\text{in}} \\ H_1 = H_1^* + \Delta H_1 \\ Q_{12} = Q_{12}^* + \Delta Q_{12} \\ H_2 = H_2^* + \Delta H_2 \\ Q_{\text{out}} = Q_{\text{out}}^* + \Delta Q_{\text{out}} \end{cases} \tag{5.20}$$

将上水箱非线性微分方程式（5.7），在平衡点 (H_1^*, Q_{in}^*) 进行泰勒级数展开，有

$$\dot{H}_1 = f_1(H_1, Q_{\text{in}}) = f_1(H_1^*, Q_{\text{in}}^*) + \frac{\partial f_1}{\partial H_1}\bigg|_{\substack{H_1=H_1^* \\ Q_{\text{in}}=Q_{\text{in}}^*}} (H_1 - H_1^*) + \frac{\partial f_1}{\partial Q_{\text{in}}}\bigg|_{\substack{H_1=H_1^* \\ Q_{\text{in}}=Q_{\text{in}}^*}} (Q_{\text{in}} - Q_{\text{in}}^*) + \cdots \tag{5.21}$$

在式（5.21）中

$$\frac{\partial f_1}{\partial H_1}\bigg|_{\substack{H_1=H_1^* \\ Q_{\text{in}}=Q_{\text{in}}^*}} = -\frac{A_{12}}{A_1} \cdot \frac{\sqrt{2g}}{2} \cdot \frac{1}{\sqrt{H_1^*}} \tag{5.22}$$

把式（5.18）代入式（5.22），得

$$\frac{\partial f_1}{\partial H_1}\bigg|_{\substack{H_1=H_1^* \\ Q_{\text{in}}=Q_{\text{in}}^*}} = -\frac{\rho g A_{12}^2}{A_1 Q_{\text{in}}^*} \tag{5.23}$$

在式（5.21）中

$$\frac{\partial f_1}{\partial Q_{\text{in}}}\bigg|_{\substack{H_1=H_1^* \\ Q_{\text{in}}=Q_{\text{in}}^*}} = \frac{1}{\rho A_1} \tag{5.24}$$

把式（5.23）和式（5.24）代入式（5.21），得

$$\dot{H}_1 = f_1(H_1, Q_{\text{in}}) = f_1(H_1^*, Q_{\text{in}}^*) - \frac{\rho g A_{12}^2}{A_1 Q_{\text{in}}^*}(H_1 - H_1^*) + \frac{1}{\rho A_1}(Q_{\text{in}} - Q_{\text{in}}^*) + \cdots \tag{5.25}$$

由式（5.17）可知 $f_1(H_1^*, Q_{\text{in}}^*) = 0$，且由式（5.20）可知 $\dot{H} = \Delta\dot{H}$，在泰勒级数展开式中忽略高阶项，将上式整理为

$$\Delta\dot{H}_1 = -\frac{\rho g A_{12}^2}{A_1 Q_{\text{in}}^*}\Delta H_1 + \frac{1}{\rho A_1}\Delta Q_{\text{in}} \tag{5.26}$$

式（5.26）是式（5.7）的线性化模型，描述了稳态时，当输入流量的偏差为 ΔQ_{in} 时，上水箱水位偏差为 ΔH_1。

类似地，将上水箱非线性微分方程式（5.8），在平衡点 (H_1^*, Q_{in}^*) 进行泰勒级数展开，有

$$Q_{12} = y_1(H_1) = y_1(H_1^*) + \frac{\mathrm{d}y_1}{\partial H_1}\bigg|_{H_1=H_1^*} (H_1 - H_1^*) + \cdots \tag{5.27}$$

将式（5.18）代入导数项中，得到式（5.27）中导数项为

$$\frac{\mathrm{d}y_1}{\partial H_1}\bigg|_{H_1=H_1^*} = \frac{\rho^2 g A_{12}^2}{Q_{\text{in}}^*} \tag{5.28}$$

忽略高阶项，得到

$$Q_{12} - y_1(H_1^*) = \frac{\rho^2 g A_{12}^2}{Q_{\text{in}}^*}(H_1 - H_1^*) \tag{5.29}$$

即

$$\Delta Q_{12} = \frac{\rho^2 g A_{12}^2}{Q_{\text{in}}^*} \Delta H_1 \tag{5.30}$$

式（5.30）是式（5.8）的线性化模型，描述了稳态时，上水箱水位偏差 ΔH_1 引起的上水箱输出流量的偏差 ΔQ_{12}。

进一步，将下水箱动态微分方程式（5.15），在平衡点 (H_1^*, H_2^*) 进行泰勒级数展开，有

$$\dot{H}_2 = f_2(H_1^*, H_2^*) + \frac{\partial f_2}{\partial H_1}\bigg|_{\substack{H_1=H_1^* \\ H_2=H_2^*}} (H_1 - H_1^*) + \frac{\partial f_2}{\partial H_2}\bigg|_{\substack{H_1=H_1^* \\ H_2=H_2^*}} (H_2 - H_2^*) + \cdots \tag{5.31}$$

把式（5.18）和式（5.19）代入导数项，得到式（5.31）中导数项为

$$\frac{\partial f_2}{\partial H_1}\bigg|_{\substack{H_1=H_1^* \\ H_2=H_2^*}} = \frac{\rho g A_{12}^2}{A_2 Q_{\text{in}}^*} \tag{5.32}$$

$$\frac{\partial f_2}{\partial H_2}\bigg|_{\substack{H_1=H_1^* \\ H_2=H_2^*}} = -\frac{\rho g A_{\text{out}}^2}{A_2 Q_{\text{in}}^*} \tag{5.33}$$

把式（5.32）和式（5.33）代入式（5.31），整理得

$$\Delta \dot{H}_2 = \frac{\rho g A_{12}^2}{A_2 Q_{\text{in}}^*} \Delta H_1 - \frac{\rho g A_{\text{out}}^2}{A_2 Q_{\text{in}}^*} \Delta H_2 \tag{5.34}$$

式（5.34）是式（5.15）的线性化模型，描述了稳态时，下水箱水位偏差 ΔH_2。

将下水箱动态微分方程式（5.16），在平衡点 (H_1^*, H_2^*) 进行泰勒级数展开，有

$$Q_{\text{out}} = y_2(H_2^*) + \frac{\mathrm{d}y_2}{\mathrm{d}H_2}\bigg|_{H_2=H_2^*} (H_2 - H_2^*) + \cdots \tag{5.35}$$

把式（5.19）代入导数项中，得到式（5.35）中导数项为

$$\frac{\mathrm{d}y_2}{\mathrm{d}H_2}\bigg|_{H_2=H_2^*} = \frac{\rho^2 g A_{\text{out}}^2}{Q_{\text{in}}^*} \tag{5.36}$$

把式（5.36）代入式（5.35），忽略高阶项，得

$$\Delta Q_{\text{out}} = \frac{\rho^2 g A_{\text{out}}^2}{Q_{\text{in}}^*} \Delta H_2 \tag{5.37}$$

式（5.37）是式（5.16）的线性化模型，描述了稳态时，下水箱出水流量偏差 ΔQ_{out}。

至此，得到双水箱液位系统的线性模型

$$\begin{cases} \Delta \dot{H}_1 = -\dfrac{\rho g A_{12}^2}{A_1 Q_{\text{in}}^*} \Delta H_1 + \dfrac{1}{\rho A_1} \Delta Q_{\text{in}} \\ \Delta Q_{12} = \dfrac{\rho^2 g A_{12}^2}{Q_{\text{in}}^*} \Delta H_1 \\ \Delta \dot{H}_2 = \dfrac{\rho g A_{12}^2}{A_2 Q_{\text{in}}^*} \Delta H_1 - \dfrac{\rho g A_{\text{out}}^2}{A_2 Q_{\text{in}}^*} \Delta H_2 \\ \Delta Q_{\text{out}} = \dfrac{\rho^2 g A_{\text{out}}^2}{Q_{\text{in}}^*} \Delta H_2 \end{cases} \tag{5.38}$$

5.5.2 传递函数模型

在零初始条件下，对式（5.38）进行拉普拉斯变换，得到传递函数

$$
\begin{cases}
s\Delta H_1(s) = -\dfrac{\rho g A_{12}^2}{A_1 Q_{\text{in}}^*}\Delta H_1(s) + \dfrac{1}{\rho A_1}\Delta Q_{\text{in}}(s) \Rightarrow \dfrac{\Delta H_1(s)}{\Delta Q_{\text{in}}(s)} = \dfrac{\dfrac{1}{\rho A_1}}{s + \dfrac{\rho g A_{12}^2}{A_1 Q_{\text{in}}^*}} \\
\Delta Q_{12}(s) = \dfrac{\rho^2 g A_{12}^2}{Q_{\text{in}}^*}\Delta H_1(s) \Rightarrow \dfrac{\Delta Q_{12}(s)}{\Delta Q_{\text{in}}(s)} = \dfrac{\dfrac{\rho g A_{12}^2}{A_1 Q_{\text{in}}^*}}{s + \dfrac{\rho g A_{12}^2}{A_1 Q_{\text{in}}^*}} \\
s\Delta H_2 = \dfrac{\rho g A_{12}^2}{A_2 Q_{\text{in}}^*}\Delta H_1 - \dfrac{\rho g A_{\text{out}}^2}{A_2 Q_{\text{in}}^*}\Delta H_2 \Rightarrow \dfrac{\Delta H_2(s)}{\Delta Q_{\text{in}}(s)} = \dfrac{\dfrac{g A_{12}^2}{A_1 A_2 Q_{\text{in}}^*}}{\left(s + \dfrac{\rho g A_{12}^2}{A_1 Q_{\text{in}}^*}\right)\left(s + \dfrac{\rho g A_{\text{out}}^2}{A_2 Q_{\text{in}}^*}\right)} \\
\Delta Q_{\text{out}} = \dfrac{\rho^2 g A_{\text{out}}^2}{Q_{\text{in}}^*}\Delta H_2 \Rightarrow \dfrac{\Delta Q_{\text{out}}(s)}{\Delta Q_{\text{in}}(s)} = \dfrac{\dfrac{\rho^2 g A_{\text{out}}^2}{Q_{\text{in}}^*}\cdot\dfrac{g A_{12}^2}{A_1 A_2 Q_{\text{in}}^*}}{\left(s + \dfrac{\rho g A_{12}^2}{A_1 Q_{\text{in}}^*}\right)\left(s + \dfrac{\rho g A_{\text{out}}^2}{A_2 Q_{\text{in}}^*}\right)}
\end{cases}
\tag{5.39}
$$

令

$$
\begin{cases}
k_1 = \dfrac{1}{\rho A_1} \\
\Omega_1 = \dfrac{\rho g A_{12}^2}{A_1 Q_{\text{in}}^*} \\
\Omega_2 = \dfrac{\rho g A_{\text{out}}^2}{A_2 Q_{\text{in}}^*} \\
k_2 = \dfrac{1}{\rho A_2} \\
k_3 = \rho A_2
\end{cases}
\tag{5.40}
$$

则式（5.39）整理为

$$
\begin{cases}
\dfrac{\Delta H_1(s)}{\Delta Q_{\text{in}}(s)} = \dfrac{k_1}{s + \Omega_1} \\
\dfrac{\Delta Q_{12}(s)}{\Delta Q_{\text{in}}(s)} = \dfrac{\Omega_1}{s + \Omega_1} \\
\dfrac{\Delta H_2(s)}{\Delta Q_{\text{in}}(s)} = \dfrac{k_2 \Omega_1}{(s + \Omega_1)(s + \Omega_2)} \\
\dfrac{\Delta Q_{\text{out}}(s)}{\Delta Q_{\text{in}}(s)} = \dfrac{k_2 k_3 \Omega_1 \Omega_2}{(s + \Omega_1)(s + \Omega_2)}
\end{cases}
\tag{5.41}
$$

式（5.41）描述了阀 1 管道上的输入流量变化量 $\Delta Q_{\text{in}}(s)$ 对上水箱水位 $\Delta H_1(s)$ 和输出流量 $\Delta Q_{12}(s)$，以及下水箱水位 $\Delta H_2(s)$ 和输出流量 $\Delta Q_{\text{out}}(s)$ 的影响。

5.6 求解与仿真分析

此节中考虑在水箱中描述流体流动的运动方程的两种解，即解析解和数值解。一般来说，难以获得非线性方程的解析解，因此只在线性模型下讨论解析解。然而，使用数值积分方法，可以得到两个非线性模型的解。通过计算机仿真，对解析解进行验证，以及对线性模型的解和非线性模型的解进行比较。

5.6.1 解析解

式（5.41）表明，水箱系统是线性时不变模型，我们可以获得阶跃输入的响应。

阶跃输入为

$$\Delta Q_{\text{in}}(s) = q_0 / s \tag{5.42}$$

式中，q_0 是阶跃输入幅值，初始条件是 $\Delta H_1(0)=0$，$\Delta Q_{12}(0)=0$，$\Delta H_2(0)=0$，$\Delta Q_{\text{out}}(0)=0$。根据式（5.41）中给的传递函数，对上水箱的输出，有

$$\Delta Q_{12}(s) = \frac{q_0\Omega_1}{s(s+\Omega_1)} \tag{5.43}$$

由部分分式展开得

$$\Delta Q_{12}(s) = \frac{-q_0}{s+\Omega_1} + \frac{q_0}{s} \tag{5.44}$$

然后，由拉普拉斯反变换得

$$\Delta Q_{12}(t) = -q_0 \mathrm{e}^{-\Omega_1 t} + q_0 \tag{5.45}$$

在稳定状态下，输出流量与平衡值的偏差等于输入流量与平衡值的偏差。通过检测式（5.41）中的变量 Ω_1，发现输出端口 A_2 开放得越大，系统达到稳定状态就越快，换句话说，Ω_1 值越大，这个 $\mathrm{e}^{-\Omega_1 t}$ 的指数项就衰减的越快，达到稳定状态也就越快。

同理，对于上水箱的水位，有

$$\Delta H_1(s) = \frac{-q_0 k_1}{\Omega_1}\left(\frac{1}{s+\Omega} - \frac{1}{s}\right) \tag{5.46}$$

进行拉普拉斯反变换得

$$\Delta H_1(t) = \frac{-q_0 k_1}{\Omega_1}(\mathrm{e}^{-\Omega_1 t} - 1) \tag{5.47}$$

由于阶跃输入幅值是 q_0，使得水位的稳态变化为

$$\Delta H_{1ss} = \frac{q_0 k_1}{\Omega_1} \tag{5.48}$$

同理，对于下水箱的水位，有

$$\Delta H_2(s) = \frac{q_0 k_2 \Omega_1}{s(s+\Omega_1)(s+\Omega_2)} \tag{5.49}$$

其拉普拉斯反变换得

$$\Delta H_2(t) = q_0 k_2 \Omega_1\left(\frac{1}{\Omega_1\Omega_2} + \frac{1}{\Omega_1\Omega_2(\Omega_1-\Omega_2)}(\Omega_2 \mathrm{e}^{-\Omega_1 t} - \Omega_1 \mathrm{e}^{-\Omega_2 t})\right) \tag{5.50}$$

由于阶跃输入幅值是 q_0，使得水位的稳态变化为

$$\Delta H_{2\mathrm{ss}} = \frac{q_0 k_2}{\Omega_2} \tag{5.51}$$

对于下水箱的输出流量，有

$$\Delta Q_{\mathrm{out}}(t) = q_0 k_2 k_3 \Omega_1 \Omega_2 \left(\frac{1}{\Omega_1 \Omega_2} + \frac{1}{\Omega_1 \Omega_2 (\Omega_1 - \Omega_2)} (\Omega_2 \mathrm{e}^{-\Omega_1 t} - \Omega_1 \mathrm{e}^{-\Omega_2 t}) \right) \tag{5.52}$$

由于阶跃输入幅值是 q_0，使得输出流量的稳态变化为

$$\Delta Q_{\mathrm{outss}} = q_0 k_2 k_3 \tag{5.53}$$

5.6.2 仿真分析

使用表 5.1 给出的常数，由式（5.7）和式（5.15），该非线性模型如下

$$\begin{cases} \dot{H}_1 = -\dfrac{A_{12}\sqrt{2g}}{A_1}\sqrt{H_1} + \dfrac{1}{\rho A_1} Q_{\mathrm{in}} \\ \dot{H}_2 = \dfrac{A_{12}\sqrt{2g}}{A_2}\sqrt{H_1} - \dfrac{A_{\mathrm{out}}\sqrt{2g}}{A_2}\sqrt{H_2} \end{cases} \tag{5.54}$$

当 $Q_{\mathrm{in}}^* = 556.32\mathrm{kg/s}$，可以对式（5.54）的模型进行数值积分，求得时间函数 $H_1(t)$ 和 $H_2(t)$。

分以下两种情况讨论：

1．当 $H_1(0) = 0.5\mathrm{m}$，$H_2(0) = 0\mathrm{m}$，$Q_{\mathrm{in}}(t) = Q_{\mathrm{in}}^*$ 时，非线性模型式（5.54）中，上下水箱液位数值的变化。

系统响应如图 5.6.1 所示。

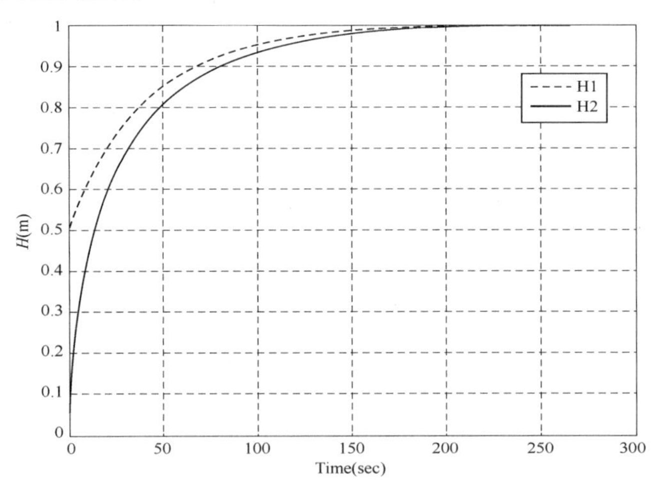

图 5.6.1　双容水箱液位系统响应

从图 5.6.1 可以看出，下水箱液位的响应呈现出滞后特性，这与传递函数描述的特性相符。在 $Q_{\mathrm{in}}(t) = Q_{\mathrm{in}}^*$ 时，系统能够达到稳态，此时 $H_1^* = H_2^* = 1\mathrm{m}$。程序和结果见图 5.6.2。

```
%Main
clc;
clear all;
```

图 5.6.2　MATLAB 程序及其结果

```
global RHO G A1 A12 A2 Aout Qinstar Hstar
RHO=1000; G=9.8;
A1=4*pi; A12=0.04*pi;
Qinstar=556.32; Hstar=1;
A2=1*pi; Aout=0.04*pi;
to=0; tf=300;
H1_0=0.5; H2_0=0;
[t,H]=ode45('d_H',[to tf],[H1_0 H2_0]);
figure(1);
plot(t,H(:,1),'k--');
hold on
plot(t,H(:,2),'k-');
grid
xlabel('Time(sec)'),ylabel('H(m)');
```

函数文件 d_H ()为

```
function D_H=d_H(t,H)
global RHO G A1 A12 A2 Aout
Qin=qin(t);
D_H=zeros(2,1);
D_H(1)=-(A12*sqrt(2*G*H(1)))/A1+Qin/(RHO*A1);
D_H(2)=(A12*sqrt(2*G*H(1)))/A2-(Aout*sqrt(2*G*H(2)))/A2;
end
```

函数文件 qin()为

```
function Qin=qin(t)
global Qinstar
Qin=Qinstar
```

图 5.6.2　MATLAB 程序及其结果（续）

2．当输入为 $Q_{\text{in}}(t)=Q_{\text{in}}^{*}+\Delta Q_{\text{in}}$ 时，系统处于 $H_1(0)=H^*$，$H_2(0)=H^*$ 状态时，非线性模型式（5.54）和线性模型式（5.38）响应的偏差分析。

输出响应曲线如图 5.6.3 所示。令 $\Delta Q_{\text{in}}(t)=10\text{kg/s}$，则其拉普拉斯变换为 10 倍单位阶跃信号。系统仿真程序和结果如图 5.6.4 所示。

从图 5.6.3（a）和图 5.6.3（b）可以看出，经过线性化处理得到的传递函数，与原始非线性模型之间仅存在很小的偏差。对上水箱而言，$\Delta Q_{\text{in}}(t)$ 造成的稳态值液位变化。非线性模型的稳态值为 $\Delta H_{1\text{ss}}=3.6158\text{cm}$，线性模型的稳态值为 $\Delta H_{1\text{ss}}=3.5901\text{cm}$；对下水箱而言，非线性模型的液位变化的稳态值 $\Delta H_{2\text{ss}}=3.6140\text{cm}$，线性模型的液位变化的稳态值 $\Delta H_{1\text{ss}}=3.5886\text{cm}$。该值与式（5.48）和式（5.51）解析求得的线性模型的稳态值一致。因此，对非线性模型，在其平衡点处进行线性化处理，是一种简化模型的有效方法。

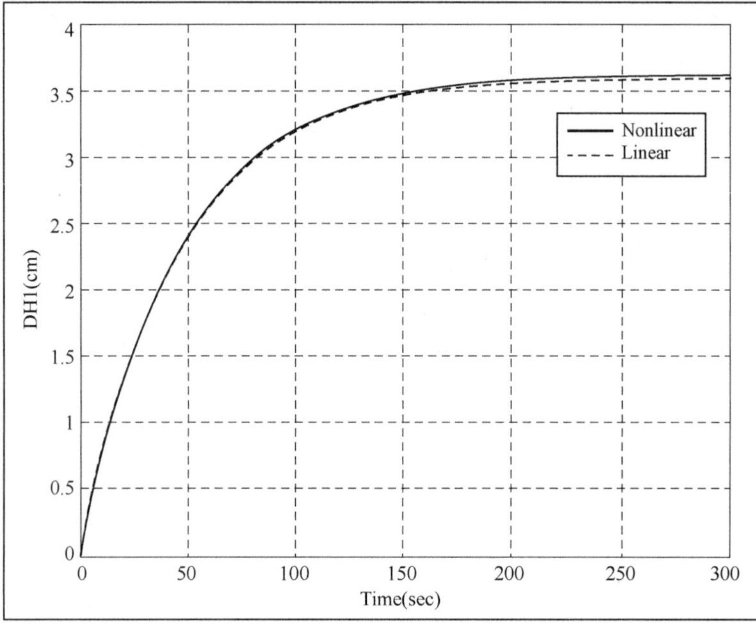

（a）

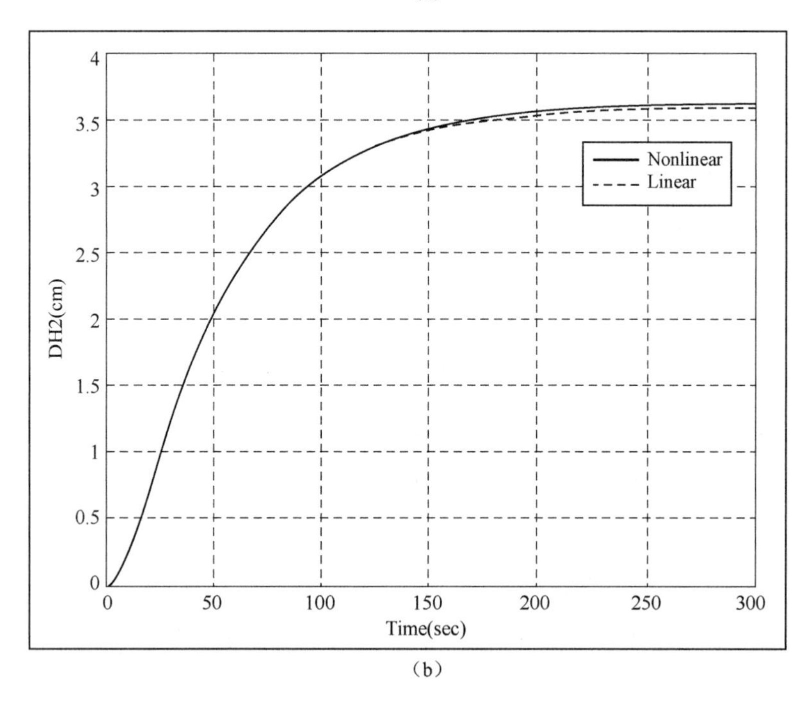

（b）

图 5.6.3 线性和非线性模型阶跃响应

```
%Main
clc;
clear all;
global RHO G A1 A12 A2 Aout Qinstar Hstar
RHO=1000; G=9.8;
A1=4*pi; A12=0.04*pi;
Qinstar=556.32; Hstar=1;
```

```
A2=1*pi; Aout=0.04*pi;
to=0; tf=300;
H1_0=Hstar;
H2_0=Hstar;
[t,H]=ode45('d_H2',[to tf],[H1_0 H2_0]);
DH_1=H(:,1)-Hstar;
DH_2=H(:,2)-Hstar;
%%%%%%%%%%%%%%%%%%%%%%%%%%%%%%%%%%%%%%%%
t_line=[0:0.1:300];
k1=1/(RHO*A1);
Omega1=A12^2*RHO*G/(A1*Qinstar);
k2=1/(RHO*A2);
Omega2=Aout^2*RHO*G/(A2*Qinstar);
num1=[k1];
den1=[1 Omega1];
DH_1_line=10*step(num1,den1,t_line);
num2=[k2*Omega1];
den2=conv([1,Omega1],[1,Omega2]);
DH_2_line=10*step(num2,den2,t_line);
%%%%%%%%%%%%%%%%%%%%%%%%%%%%%%%%%%%%%%%%%%
figure(1);
plot(t,100*DH_1,'k-');
hold on
plot(t_line,100*DH_1_line,'k--');
grid on;
xlabel('Time(sec)'),ylabel('DH1(cm)');
figure(2);
plot(t,100*DH_2,'k-');
hold on
plot(tLine,100*DH_2_line,'k--');
grid on;
xlabel('Time(sec)'),ylabel('DH2(cm)');
```

其中，函数文件 d_H2()为

```
function D_H=d_H2(t,H)
global RHO G A1 A12 A2 Aout
Qin=qin2(t);
D_H=zeros(2,1);
D_H(1)=-(A12*sqrt(2*G*H(1)))/A1+Qin/(RHO*A1);
D_H(2)=(A12*sqrt(2*G*H(1)))/A2-(Aout*sqrt(2*G*H(2)))/A2;
end
```

函数文件 qin2()为

```
function Qin=qin2(t)
global Qinstar
```

图 5.6.4 MATLAB 程序及其结果

```
D_Qin=10;
Qin=Qinstar+D_Qin;
```

```
DH_1_final =

    3.6158

DH1_line_final =

    3.5901

DH_2_final =

    3.6140

DH_2_line_final =

    3.5886
```

图 5.6.4 MATLAB 程序及其结果（续）

本章小结

（1）依据系统的物理特性进行了合理地假设；
（2）推导了系统的非线性动态微分方程模型，并对其线性化；
（3）推导了系统的传递函数，利用拉普拉斯反变换得到各传递函数的解析解；
（4）通过 MATLAB 分析了系统液位特性，实现了非线性和线性模型的响应分析。

Chapter 5
Modeling of a Double-Tank System

Chapters 3 and 4 focus on the use of MATLAB software to simulate the control system of a given model. On this basis, this chapter centers on the process of establishing the mathematical model of the physical system. The double-tank system is taken as an example to establish its nonlinear dynamic differential equation model, which is then linearized to obtain the transfer function model. Finally, the nonlinear and linear models are both analyzed briefly.

5.1 Introduction to Modeling Methods

One of the key steps in control system design is mathematical modeling of the controlled object. This chapter discusses the establishment of a mathematical model for a double-tank system.

The steps of modeling the control system are as follows:

1. Define the system and its composition.
2. Make the necessary assumptions and formulate the physical equivalent mathematical expression.
3. Derive the differential equation model of the system.
4. Identify the input and output variables.
5. Solve the equations for the output variables.
6. Examine the solutions and the assumptions.

The core of this chapter is to establish a mathematical model of the physical system. The starting point is to determine the composition of the control system. Since the control system consists of multiple parts such as controllers, actuators and controlled objects, it is necessary to consider more elements while determining the composition of the system. This chapter only considers the process of obtaining the model of the controlled object. The issues related to the selection of the controller and the actuator and modeling are not discussed. By taking a double-tank system as an example, the physical process is deduced, a nonlinear model is obtained, and a linear model is obtained through linearization. Finally, MATLAB simulation is used for analysis.

5.2 A Double-Tank System

The fluid height is an important parameter in production. There is great demand for its control. For example, the fluid height in oil tanks need to be measured to calculate the product output and raw material consumption; chemical reaction towers need to maintain a certain depth of liquid and achieve higher productivity. As a common controlled object in industrial production, the double-tank system is composed of two single water tanks with self-balancing capacity connected up and down in series. It is usually required to control the fluid height of the lower tank as the controlled variable, and choose the water flow rate of the upper water tank as a given amount. Double-tank system generally has second-order characteristics.

The double-tank system is shown in Figure 5.2.1. The upper and lower tanks are connected in series to form a fluid height system. The water is supplied to the upper tank through the pipe controlled by Valve 1, and the water in the upper tank is discharged into the lower tank by the pipe controlled by Valve 2, and finally flows out of the pipe controlled by Valve 3.

In Figure 5.2.1, Q_{in} , Q_{12} , and Q_{out} are the instantaneous water flow for each pipe, respectively. This chapter uses Q_{in} as a system input to study the variation in water height H_1, H_2 and water flow rates Q_{12}, Q_{out} caused by input. The variables involved are fluid velocity V(m/s), water height H (m), and pressure P (N/m^2).

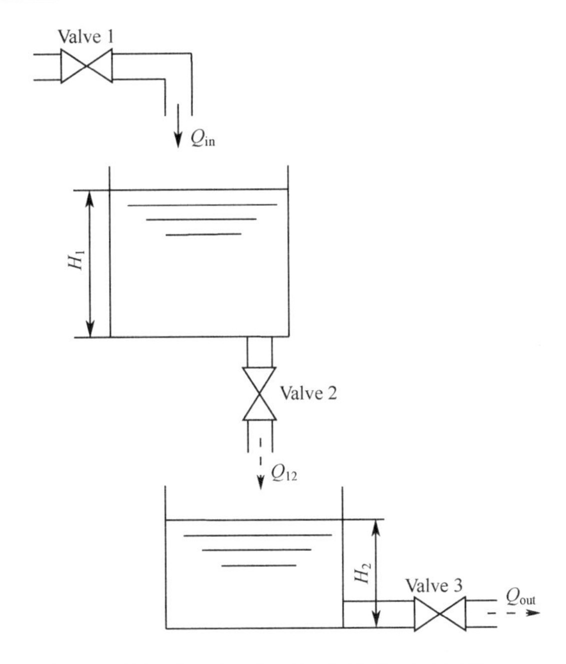

Figure 5.2.1 The Double-Tank Fluid Height System

5.3 Assumptions

In general, while modeling a physical system, it is necessary to make appropriate assumptions based on its physical characteristics, thereby reducing the complexity of the mathematical model and improving the feasibility of modeling. This chapter models fluids and needs to make the following assumptions.

5.3.1 Compressibility

We assume that the water in the tank is uncompressed, so it has a constant density ρ (kg/m^3). In practice, to a certain extent, all liquids are compressible, and the compressibility coefficient k is a measure of liquid compressibility. A smaller value of k indicates less compressibility. Air has a compressibility coefficient $k_{\text{air}} = 0.98\text{m}^2/\text{N}$, and water has a compressibility $k_{\text{water}} = 4.9\times10^{-10}\text{m}^2/\text{N} = 50\times10^{-6}\text{atm}^{-1}$. In this example it can be assumed that the water is uncompressed.

5.3.2 Viscosity

We assume that the fluid is in motion. The initial flow rates of adjacent layers of the liquid are different. The exchange of internal friction between layers and momentum is called viscosity. Solids are more viscous than fluids and fluids are more viscous than gases. The coefficient of viscosity, μ (Nsec/m^2), is used to measure the degree of viscosity . A larger viscosity coefficient means higher viscosity. Under the standard condition of 20℃, the viscosity coefficient of air is

$$\mu_{\text{air}} = 0.178\times10^{-4}\text{N s/ m}^2$$

the viscosity coefficient of water is

$$\mu_{H_2O} = 1.054 \times 10^{-3} \text{N s/ m}^2$$

The water is about 60 times more viscous than air. Viscosity depends on temperature rather than pressure. For low-viscosity liquids, such as water, only the boundary layer friction is significant. Considering the thin boundary layer on the inner layer of the tank and the output pipe, the viscosity can be ignored and the fluid is called to be in-viscid.

5.3.3 Irrotational Flow

If each fluid element at each point in the flow has no net angular velocity about that point, the flow is termed irrotational. Here the water in the tank is assumed irrotational.

5.3.4 Steady Flow

If the velocity at each point is constant in the liquid and does not change over time, the flow is steady. Steady state can be achieved at low flow rates. In this case, the tank outlet area is small and the water flow in the tank is slow, so we can assume that the water flow in the tank is steady.

5.4 Differential Equations of Motion

The mass of water in the upper tank at any given time is

$$m_1 = \rho A_1 H_1 \tag{5.1}$$

where A_1 is the area of the tank, ρ is the water density, and H_1 is the water height in the upper tank.

In the following formulas, Subscript *1* indicates the quantities of the upper tank, Subscript *2* indicates the quantities of the lower tank, Subscript *in* indicates the quantities of Valve 1 pipe or input pipe, and Subscript *12* indicates the quantities of Valve 2 pipe. Subscript out indicates the quantities of Valve 3, i.e. the output pipe.

Take the time derivative of Equation(5.1) to obtain the instantaneous change rate of water quality in the upper tank

$$\dot{m}_1 = Q_{in} - Q_{12} = \rho A_1 \dot{H}_1 \tag{5.2}$$

That is, the change of water stored in the container is equal to the amount that the input minus the output. In the formula, Q_{12} is the flow rate controlled by Valve 2, which is expressed as a function of area A_{12} and flow rate V_{12}:

$$Q_{12} = \rho A_{12} V_{12} \tag{5.3}$$

Substituting Equation (5.3) into Equation (5.2) to yield

$$\rho A_1 \dot{H}_1 = Q_{in} - \rho A_{12} V_{12} \tag{5.4}$$

According to the Bernoulli equation, for the upper tank, we can get

$$\frac{1}{2}\rho V_{in}^2 + P_1 + \rho g H_1 = \frac{1}{2}\rho V_{12}^2 + P_{12} \tag{5.5}$$

In the equation, V_{in} is the relocity of water which flows down from at which the uppermost layer in the water tank. P_1 and P_{12} are the respective pressures at the input and output ports of the upper tank, so we have $P_1 = P_{12}$, which is equal to the atmospheric pressure. When the water in the tanks changes slowly, the momentum conservation formula can be applied at every minute time interval:

$$A_1 V_{\text{in}} = A_{12} V_{12}$$

Because $A_1 >> A_{12}$, so $V_{\text{in}} << V_{12}$, V_{in} in Equation (5.5) can be ignored to get

$$V_{12} = \sqrt{2gH_1} \tag{5.6}$$

The constants for the double-tank system are listed in Table 5.1.

Table 5.1 Physical Constants of Water Tank

ρ (kg/m^3)	g (m/s^2)	A_1 (m^2)	A_{12} (m_2)	A_2 (m_2)	A_{out} (m_2)
1000	9.8	4π	0.04π	π	0.04π
H^* (m)	Q_{in}^* (kg / s)				
1	556.32				

Substituting Equation (5.6) into (5.4) to get $\dot{H}_1$

$$\dot{H}_1 = -\frac{A_{12}\sqrt{2g}}{A_1}\sqrt{H_1} + \frac{1}{\rho A_1} Q_{\text{in}} = f_1(H_1, Q_{\text{in}}) \tag{5.7}$$

Substituting Equation (5.6) into (5.3) to get the flow rate of the upper tank Q_{12}

$$Q_{12} = (\rho A_{12}\sqrt{2g})\sqrt{H_1} = y_1(H_1) \tag{5.8}$$

Equation (5.7) and (5.8) are the dynamic differential equations for the upper tank. It can be seen that the above equation is a first-order, nonlinear, ordinary differential equation. The nonlinearity comes from $\sqrt{H_1}$. Define Equation (5.7) and (5.8) to have function forms $f_1(H_1, Q_{\text{in}})$ and $y_1(H_1)$ respectively.

Similarly, analyze the lower tank and we get

$$m_2 = \rho A_2 H_2 \tag{5.9}$$

Taking the derivative of Equation (5.9), we have

$$\dot{m}_2 = Q_{12} - Q_{\text{out}} = \rho A_2 \dot{H}_2 \tag{5.10}$$

Because

$$Q_{\text{out}} = \rho A_{\text{out}} V_{\text{out}} \tag{5.11}$$

Substituting Equation (5.8) and (5.11) into Equation (5.10) and we get

$$\rho A_2 \dot{H}_2 = \rho A_{12}\sqrt{2gH_1} - \rho A_{\text{out}} V_{\text{out}} \tag{5.12}$$

According to the Bernoulli equation, for the lower tank, we can get

$$\frac{1}{2}\rho V_{12}^2 + P_{12} + \rho g H_2 = \frac{1}{2}\rho V_{\text{out}}^2 + P_{\text{out}} \tag{5.13}$$

Similarly, because $P_{12} = P_{\text{out}}$, $A_2 >> A_{\text{out}}$, so $V_{12} << V_{\text{out}}$, V_{12} in Equation (5.13) can be ignored to get

$$V_{\text{out}} = \sqrt{2gH_2} \tag{5.14}$$

Substituting Equation (5.14) into Equation (5.12), we find $\dot{H}_2$:

$$\dot{H}_2 = \frac{A_{12}\sqrt{2g}}{A_2}\sqrt{H_1} - \frac{A_{out}\sqrt{2g}}{A_2}\sqrt{H_2} = f_2(H_1, H_2) \tag{5.15}$$

Substituting Equation (5.14) into Equation (5.11), we get

$$Q_{out} = (\rho A_{out}\sqrt{2g})\sqrt{H_2} = y_2(H_2) \tag{5.16}$$

Equation (5.15) and (5.16) are the dynamic differential equations for the lower tank. The above equations are a first-order, nonlinear, ordinary differential equation. The nonlinearity comes from $\sqrt{H_1}$ and $\sqrt{H_2}$. Define Equation (5.15) and (5.16) to have function forms $f_2(H_1, H_2)$ and $y_2(H_2)$ respectively.

In summary, a nonlinear differential equation model for a double-tank system is obtained

$$\begin{cases} \dot{H}_1 = -\dfrac{A_{12}\sqrt{2g}}{A_1}\sqrt{H_1} + \dfrac{1}{\rho A_1}Q_{in} \\ Q_{12} = (\rho A_{12}\sqrt{2g})\sqrt{H_1} \\ \dot{H}_2 = \dfrac{A_{12}\sqrt{2g}}{A_2}\sqrt{H_1} - \dfrac{A_{out}\sqrt{2g}}{A_2}\sqrt{H_2} \\ Q_{out} = (\rho A_{out}\sqrt{2g})\sqrt{H_2} \end{cases}$$

5.5 Linearization and Transfer Functions

5.5.1 Linearization

From the above derivation , we have obtained the dynamic differential equation of the upper water tank—Equation (5.7) and Equation (5.8)—and the dynamic differential equation of the lower water tank—Equation (5.15) and Equation (5.16). However, they are all nonlinear differential equations. In order to simplify models and analysis, they need to be linearized.

When the double-tank system is in equilibrium, that is, enough water enters the tank to make up for the water leaving the exit, we obtain

$$\begin{cases} \dot{H}_1 = f_1(H_1^*, Q_{in}^*) = 0 \\ \dot{H}_2 = f_2(H_1^*, H_2^*) = 0 \end{cases} \tag{5.17}$$

Where the equilibrium point for the upper tank is (H_1^*, Q_{in}^*), and that for the lower tank is (H_1^*, H_2^*). According to Equation (5.7), at the equilibrium point, we have

$$\sqrt{H_1^*} = \frac{Q_{in}^*}{\rho A_{12}\sqrt{2g}} \tag{5.18}$$

Substituting Equation (5.18) into Equation (5.15), at the equilibrium point we obtain

$$\sqrt{H_2^*} = \frac{Q_{in}^*}{\rho A_{out}\sqrt{2g}} \tag{5.19}$$

Each variable's instantaneous value can be represented by the equilibrium point value and a small deviation Δ from the equilibrium value

$$\begin{cases} Q_{\text{in}} = Q_{\text{in}}^* + \Delta Q_{\text{in}} \\ H_1 = H_1^* + \Delta H_1 \\ Q_{12} = Q_{12}^* + \Delta Q_{12} \\ H_2 = H_2^* + \Delta H_2 \\ Q_{\text{out}} = Q_{\text{out}}^* + \Delta Q_{\text{out}} \end{cases} \tag{5.20}$$

Using the Taylor series expansion about the equilibrium point (H_1^*, Q_{in}^*) for the nonlinear differential equation of the upper tank Equation (5.7), we have

$$\dot{H}_1 = f_1(H_1, Q_{\text{in}}) = f_1(H_1^*, Q_{\text{in}}^*) + \frac{\partial f_1}{\partial H_1}\Big|_{\substack{H_1 = H_1^* \\ Q_{\text{in}} = Q_{\text{in}}^*}} (H_1 - H_1^*) + \frac{\partial f_1}{\partial Q_{\text{in}}}\Big|_{\substack{H_1 = H_1^* \\ Q_{\text{in}} = Q_{\text{in}}^*}} (Q_{\text{in}} - Q_{\text{in}}^*) + \cdots \tag{5.21}$$

In Equation (5.21),

$$\frac{\partial f_1}{\partial H_1}\Big|_{\substack{H_1 = H_1^* \\ Q_{\text{in}} = Q_{\text{in}}^*}} = -\frac{A_{12}}{A_1} \cdot \frac{\sqrt{2g}}{2} \cdot \frac{1}{\sqrt{H_1^*}} \tag{5.22}$$

Substituting Equation (5.18) into Equation (5.22), we get

$$\frac{\partial f_1}{\partial H_1}\Big|_{\substack{H_1 = H_1^* \\ Q_{\text{in}} = Q_{\text{in}}^*}} = -\frac{\rho g A_{12}^2}{A_1 Q_{\text{in}}^*} \tag{5.23}$$

In Equation (5.21),

$$\frac{\partial f_1}{\partial Q_{\text{in}}}\Big|_{\substack{H_1 = H_1^* \\ Q_{\text{in}} = Q_{\text{in}}^*}} = \frac{1}{\rho A_1} \tag{5.24}$$

Substituting Equation (5.23) and (5.24) into Equation (5.21), we get

$$\dot{H}_1 = f_1(H_1, Q_{\text{in}}) = f_1(H_1^*, Q_{\text{in}}^*) - \frac{\rho g A_{12}^2}{A_1 Q_{\text{in}}^*}(H_1 - H_1^*) + \frac{1}{\rho A_1}(Q_{\text{in}} - Q_{\text{in}}^*) + \cdots \tag{5.25}$$

Frow Equation (5.17), we can see $f_1(H_1^*, Q_{\text{in}}^*) = 0$, and from Equation (5.20), we can see $\dot{H} = \Delta\dot{H}$. In the Taylor series expansion, if we ignore high-order terms, then the formula will be turned into

$$\Delta\dot{H}_1 = -\frac{\rho g A_{12}^2}{A_1 Q_{\text{in}}^*}\Delta H_1 + \frac{1}{\rho A_1}\Delta Q_{\text{in}} \tag{5.26}$$

Equation (5.26) is the linearized model of Equation (5.7), which describes that, when the deviation of the input flow is ΔQ_{in} in the steady-state, the deviation of water height in the upper tank is ΔH_1.

Similarly, using the Taylor series expansion at the equilibrium point (H_1^*, Q_{in}^*) for the upper tank nonlinear differential equation, Equation (5.8), we get

$$Q_{12} = y_1(H_1) = y_1(H_1^*) + \frac{\mathrm{d}y_1}{\partial H_1}\Big|_{H_1 = H_1^*} (H_1 - H_1^*) + \cdots \tag{5.27}$$

Substituting Formula (5.18) into this derivative term yields the derivative item in Equation (5.27) as

$$\frac{\mathrm{d}y_1}{\partial H_1}\Big|_{H_1 = H_1^*} = \frac{\rho^2 g A_{12}^2}{Q_{\text{in}}^*} \tag{5.28}$$

Ignoring high-order terms yields

$$Q_{12} - y_1(H_1^*) = \frac{\rho^2 g A_{12}^2}{Q_{\text{in}}^*}(H_1 - H_1^*) \tag{5.29}$$

which is

$$\Delta Q_{12} = \frac{\rho^2 g A_{12}^2}{Q_{\text{in}}^*}\Delta H_1 \tag{5.30}$$

Equation (5.30) is the linearized model Equation (5.8), which describes the deviation of the upper tank output flow ΔQ_{12} caused by the deviation of the upper tank water height ΔH_1 during steady state.

Furthermore, using the Taylor series expansion about the equilibrium point (H_1^*, H_2^*) for the nonlinear differential equation of the lower tank Equation (5.15), we have

$$\dot{H}_2 = f_2(H_1^*, H_2^*) + \frac{\partial f_2}{\partial H_1}\Big|_{\substack{H_1=H_1^* \\ H_2=H_2^*}} (H_1 - H_1^*) + \frac{\partial f_2}{\partial H_2}\Big|_{\substack{H_1=H_1^* \\ H_2=H_2^*}} (H_2 - H_2^*) + \cdots \tag{5.31}$$

Substituting Equation (5.18) and (5.19) into the derivative term yields the derivative item in Equation (5.31) as

$$\frac{\partial f_2}{\partial H_1}\Big|_{\substack{H_1=H_1^* \\ H_2=H_2^*}} = \frac{\rho g A_{12}^2}{A_2 Q_{\text{in}}^*} \tag{5.32}$$

$$\frac{\partial f_2}{\partial H_2}\Big|_{\substack{H_1=H_1^* \\ H_2=H_2^*}} = -\frac{\rho g A_{\text{out}}^2}{A_2 Q_{\text{in}}^*} \tag{5.33}$$

Substituting Equation (5.32) and (5.33) into Equation (5.31), we get

$$\Delta \dot{H}_2 = \frac{\rho g A_{12}^2}{A_2 Q_{\text{in}}^*}\Delta H_1 - \frac{\rho g A_{\text{out}}^2}{A_2 Q_{\text{in}}^*}\Delta H_2 \tag{5.34}$$

Equation (5.34) is the linearized model of the Equation (5.15), which describes the deviation of the water height ΔH_2 in the lower tank at steady state.

Using the Taylor series expansion about the equilibrium point (H_1^*, H_2^*) for the nonlinear differential equation of the lower tank Equation (5.16), we have

$$Q_{\text{out}} = y_2(H_2^*) + \frac{\mathrm{d}y_2}{\mathrm{d}H_2}\Big|_{H_2=H_2^*} (H_2 - H_2^*) + \cdots \tag{5.35}$$

Substituting Equation (5.19) into the derivative term yields the derivative item in Equation (5.35) as

$$\frac{\mathrm{d}y_2}{\mathrm{d}H_2}\Big|_{H_2=H_2^*} = \frac{\rho^2 g A_{\text{out}}^2}{Q_{\text{in}}^*} \tag{5.36}$$

Substituting Equation (5.36) into Equation (5.35) and ignoring the high-order term , we have

$$\Delta Q_{\text{out}} = \frac{\rho^2 g A_{\text{out}}^2}{Q_{\text{in}}^*}\Delta H_2 \tag{5.37}$$

Equation (5.37) is the linearized model of the Equation (5.16). It describes the deviation of the discharge flow rate ΔQ_{out} of the lower tank in the steady state.

In summary, the linear model of the double-tank system is obtained as

$$\begin{cases} \Delta\dot{H}_1 = -\dfrac{\rho g A_{12}^2}{A_1 Q_{\text{in}}^*}\Delta H_1 + \dfrac{1}{\rho A_1}\Delta Q_{\text{in}} \\ \Delta Q_{12} = \dfrac{\rho^2 g A_{12}^2}{Q_{\text{in}}^*}\Delta H_1 \\ \Delta\dot{H}_2 = \dfrac{\rho g A_{12}^2}{A_2 Q_{\text{in}}^*}\Delta H_1 - \dfrac{\rho g A_{\text{out}}^2}{A_2 Q_{\text{in}}^*}\Delta H_2 \\ \Delta Q_{\text{out}} = \dfrac{\rho^2 g A_{\text{out}}^2}{Q_{\text{in}}^*}\Delta H_2 \end{cases} \tag{5.38}$$

5.5.2 Transfer Function Model

For Equation (5.38), taking the Laplace transform under zero initial conditions yields the transfer function

$$\begin{cases} s\Delta H_1(s) = -\dfrac{\rho g A_{12}^2}{A_1 Q_{\text{in}}^*}\Delta H_1(s) + \dfrac{1}{\rho A_1}\Delta Q_{\text{in}}(s) \Rightarrow \dfrac{\Delta H_1(s)}{\Delta Q_{\text{in}}(s)} = \dfrac{\dfrac{1}{\rho A_1}}{s + \dfrac{\rho g A_{12}^2}{A_1 Q_{\text{in}}^*}} \\ \Delta Q_{12}(s) = \dfrac{\rho^2 g A_{12}^2}{Q_{\text{in}}^*}\Delta H_1(s) \Rightarrow \dfrac{\Delta Q_{12}(s)}{\Delta Q_{\text{in}}(s)} = \dfrac{\dfrac{\rho g A_{12}^2}{A_1 Q_{\text{in}}^*}}{s + \dfrac{\rho g A_{12}^2}{A_1 Q_{\text{in}}^*}} \\ s\Delta H_2 = \dfrac{\rho g A_{12}^2}{A_2 Q_{\text{in}}^*}\Delta H_1 - \dfrac{\rho g A_{\text{out}}^2}{A_2 Q_{\text{in}}^*}\Delta H_2 \Rightarrow \dfrac{\Delta H_2(s)}{\Delta Q_{\text{in}}(s)} = \dfrac{\dfrac{g A_{12}^2}{A_1 A_2 Q_{\text{in}}^*}}{\left(s + \dfrac{\rho g A_{12}^2}{A_1 Q_{\text{in}}^*}\right)\left(s + \dfrac{\rho g A_{\text{out}}^2}{A_2 Q_{\text{in}}^*}\right)} \\ \Delta Q_{\text{out}} = \dfrac{\rho^2 g A_{\text{out}}^2}{Q_{\text{in}}^*}\Delta H_2 \Rightarrow \dfrac{\Delta Q_{\text{out}}(s)}{\Delta Q_{\text{in}}(s)} = \dfrac{\dfrac{\rho^2 g A_{\text{out}}^2}{Q_{\text{in}}^*} \cdot \dfrac{g A_{12}^2}{A_1 A_2 Q_{\text{in}}^*}}{\left(s + \dfrac{\rho g A_{12}^2}{A_1 Q_{\text{in}}^*}\right)\left(s + \dfrac{\rho g A_{\text{out}}^2}{A_2 Q_{\text{in}}^*}\right)} \end{cases} \tag{5.39}$$

Let

$$\begin{cases} k_1 = \dfrac{1}{\rho A_1} \\ \Omega_1 = \dfrac{\rho g A_{12}^2}{A_1 Q_{\text{in}}^*} \\ \Omega_2 = \dfrac{\rho g A_{\text{out}}^2}{A_2 Q_{\text{in}}^*} \\ k_2 = \dfrac{1}{\rho A_2} \\ k_3 = \rho A_2 \end{cases} \tag{5.40}$$

Equation (5.39) is then organized into

$$
\begin{cases}
\dfrac{\Delta H_1(s)}{\Delta Q_{\text{in}}(s)} = \dfrac{k_1}{s+\Omega_1} \\
\dfrac{\Delta Q_{12}(s)}{\Delta Q_{\text{in}}(s)} = \dfrac{\Omega_1}{s+\Omega_1} \\
\dfrac{\Delta H_2(s)}{\Delta Q_{\text{in}}(s)} = \dfrac{k_2\Omega_1}{(s+\Omega_1)(s+\Omega_2)} \\
\dfrac{\Delta Q_{\text{out}}(s)}{\Delta Q_{\text{in}}(s)} = \dfrac{k_2 k_3 \Omega_1 \Omega_2}{(s+\Omega_1)(s+\Omega_2)}
\end{cases}
\tag{5.41}
$$

Equation (5.41) describes the effect of the input flow rate $\Delta Q_{\text{in}}(s)$ on the water height in the upper tank $\Delta H_1(s)$, the output flow rate $\Delta Q_{12}(s)$, the water height in the lower tank $\Delta H_2(s)$ and output flow rate $\Delta Q_{\text{out}}(s)$.

5.6 Solution and Simulation Analysis

This section aims to find two solutions to describe the equation of motion describing a fluid flow in the tank, namely an analytical solution and a numerical solution. In general, it is difficult to obtain the analytical solution of the nonlinear equation of motion, so the analytical solution is only discussed in the linear model. However, solutions to both nonlinear model can be obtained by means of numerical integration method. Through computer simulation, the analytical solution is verified and the solutions obtained from linear models can be compared with the solutions obtained from the nonlinear model.

5.6.1 Analytical Solution

Equation (5.41) shows that the tank system is a linear time-invariant model, and we can obtain the response to the step input.

Let the step input

$$\Delta Q_1(s) = q_0 / s \tag{5.42}$$

Where q_0 is the magnitude of the step input, and the initial condition is $\Delta H_1(0)=0$, $\Delta Q_{12}(0)=0$, $\Delta H_2(0)=0$, $\Delta Q_{\text{out}}(0)=0$. According to the transfer function given in Equation (5.41), the output of the upper tank is

$$\Delta Q_{12}(s) = \frac{q_0\Omega_1}{s(s+\Omega_1)} \tag{5.43}$$

The partial fraction expansion yields

$$\Delta Q_{12}(s) = \frac{-q_0}{s+\Omega_1} + \frac{q_0}{s} \tag{5.44}$$

Then, taking the Laplace inverse transform yields

$$\Delta Q_{12}(t) = -q_0 e^{-\Omega_1 t} + q_0 \tag{5.45}$$

We see that in the steady-state, the deviation of the output mass flow rate from the equilibrium value is equal to the deviation of the input mass flow rate from the equilibrium value. By examining the variable Ω_1 in Eq.(5.41), we find that the larger the output port opening, A_2, the faster the system reaches steady-state. In other words, as Ω_1 gets larger, the exponential term $e^{-\Omega_1 t}$ vanishes more quickly, and steady-state is reached faster.

Similarly, for the water height in the upper tank, we have

$$\Delta H_1(s) = \frac{-q_0 k_1}{\Omega_1}\left(\frac{1}{s+\Omega} - \frac{1}{s}\right) \tag{5.46}$$

Taking Laplace inverse transforms yields

$$\Delta H_1(t) = \frac{-q_0 k_1}{\Omega_1}(e^{-\Omega_1 t} - 1) \tag{5.47}$$

Due to the step input amplitude q_0, the steady-state change in water height is

$$\Delta H_{1ss} = \frac{q_0 k_1}{\Omega_1} \tag{5.48}$$

For the water height in the lower tank, we have

$$\Delta H_2(s) = \frac{q_0 k_2 \Omega_1}{s(s+\Omega_1)(s+\Omega_2)} \tag{5.49}$$

Taking Laplace inverse transforms yields

$$\Delta H_2(t) = q_0 k_2 \Omega_1\left(\frac{1}{\Omega_1\Omega_2} + \frac{1}{\Omega_1\Omega_2(\Omega_1-\Omega_2)}(\Omega_2 e^{-\Omega_1 t} - \Omega_1 e^{-\Omega_2 t})\right) \tag{5.50}$$

Due to the step input amplitude q_0, the steady-state change in water height is

$$\Delta H_{2ss} = \frac{q_0 k_2}{\Omega_2} \tag{5.51}$$

For the output flow in the lower tank, we have

$$\Delta Q_{out}(t) = q_0 k_2 k_3 \Omega_1 \Omega_2\left(\frac{1}{\Omega_1\Omega_2} + \frac{1}{\Omega_1\Omega_2(\Omega_1-\Omega_2)}(\Omega_2 e^{-\Omega_1 t} - \Omega_1 e^{-\Omega_2 t})\right) \tag{5.52}$$

Due to the step input amplitude q_0, the steady-state change in output flow is

$$\Delta Q_{out\,ss} = q_0 k_2 k_3 \tag{5.53}$$

5.6.2 Simulation Analysis

Using the constants given in Table 5.1, from Equation (5.7) and Equation (5.15), the nonlinear model is as follows

$$\begin{cases} \dot{H}_1 = -\dfrac{A_{12}\sqrt{2g}}{A_1}\sqrt{H_1} + \dfrac{1}{\rho A_1}Q_{in} \\ \dot{H}_2 = \dfrac{A_{12}\sqrt{2g}}{A_2}\sqrt{H_1} - \dfrac{A_{out}\sqrt{2g}}{A_2}\sqrt{H_2} \end{cases} \tag{5.54}$$

With $Q_{in}^* = 556.32\text{kg/s}$, we can obtain the time function $H_1(t)$ and $H_2(t)$ by numerically integrating the model of Equation (5.54).

There are two cases:

1.When $H_1(0) = 0.5\text{m}$, $H_2(0) = 0\text{m}$, $Q_{in}(t) = Q_{in}^*$, the water height changes of the upper tank and the lower tank in the nonlinear model Equation (5.54).

The system response is shown in Figure 5.6.1.

We can see from Figure 5.6.1, the water height response of the lower tank exhibits a hysteresis characteristic that matches the characteristics described by its transfer function. When $Q_{\text{in}}(t)=Q_{\text{in}}^*$, system can reach steady state, and then $H_1^*=H_2^*=1\text{m}$. The script and results are shown in Figure 5.6.2.

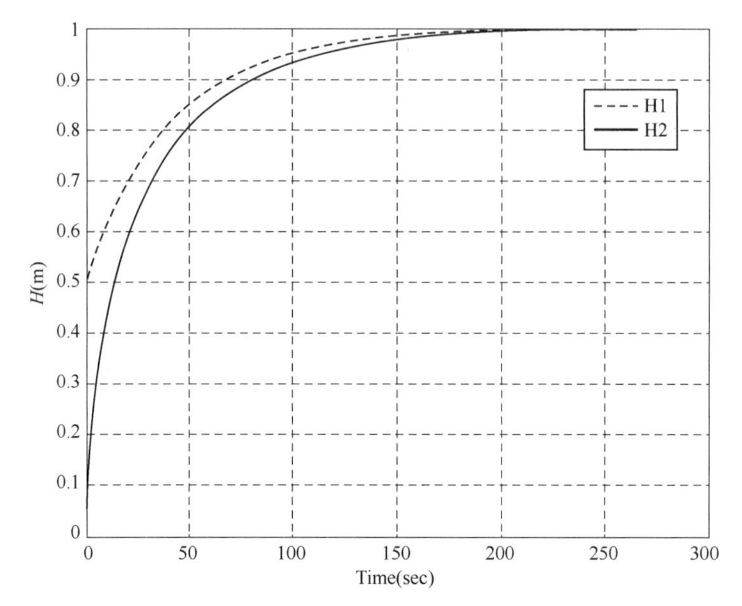

Figure 5.6.1 Water Height Response of Double-Tank System

```
%Main
clc;
clear all;
global RHO G A1 A12 A2 Aout Qinstar Hstar
RHO=1000; G=9.8;
A1=4*pi; A12=0.04*pi;
Qinstar=556.32; Hstar=1;
A2=1*pi; Aout=0.04*pi;
to=0; tf=300;
H1_0=0.5; H2_0=0;
[t,H]=ode45('d_H',[to tf],[H1_0 H2_0]);
figure(1);
plot(t,H(:,1),'k--');
hold on
plot(t,H(:,2),'k-');
grid
xlabel('Time(sec)'),ylabel('H(m)');
```

Figure 5.6.2 MATLAB Script and Its Results

Function file d_H() is:

```
function D_H=d_H(t,H)
global RHO G A1 A12 A2 Aout
Qin=qin(t);
D_H=zeros(2,1);
D_H(1)= - (A12*sqrt(2*G*H(1)))/A1+Qin/(RHO*A1);
D_H(2)=(A12*sqrt(2*G*H(1)))/A2- (Aout*sqrt(2*G*H(2)))/A2;
end
```

Function file qin() is:

```
function Qin=qin(t)
global Qinstar
Qin=Qinstar
```

Figure 5.6.2 MATLAB Script and Its Results (continued)

2. When the input is $Q_{\text{in}}(t)=Q_{\text{in}}^{*}+\Delta Q_{\text{in}}$ and the system is in steady state with $H_1(0)=H^{*}$, $H_2(0)=H^{*}$, the response deviation of both the nonlinear model Equation (5.54) and linear model Equation (5.38).

The output response curve is shown in Figure 5.6.3. Let $\Delta Q_{\text{in}}(t)=10\text{kg}/\text{sec}$, its Laplace transform is 10 times the unit step signal. The system simulation script and results are shown in Figure 5.6.4.

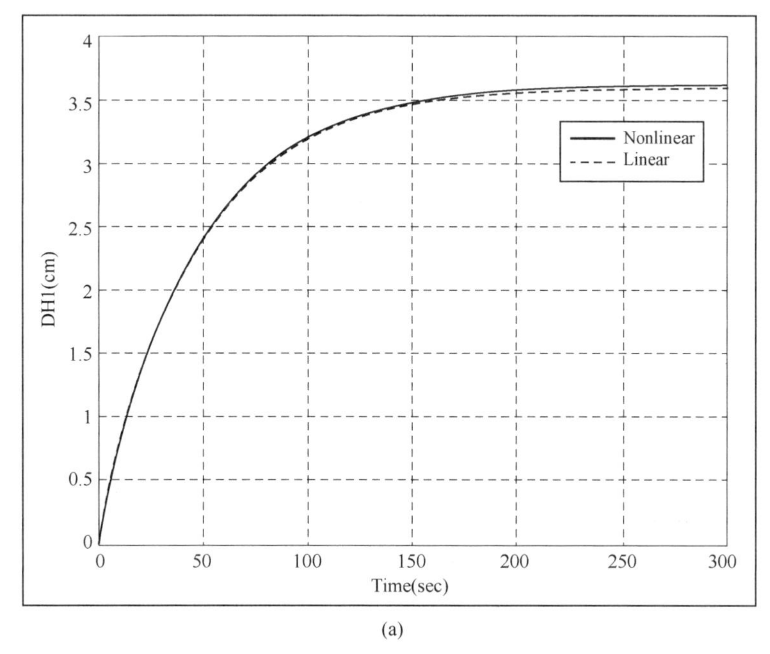

(a)

Figure 5.6.3 Step Response of both Linear and Nonlinear Models

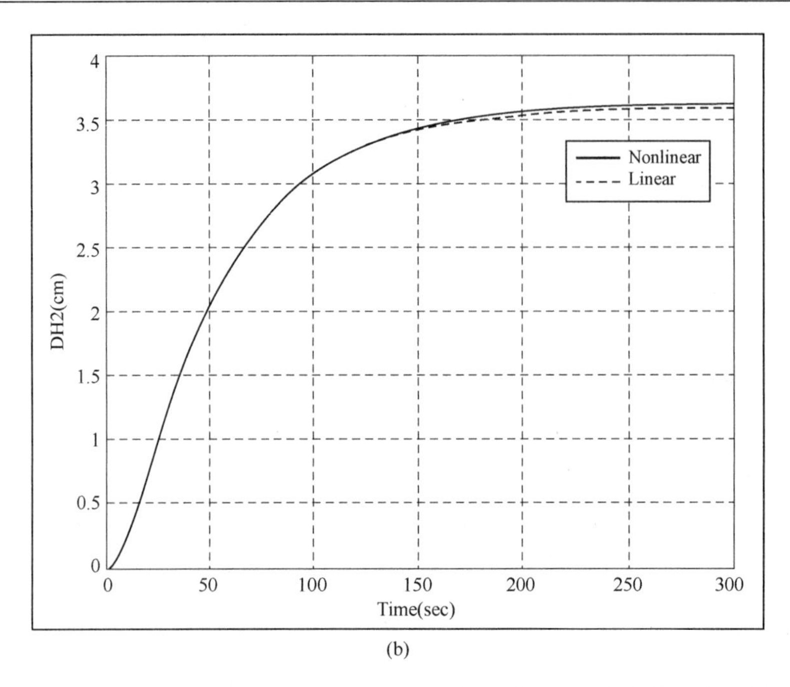

(b)

Figure 5.6.3 Step Response of both Linear and Nonlinear Models（continued）

```
%Main
clc;
clear all;
global RHO G A1 A12 A2 Aout Qinstar Hstar
RHO=1000; G=9.8;
A1=4*pi; A12=0.04*pi;
Qinstar=556.32; Hstar=1;
A2=1*pi; Aout=0.04*pi;
to=0; tf=300;
H1_0=Hstar;
H2_0=Hstar;
[t,H]=ode45('d_H2',[to tf],[H1_0 H2_0]);
DH_1=H(:,1)-Hstar;
DH_2=H(:,2)-Hstar;
%%%%%%%%%%%%%%%%%%%%%%%%%%%%%%%%%%%%%%%
t_line=[0:0.1:300];
k1=1/(RHO*A1);
Omega1=A12^2*RHO*G/(A1*Qinstar);
k2=1/(RHO*A2);
Omega2=Aout^2*RHO*G/(A2*Qinstar);
num1=[k1];
den1=[1 Omega1];
DH_1_line=10*step(num1,den1,t_line);
num2=[k2*Omega1];
den2=conv([1,Omega1],[1,Omega2]);
```

Figure 5.6.4 MATLAB Script and Its Results

```
DH_2_line=10*step(num2,den2,t_line);
```

```
%%%%%%%%%%%%%%%%%%%%%%%%%%%%%%%%%%%%%%%%
figure(1);
plot(t,100*DH_1,'k-');
hold on
plot(t_line,100*DH_1_line,'k--');
grid on;
xlabel('Time(sec)'),ylabel('DH1(cm)');
figure(2);
plot(t,100*DH_2,'k-');
hold on
plot(tLine,100*DH_2_line,'k--');
grid on;
xlabel('Time(sec)'),ylabel('DH2(cm)');
```

Function file d_H2() is

```
function D_H=d_H2(t,H)
global RHO G A1 A12 A2 Aout
Qin=qin2(t);
D_H=zeros(2,1);
D_H(1)=-(A12*sqrt(2*G*H(1)))/A1+Qin/(RHO*A1);
D_H(2)=(A12*sqrt(2*G*H(1)))/A2-(Aout*sqrt(2*G*H(2)))/A2;
end
```

Function file qin2() is

```
function Qin=qin2(t)
global Qinstar
D_Qin=10;
Qin=Qinstar+D_Qin;
```

```
DH_1_final =

    3.6158

DH1_line_final =

    3.5901

DH_2_final =

    3.6140

DH_2_line_final =

    3.5886
```

Figure 5.6.4 MATLAB Script and Its Results (continued)

It can be seen from Figure 5.6.3(a) and Figure 5.6.3(b) that there exists small deviation between the transfer function obtained by linearization and the original nonlinear model. For the upper tank, $\Delta Q_{\text{in}}(t)$ caused the steady-state value of the water height change, for nonlinear model it is $\Delta H_{1\text{ss}} = 3.6158\text{cm}$, for linear model it is $\Delta H_{1\text{ss}} = 3.5901\text{cm}$. For the lower tank, the steady-state value of the water height change is $\Delta H_{2\text{ss}} = 3.6140\text{cm}$ for the nonlinear model and $\Delta H_{1\text{ss}} = 3.5886\text{cm}$ for the linear model. This value is consistent with the steady-state value of the linear model obtained by Analytical Equation (5.48) and Equation (5.51). Therefore, linearizing the nonlinear model at its equilibrium is an effective method to simplify the model.

Summary

1. Rational assumptions are conducted based on the physical characteristics of the system.

2. Derivation of the nonlinear dynamic differential equation model of the system is derived and linearized.

3. The transfer function of the system is derived through, the analytical solutions of each transfer function by means of Laplace inverse transform.

4. The liquid depth characteristics of the system and the nonlinear/linear models are analyzed through MATLAB.

生词注解

controlled	v. 控制，约束
	adj. 受控制的
nonlinear	adj. 非线性的
boiler	n. 锅炉，烧水器
instantaneous	adj. 瞬间的，即时的
compressibility	n. 压缩性，压缩系数
coefficient	n. 系数
	adj. 合作的
viscosity	n. 黏性
irrotational	adj. 无旋涡的
linearization	n. 线性化
derivative	n. 衍生物，派生物，导数
	adj. 派生的
equilibrium	n. 均衡
deviation	n. 偏差
analytical solution	n. 解析解
invariant	n. 不变量
	adj. 不变的
hysteresis	n. 磁滞现象

第6章 控制系统的稳定性

本章介绍控制系统的稳定性分析。在控制系统的分析、改造和设计过程中，对系统性能的要求包括稳定性、快速性和准确性。其中，稳定性是前提。本章将介绍劳斯稳定判据的MATLAB实现，并通过一个实际的例子——人骑行自行车的稳定控制来展开对稳定性的分析和讨论。

6.1 稳定性的定义和判定

当设计和分析控制系统时，最重要的是要考虑稳定性。从实用的观点看，不稳定的闭环反馈系统没有多大价值。除了个别情况，通常要求所设计的控制系统是闭环稳定的。

通常，使用“稳定的”或“不稳定的”来描述一个闭环反馈系统，这里指的是绝对稳定性。具有绝对稳定性的系统称为稳定系统。若一个闭环系统是稳定的，则可以利用相对稳定性来进一步衡量其稳定程度。

6.1.1 系统稳定性

对线性连续系统而言，设其闭环传递函数为

$$\Phi(s)=\frac{C(s)}{R(s)}=\frac{b_0 s^m + b_1 s^{m-1} + \cdots + b_{m-1}s + b_m}{a_0 s^n + a_1 s^{n-1} + \cdots a_{n-1}s + a_n} \tag{6.1}$$

式中，$D(s)=a_0 s^n + a_1 s^{n-1} + \cdots + a_{n-1}s + a_n$ 为系统特征多项式；$D(s)=a_0 s^n + a_1 s^{n-1} + \cdots + a_{n-1}s + a_n = 0$ 为系统特征方程，其解称为系统闭环极点，也称系统特征方程的根。

对于连续时间系统，确定一个系统是否稳定（绝对稳定）的方法是，判断该系统闭环传递函数的所有极点（即闭环特征根）是否均位于 s 左半平面。

对于状态空间模型，判断其稳定性就是判断系统矩阵 $\boldsymbol{A}$ 的特征根是否均位于 s 左半平面。若所有极点（或特征根）均位于 s 左半平面，则系统绝对稳定，就可以进一步通过极点（或特征根）的相对位置来判断相对稳定性。

对于离散时间系统，如果系统全部极点都位于 z 平面的单位圆内，则系统是稳定的。

若系统在有界输入的作用下其响应的幅度也是有界的，则这个系统是稳定的，称为BIBO（Bounded-Input Bounded-Output）稳定性。

一个系统 BIBO 稳定的充分必要条件是系统传递函数的所有极点均有负的实部。

如果所有极点不是都位于 s 左半平面，则系统是不稳定的。如果特征方程有一对共轭复根在虚轴（jw 轴）上，而其他根均位于 s 左半平面，则系统在有界的输入下，其稳态输出保持等幅振荡；当输入为正弦波，且频率等于虚轴上根的幅值时，其输出才会变成无界的。这样的系统称为临界稳定系统。对于不稳定系统，特征方程至少有一个根位于 s 右半平面，在这种情况下，系统的输出对任何输入都是不稳定的。

系统极点在复平面的位置对系统的稳定性是至关重要的。众所周知，系统极点的位置也会影响系统的性能，可以通过超调量、调整时间和峰值时间等性能指标表现出来。对于判定系统的稳定性来说，没必要知道系统极点在左半平面的具体位置，只需知道系统的 所有极点均位于左半平面即可。若系统在虚轴上有一对共轭极点，如 $s_{1,2}=\pm 5\mathrm{j}$，则系统是临界稳定的；而若系统在此有两对共轭极点，如 $s_{1,2}=\pm 4\mathrm{j}$ 和 $s_{3,4}=\pm 5\mathrm{j}$，则系统是不稳定的。

除了根据定义直接判断稳定性，还可以利用其他稳定性判据进行判断。本章将介绍劳斯稳定判据的使用。

若连续时间系统的全部零极点都位于 s 左半平面；或若离散时间系统的全部零极点都位于 z 平面单位圆内，则系统是最小相位系统。

6.1.2　MATLAB 中稳定判断的函数

根据稳定性的定义可知，判断稳定的最直接方法就是计算出所有的闭环特征方程的根，根据其在 s 平面的位置进行判断。MATLAB 提供了相关的函数，其说明见表 6.1。

表 6.1　判断稳定的相关函数

函　数	说　明
E=eig(G)	求取矩阵特征根，系统的模型 G 可以是传递函数，状态和零极点模型，也可以是连续或离散的
P=pole(G) Z=zero(G)	分别求系统的极点和零点，G 是已定义的系统模型
[p,z]=pzmap(sys)	求系统的极点和零点，sys 是已定义的系统模型
r=roots(P)	求特征方程的根，P 是按降幂排列的闭环系统特征多项式的系数向量
ind=find(X)	求取满足条件 X 的下标向量，用列向量表示

6.1.3　稳定判断实例

例 6.1　已知系统闭环传递函数为 $\Phi(s)=\dfrac{s^3+s^2+4}{s^6+12s^5+17s^4+22s^3+31s^2+16s+12}$，请用 MATLAB 判定系统稳定性。

例 6.1 的程序和结果见图 6.1.1。

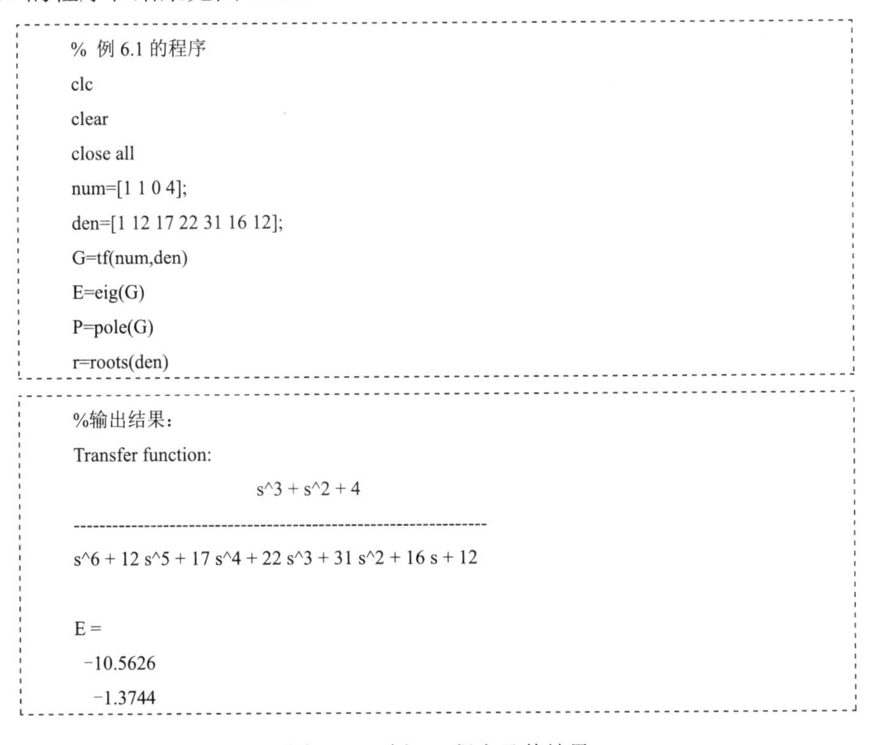

```
% 例 6.1 的程序
clc
clear
close all
num=[1 1 0 4];
den=[1 12 17 22 31 16 12];
G=tf(num,den)
E=eig(G)
P=pole(G)
r=roots(den)
```

```
%输出结果：
Transfer function:
                    s^3 + s^2 + 4
-----------------------------------------------------------
s^6 + 12 s^5 + 17 s^4 + 22 s^3 + 31 s^2 + 16 s + 12

E =
 -10.5626
  -1.3744
```

图 6.1.1　例 6.1 程序及其结果

```
    0.2265 + 1.1416i
    0.2265 - 1.1416i
   -0.2580 + 0.7373i
   -0.2580 - 0.7373i

P=
 -10.5626
  -1.3744
    0.2265 + 1.1416i
    0.2265 - 1.1416i
   -0.2580 + 0.7373i
   -0.2580 - 0.7373i

r =
 -10.5626
  -1.3744
    0.2265 + 1.1416i
    0.2265 - 1.1416i
   -0.2580 + 0.7373i
   -0.2580 - 0.7373i
```

图 6.1.1　例 6.1 程序及其结果（续）

从结果可知，系统有两个正实部的闭环特征根，系统不稳定。此外，eig(G)，pole(G)，roots(den)函数在此例中的作用是一样的，都是用来求系统的闭环特征根，可根据需要选择使用。

例 6.2　系统闭环传递函数为

$\varPhi(s)=\dfrac{3s^3+16s^2+41s+28}{s^6+14s^5+110s^4+528s^3+1494s^2+2117s+112}$，判断系统的稳定性，以及系统是否为最小相位系统。

程序见图 6.1.3，其输出结果显示了系统的零极点和增益，得到系统稳定且为最小相位系统的结论，输出零极点分布见图 6.1.2。

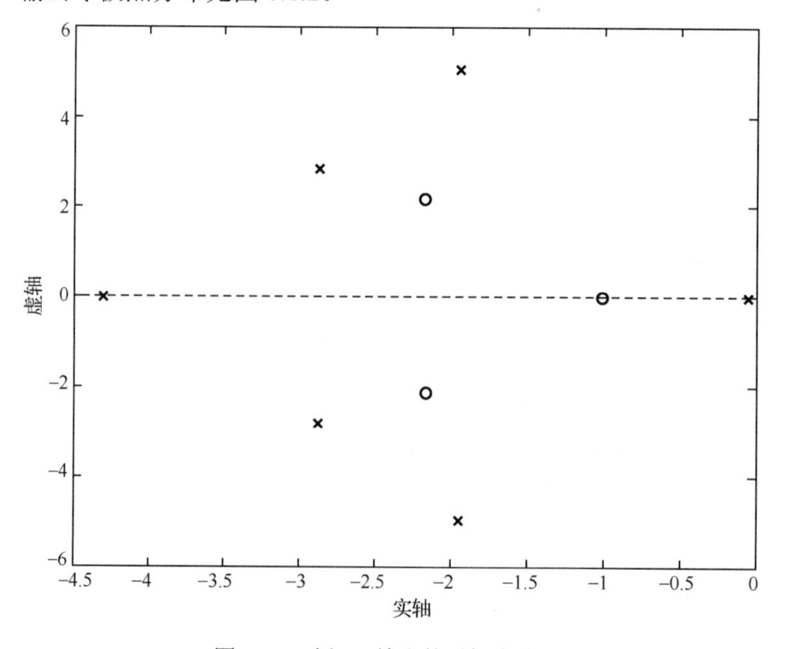

图 6.1.2　例 6.2 输出的零极点分布

```
%例 6.2 程序
clc;
clear;
close all;
num=[3 16 41 28];
den=[1 14 110 528 1494 2117 112];
[z,p,k]=tf2zp(num,den)
%检验零点的实部；求实部大于零的零点下标
ind_z=find(real(z)>0);
%求实部大于零的零点个数
m=length(ind_z);
%检验极点的实部；求实部大于零的极点下标
ind_p=find(real(p)>0);
%求实部大于零的极点个数
n=length(ind_p);
%判稳
if(n>0)
    disp('the system is unstable')
    disp('the unstable pole is:')
    disp(p(ind_p))
else
    disp('the system is stable')
    if (m>0) %判断是否为最小相位系统
        disp('the system is a nonminimal phase one')
    else
        disp('the system is a minimal phase one')
    end
end
%绘制零极点图
pzmap(p,z)
```

```
%例 6.2 的输出结果
z =


  -2.1667 + 2.1538i
  -2.1667 - 2.1538i
  -1.0000


p =


  -1.9474 + 5.0282i
  -1.9474 - 5.0282i
  -4.2998
  -2.8752 + 2.8324i
  -2.8752 - 2.8324i
  -0.0550
k =
     3


the system is stable
the system is a minimal phase one
```

图 6.1.3　例 6.2 程序及其结果

例 6.3 已知某系统的模型为

$$\dot{x}=\begin{bmatrix}1&2&-1&2\\2&6&3&0\\4&7&-8&-5\\7&2&1&6\end{bmatrix}x+\begin{bmatrix}-1\\0\\0\\1\end{bmatrix}u$$

$$y=[-2\quad 5\quad 6\quad 1]x+7u$$

试判断系统的稳定性及系统是否为最小相位系统。

程序见图 6.1.5，其输出结果显示了系统的零极点和增益，得出结论，系统是不稳定的。输出的零极点分布见图 6.1.4。

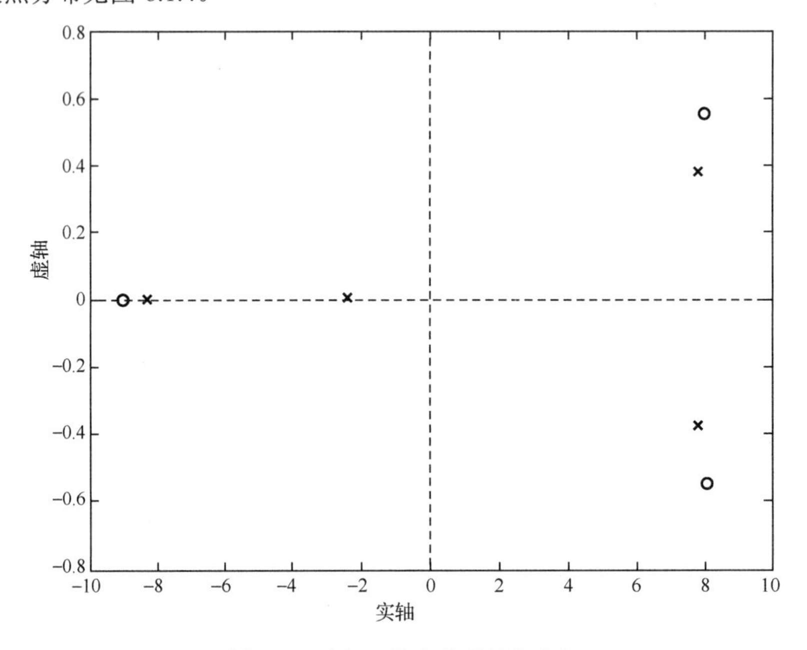

图 6.1.4 例 6.3 输出的零极点分布

```
%例 6.3 程序
clc
clear
close all
a=[1 2 -1 2;2 6 3 0;4 7 -8 -5;7 2 1 6];
b=[-1 0 0 1]';
c=[-2 5 6 1];d=7;
[z,p,k]=ss2zp(a,b,c,d)
ind_z=find(real(z)>0);
m=length(ind_z);
ind_p=find(real(p)>0);
n=length(ind_p);
if(n>0)
    disp('the system is unstable')
```

图 6.1.5 例 6.3 程序及其结果

```
        disp('the unstable pole is:')
        disp(p(ind_p))
else
        disp('the system is stable')
        if(m>0)
              disp('the system is a nonminimal phase one')
        else
              disp('the system is a minimal phase one')
        end
end
pzmap(p,z)
```

图 6.1.5　例 6.3 程序及其结果（续）

6.2　劳斯稳定判据

劳斯稳定判据是用来判断，在一个特征方程式中是否存在不稳定根，而不必求解该方程式。该稳定判据只能用在具有有限项的特征方程式中。当把该判据用到控制系统时，根据特征方程的系数，可以直接判断系统的稳定性。

劳斯稳定判据的方法是判断系统特征方程的根有几个具有正实部（即有几个特征根位于 s 右半平面）。

特征方程式为

$$a_0 s^n + a_1 s^{n-1} + \cdots + a_{n-1} s + a_n = 0 \tag{6.2}$$

式中，系数为实数（$a_n \neq 0$）。

劳斯稳定判据说明：式（6.2）中具有正实部根的个数，等于劳斯阵列中第一列系数符号的改变次数。应当指出，第一列中各项系数的精确值没有必要知道，只需要知道它们的符号。式（6.2）的所有根都位于 s 左半平面的充分必要条件是全部系数都是正值，并且劳斯阵列第一列中的所有项都具有正号。

有一个三阶特征方程式为

$$a_0 s^3 + a_1 s^2 + a_2 s + a_3 = 0$$

式中所有系数均为正数。对上式应用劳斯稳定判据，劳斯阵列为

$$\begin{array}{lll} s^3 & a_0 & a_2 \\ s^2 & a_1 & a_3 \\ s^1 & \dfrac{a_1 a_2 - a_0 a_3}{a_1} & \\ s^0 & a_3 & \end{array}$$

显然，所有根都具有负实部的条件是：$a_1 a_2 > a_0 a_3$。

例 6.4　某系统的特征方程为

$$3s^4 + 10s^3 + 5s^2 + s + 2 = 0$$

应用劳斯稳定判据判断该系统是否稳定。

建立劳斯阵列，有

$$
\begin{array}{llll}
s^4 & 3 & 5 & 2 \\
s^3 & 10 & 1 & 0 \\
s^2 & \dfrac{47}{10} & 2 & \\
s^1 & -\dfrac{153}{47} & & \\
s^0 & 2 & &
\end{array}
$$

在此，首列元素出现了两次符号变化，这说明系统特征方程式还有两个根位于 s 右半平面，因此系统是不稳定的。

利用 MATLAB，通过调用 roots 函数直接求解特征方程的根，可以验证用劳斯稳定判据得到的结果，如图 6.2.1 所示。

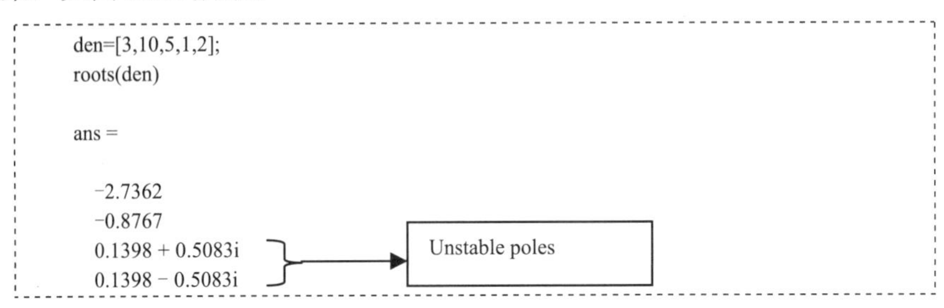

```
den=[3,10,5,1,2];
roots(den)

ans =

  -2.7362
  -0.8767
  0.1398 + 0.5083i
  0.1398 - 0.5083i
```


图 6.2.1　求解特征方程的根

6.3　人控制的自行车

本节将以人控制的自行车为例，说明实现系统稳定性的控制器设计。自行车在日常生活中使用非常广泛，给人们带来了便利。骑行自行车时，需保持一定的姿态、速度和稳定行驶。

在系统中，被控对象是自行车，控制器是人脑，眼睛、耳朵作为传感器，通过视听获得信号并传递给大脑，大脑形成控制决策。人的躯干和四肢执行该决策，调整自行车行驶的姿态和速度，保持稳定（见图 6.3.1）。

图 6.3.1　人控制的自行车

本章只讨论控制器参数设计，被控对象建模和控制器选型不作为讨论内容，劳斯稳定判据被用来完成系统未定参数的设计。

6.3.1　控制要求及参数分析

由人控制的自行车如图 6.3.1 所示。自行车将沿直线以恒定速度 V 前进。用 φ 表示自行车对称轴平面和垂直平面之间的夹角。当自行车稳定行驶时，期望的角度 $\varphi_d(s)$

$$\varphi_d(s)=0 \tag{6.3}$$

1. 控制目标

控制自行车在垂直位置稳定行驶，控制自行车在有扰动时维持在垂直位置。

2. 控制变量

自行车偏离垂直位置的角度，即 $\varphi(s)$。

3. 设计指标

闭环系统必须稳定。

本章仅关注系统的稳定性而非其瞬态响应特征，因此控制指标只与稳定性有关。

已知自行车的模型为

$$G(s)=\frac{\Omega_2 s+\Omega_3}{s^2-\Omega_1} \tag{6.4}$$

其中，$\Omega_1=g/h$，$g=9.806\text{m/s}^2$，h 是自行车的重心离地面的高度，$\Omega_2=VL/(hw)$，$\Omega_3=V^2/(hw)$（见图 6.3.1）。自行车的前进速度用 V 表示，其轴距（两轮心之间的距离）用 w 表示。自行车的重心与其前轮轴之间的水平距离用 L 表示。

由于有极点 $s=\pm\sqrt{\Omega_1}$，所以自行车是不稳定的。

人作为控制器，该控制器为

$$G_c(s)=\frac{K_1 s+K_2}{\tau s+1} \tag{6.5}$$

控制器参数 τ，K_1 和 K_2 共同调节系统的响应。

在传递函数的建模过程中，忽略了前轮在垂直面上的旋转，且假设前进速度为恒定，因此需增加反馈环节来调整实际的前进速度。系统物理参数如表 6.2 所示。

表 6.2　物理参数

参　量	值	单　位
τ	0.5	s
Ω_1	12.56	$1/s^2$
Ω_2	1.86	$1/s^2$
Ω_3	4.66	1/s
h	0.78	m
V	2.0	m/s
L	0.8	m
w	1.1	m
K_1	0.1	
K_2	1	

反馈系统的组成结构如图 6.3.2 所示。

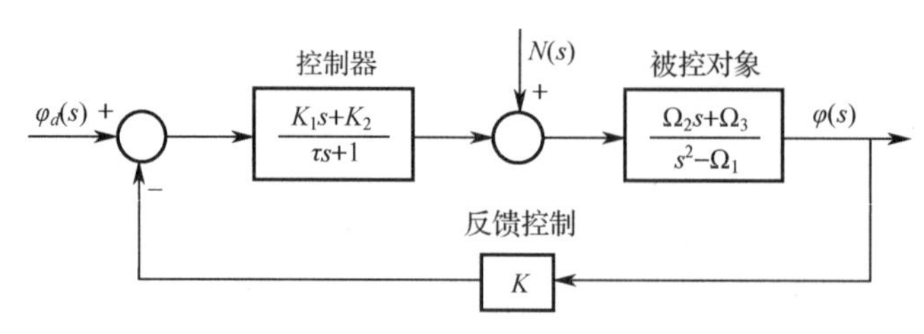

图 6.3.2 人控制的自行车系统结构框图

观察图 6.3.2 可知，被控对象由物理系统固有参数 h，L，w 以及 V 表达，控制器由参数 τ，K_1，K_2 表达。在这个例子中，假定 τ，K_1 和 K_2 给定，则需要调节的参数为反馈环节中的参数 K。

4. 关键待调参数

反馈增益 K 是关键待调参数。

需要说明的是，关键待调参数不总是在前向通道中，事实上它们可能存在于方框图中的任何子系统中。

接下来对系统的稳定性进行设计，用劳斯判据来分析闭环系统的稳定性，确定当 K 取什么值时才能使闭环系统稳定。

6.3.2 稳定性分析

闭环系统从 $\varphi_d(s)$ 到 $\varphi(s)$ 的传递函数为

$$T(s)=\frac{(K_1s+K_2)(\Omega_2s+\Omega_3)}{\Delta(s)} \tag{6.6}$$

其中

$$\Delta(s)=\tau s^3+(1+K_1\Omega_2K)s^2+(K_1\Omega_3K+K_2\Omega_2K-\tau\Omega_1)s+K_2\Omega_3K-\Omega_1 \tag{6.7}$$

特征方程为

$$\Delta(s)=0 \tag{6.8}$$

故需要解决的问题是：当 K 取什么值时，特征方程式（6.8）的所有根都在 s 左半平面？可以列出如下劳斯阵列

$$\begin{array}{c|cc} s^3 & \tau & K_1\Omega_3K+K_2\Omega_2K-\tau\Omega_1 \\ s^2 & 1+K_1\Omega_2K & K_2\Omega_3K-\Omega_1 \\ s^1 & \alpha & \\ s^0 & K_2\Omega_3K-\Omega_1 & \end{array} \tag{6.9}$$

其中

$$\alpha=\frac{(1+K_1\Omega_2K)(K_1\Omega_3K+K_2\Omega_2K-\tau\Omega_1)-\tau(K_2\Omega_3K-\Omega_1)}{1+K_1\Omega_2K} \tag{6.10}$$

为了使系统稳定，根据劳斯阵列中的第一列必须满足

$$
\begin{cases}
\tau > 0 \\
K > -\dfrac{1}{K_1\Omega_2} \\
\alpha > 0 \\
K > \dfrac{\Omega_1}{K_2\Omega_3}
\end{cases}
\tag{6.11}
$$

显然$\tau > 0$，且由于$K_1 > 0$，$\Omega_2 > 0$，因此第二个不等式也能够满足。将表 6.2 中的取值代入第 4 个不等式中，可得：$K > 2.695$。因此，只要选择$K > 2.695$，系统就能够稳定。

由于人控制的自行车系统控制目标是保持自行车直立，即希望$\varphi_d(s) = 0$，因此，无论由输入还是由扰动所导致的$\varphi_d(s) \neq 0$的情况，相应的输出都应该达到或逼近$\varphi_d(s) = 0$的状态。因此，需要选择合适的K，使其达到控制目标。

通过 MATLAB 仿真来选取K值。

首先，取$K = 2.7$，系统在单位阶跃输入下的输出响应如图 6.3.3 所示。虽然系统能够稳定，但是振荡幅度和，稳态误差都过大，不符合自行车的稳定控制要求。

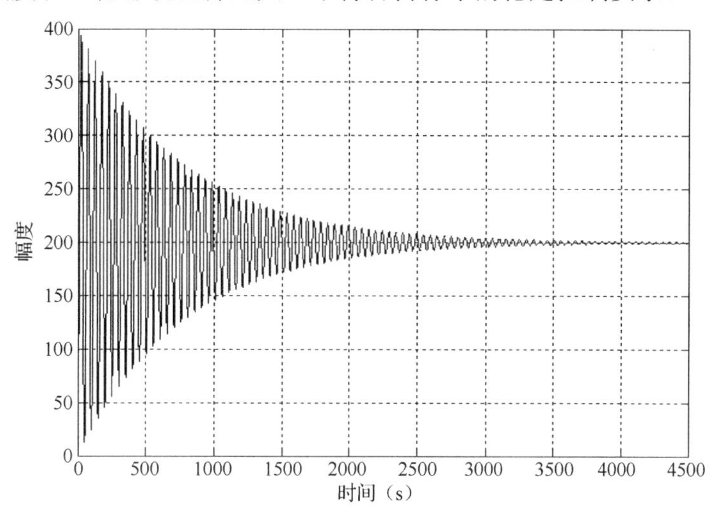

图 6.3.3　人控制的自行车系统单位阶跃响应（K=2.7）

此外，取$K = 6$，系统在单位阶跃输入下的输出响应如图 6.3.4 所示。虽然系统能够稳定，振荡幅度明显减小，但稳态误差约为输入信号幅值的 1/3，仍然过大，自行车难以在较大的倾角下持续运行，因此仍然未能达到预期的控制目标。

最后，取$K = 26$，系统在单位阶跃输入下的输出响应如图 6.3.5 所示。系统能够稳定，振荡幅度和稳态误差均较小。调节时间约为 1.2s，输出响应符合人控制的自行车稳定性要求，即达到预期控制目标。

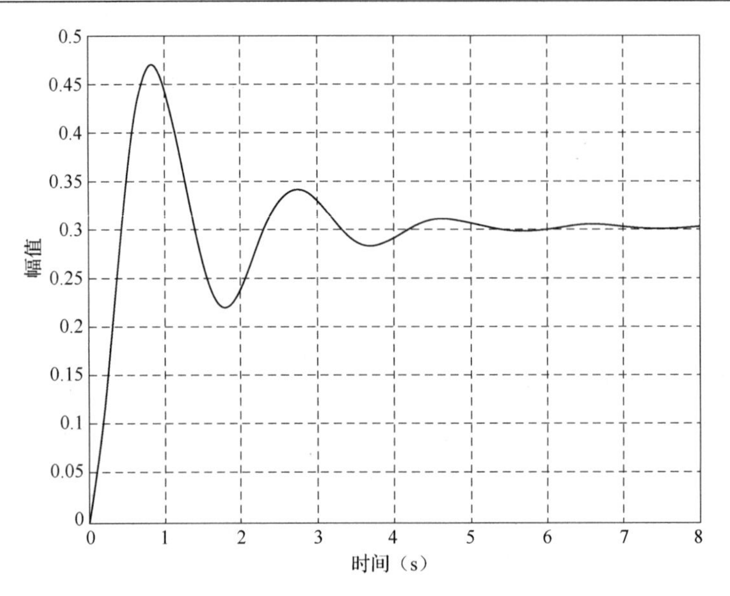

图 6.3.4 人控制的自行车系统单位阶跃响应（K=6）

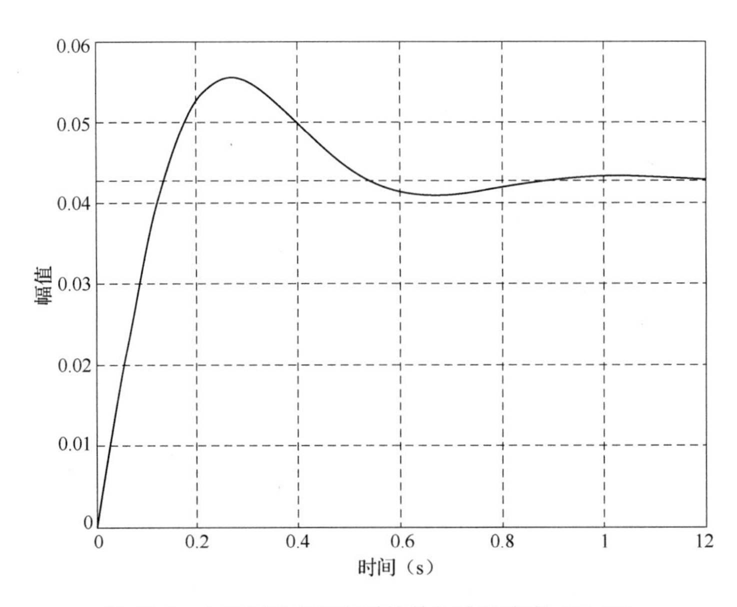

图 6.3.5 人控制的自行车系统单位阶跃响应（K=26）

MATLAB 程序见图 6.3.6。

```
%图 6.3.3、图 6.3.4 和图 6.3.5 的程序
clc;
clear;
close all;
h=0.78;v=2;L=0.8;w=1.1;g=9.8;
Omiga1=g/h;Omiga2=v*L/(h*w);Omiga3=v^2/(h*w);
tau=0.5;k1=0.1;k2=1;
```

图 6.3.6 MATLAB 程序

```
k=2.7;
%k=6;
%k=26;
nc=[k1 k2];dc=[tau 1];
nb=[Omiga2 Omiga3];db=[1 0 -Omiga1];
nf=[0 k];df=[0 1];
[na,da]=series(nc,dc,nb,db);
[num,den]=feedback(na,da,nf,df);
printsys(num,den)
figure(1);
step(num,den),grid
```

图 6.3.6　MATLAB 程序（续）

6.3.3　扰动响应分析

考虑自行车本身，即如图 6.3.7 所示，不含控制器和反馈时，系统从扰动 $N(s)$ 到输出 $\varphi(s)$ 的结构框图。

$$N(s) \rightarrow \boxed{\frac{\Omega_2 s+\Omega_3}{s^2-\Omega_1}} \rightarrow \varphi(s)$$

图 6.3.7　在扰动输入下的自行车系统结构

从扰动 $N(s)$ 到输出 $\varphi(s)$ 的传递函数为

$$T_N(s)=\frac{\varphi(s)}{N(s)}=\frac{\Omega_2 s+\Omega_3}{s^2-\Omega_1} \tag{6.12}$$

特征方程为

$$q(s)=s^2-\Omega_1=0 \tag{6.13}$$

系统极点为

$$\begin{aligned} s_1&=-\sqrt{\Omega_1} \\ s_2&=+\sqrt{\Omega_1} \end{aligned} \tag{6.14}$$

可见，系统在 s 右半平面有一个极点，因此自行车本身是不稳定的。没有控制器和反馈时，任何外界扰动都会使自行车歪倒。显然，必须对自行车进行控制。当增加控制器和反馈时，从扰动 $N(s)$ 到输出 $\varphi(s)$ 的闭环传递函数为

$$T_N(s)=\frac{(\tau s+1)(\Omega_2 s+\Omega_3)}{\Delta(s)} \tag{6.15}$$

式中的 $\Delta(s)$ 见式（6.7）。

阶跃扰动信号 $N(s)=1/s$ 的响应如图 6.3.8 所示，响应是稳定的。系统的振荡幅度和稳态误差均较小，最终当 $\varphi=0.0425\,\text{rad}$ 时，微小倾斜能够满足使自行车直立行驶的要求。MATLAB 程序见图 6.3.9。

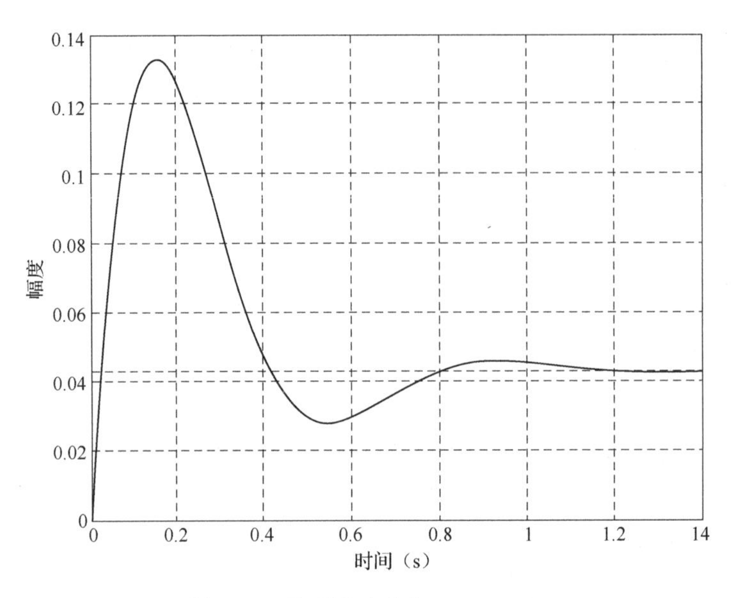

图 6.3.8　阶跃扰动响应（ $K=26$ ）

```
%图 6.3.8 程序
clc;
clear;
char cluse all;
h=0.78;v=2;L=0.8;w=1.1;g=9.8;
Omiga1=g/h;Omiga2=v*L/(h*w);Omiga3=v^2/(h*w);
tau=0.5;k1=0.1;k2=1;
k=26;
nc=[k1 k2];dc=[tau 1];
nb=[Omiga2 Omiga3];db=[1 0 -Omiga1];
nf=[0 k];df=[0 1];
[na,da]=series(nf,df,nc,dc);
[num,den]=feedback(nb,db,na,da);
printsys(num,den)
figure(2);
step(num,den,'k'),grid
```

图 6.3.9　MATLAB 程序

6.3.4　速度影响分析

众所周知，人们在骑行自行车时，速度越慢，稳定越难以控制。对此，可利用改变速度 V 的取值来加以验证。

当选择反馈增益 $K=26$ 时，不断改变前进速度，对应的特征方程的根如图 6.3.10 所示。当自行车处于低速时，如 $V=0.6\text{m/s}$ ，特征方程的根为 $r_1=-2.5391+3.6222\text{i}$ ， $r_2=-2.5391-3.6222\text{i}$ ， $r_3=0.1692$ ，存在一个 s 右半平面的根，系统不稳定；随着速度提高，当 $V=0.7\text{m/s}$ 时，特征方程的根为 $r_1=-2.5990+4.0709\text{i}$ ， $r_2=-2.5990-4.0709\text{i}$ ， $r_3=-0.1959$ ，所有的根

位于 s 左半平面，系统稳定；随着速度 V 的继续增加，特征方程的根依然位于 s 左半平面，系统稳定。MATLAB 程序见图 6.3.11。

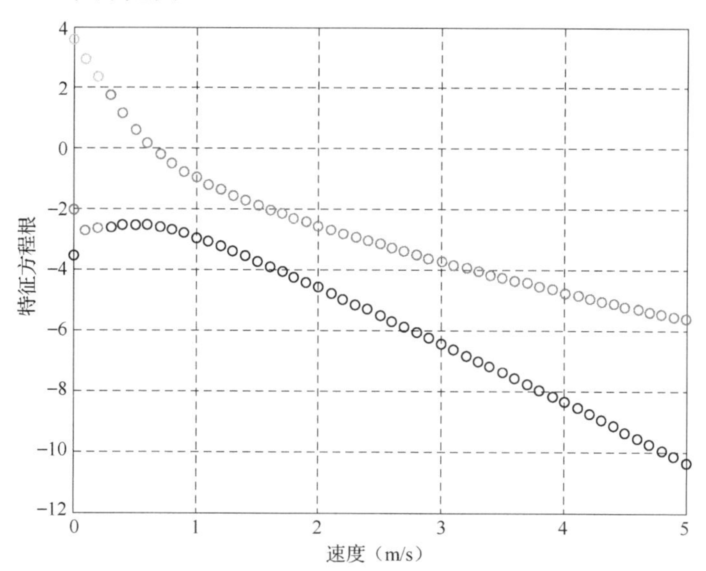

图 6.3.10 速度变化对特征方程根的影响

```
%图 6.3.10 程序
clc;
clear;
close all;
h=0.78;L=0.8;w=1.1;g=9.8;
v=[0:0.1:5];
Omiga1=g/h;
Omiga2=v*L/(h*w);
Omiga3=v.^2/(h*w);
tau=0.5;k1=0.1;k2=1;
k=26;
nc=[k1 k2];dc=[tau 1];
nf=[0 k];df=[0 1];
for i=1:length(v)
nb=[Omiga2(i) Omiga3(i)];db=[1 0 -Omiga1];

[na,da]=series(nc,dc,nb,db);
[num,den]=feedback(na,da,nf,df);
    r(:,i)=roots(den);
end
plot(v,r,'o'),grid
xlabel('Velocity(m/sec)');ylabel('Characteristic equation roots');
```

图 6.3.11 MATLAB 程序

本章小结

（1）介绍稳定的概念和劳斯稳定判据；

（2）分析人控制的自行车系统稳定性，用劳斯稳定判据确定未定的参数；

（3）通过对扰动和速度的响应分析，实现了人控制的自行车系统的稳定控制。

Chapter 6
Stability of Control System

This chapter describes the stability analysis of the control system. During the analysis, modification and design of the control system, the requirements for system performance include stability, rapidity, and accuracy, in which stability is the prerequisite. This chapter will introduce the MATLAB implementation of the Routh's Stability Criterion, and go further to analyze and discuss stability control by taking the example of riding bicycles.

6.1 Definition and Judgment of Stability

Stability is of the utmost importance in designing and analyzing control system. From a practical point of view, an unstable closed-loop feedback system is of little value. With rare exceptions, it is usually required that the designed control system is closed-loop stable.

In general, a closed-loop feedback system is described as being “stable” or “unstable” to refer to its absolute stability. Systems with absolute stability are called stable systems. If a closed-loop system is stable, then relative stability can be used to further measure its stability.

6.1.1 Stability of System

For a linear continuous system, set its closed-loop transfer function as

$$\Phi(s)=\frac{C(s)}{R(s)}=\frac{b_0s^m+b_1s^{m-1}+\cdots+b_{m-1}s+b_m}{a_0s^n+a_1s^{n-1}+\cdots+a_{n-1}s+a_n} \tag{6.1}$$

where $D(s)=a_0s^n+a_1s^{n-1}+\cdots+a_{n-1}s+a_n$ is system Characteristic Polynomial, $D(s)=a_0s^n+a_1s^{n-1}+\cdots+a_{n-1}s+a_n=0$ is system characteristic equation. The solutions are the system closed-loop poles, also called the roots of the system’s characteristic equation.

For continuous time systems, the method to determine whether a system is stable (i.e. absolutely stable) is to determine whether all the poles of the closed-loop transfer function (i.e. the roots of the characteristic equation) are located in the left-half s plane.

For the state-space model, the method to determine its stability is to determine whether the eigenvalues of the system matrix $\boldsymbol{A}$ are all located in the left-half s plane. If all poles (or eigenvalues) are located in the left-half s plane, then the system is absolutely stable, and relative stability can be further determined by means of the relative position of the poles (or eigenvalues).

For the discrete-time system, the system is stable if all poles of the system are within the unit circle of the z plane.

If the response amplitude of the system, under bounded input, is also bounded, then the system is stable, called BIBO (Bounded-Input Bounded-Output) stability.

A necessary and sufficient condition for the BIBO stability is that all the poles of the transfer function have a negative real part.

If all of the poles are not located in the left-half s plane, the system is not stable. If, in the characteristic equations, there are a pair of conjugate complex roots on the imaginary axis (jw axis) and the other roots are on the left-half s plane, then the steady-state output of the system, at the bounded input, maintains persistent oscillation; when the input is a sine wave and the frequency is equal to the amplitude of the root on the imaginary axis, its output becomes unbounded. Such a system is called a critical stability system. For an unstable system, at least one root of the characteristic equation is located in the right-half of s plane. In this case, the output of the system is unstable to any input.

The location of the poles in the negative plane are critical to stability. It is well known that the pole locations also affect the system performance, as measured by overshoot, settling time and peak time etc. To determine the system stability, it is not necessary to learn the specific locations of the poles in the left-half s plane; instead, it is necessary to know that all the poles are located in the left-half and s plane. If there is a pair of conjugate poles on the imaginary axis, such as $s_{1,2}=\pm 5j$, the system is critically stable; If there are two pairs of conjugate poles on the imaginary axis, such as $s_{1,2}=\pm 3j$ and $s_{3,4}=\pm 5j$, the system is unstable.

In addition to direct stability judgment based on definitions, other stability criteria can also be used. This chapter will introduce the Routh's Stability Criterion.

If all zero poles of the continuous-time system are located in the left-half s plane, or if all zero poles of the discrete-time system are within the unit circle of the z plane, then the system is the minimum phase system.

6.1.2 Functions for Stability Judgment in MATLAB

According to the definition of stability, the direct way to determine the stability is to calculate all the roots of closed-loop characteristic equation, and make judgment according to their positions in the s plane. MATLAB provides related functions, and their descriptions are shown in Table 6.1.

Table 6.1 Related Functions Stability Judgment

Function	Descriptions
E = eig(G)	Find the eigenvalues of the matrix, the model G of the system can be a transfer function, a state-space model and a zero-pole model, which can be continuous or discrete
P = pole(G) Z = zero(G)	Find the poles and zeros of the system respectively. G is the defined system model
[p,z] = pzmap(sys)	Find the poles and zeros of the system. sys is the defined system model
r = roots(P)	Find the root of the characteristic equation. P is the coefficient vector in descending power of the characteristic polynomial of closed-loop systems
ind = find(X)	Find the subscript vector satisfying condition X, represented by a column vector

6.1.3 Instances of Stability Judgment

Example 6.1 For a closed-loop transfer function

$$\Phi(s)=\frac{s^3+s^2+4}{s^6+12s^5+17s^4+22s^3+31s^2+16s+12},$$ judge the system stability with MATLAB.

The script and results of Example 6.1 are shown in Figure 6.1.1.

```
% MATLAB script used for Example 6.1
clc;
clear;
close all;
num=[1 1 0 4];
```

Figure 6.1.1 MATLAB Script and Results of Example 6.1

```
den=[1 12 17 22 31 16 12];
G=tf(num,den)
E=eig(G)
P=pole(G)
r=roots(den)
```

```
%Output:
Transfer function:
                    s^3 + s^2 + 4
---------------------------------------------------------
s^6 + 12 s^5 + 17 s^4 + 22 s^3 + 31 s^2 + 16 s + 12

E=
 -10.5626
  -1.3744
   0.2265 + 1.1416i
   0.2265 - 1.1416i
  -0.2580 + 0.7373i
  -0.2580 - 0.7373i

P=
 -10.5626
  -1.3744
   0.2265 + 1.1416i
   0.2265 - 1.1416i
  -0.2580 + 0.7373i
  -0.2580 - 0.7373i

r =
 -10.5626
  -1.3744
   0.2265 + 1.1416i
   0.2265 - 1.1416i
  -0.2580 + 0.7373i
  -0.2580 - 0.7373i
```

Figure 6.1.1　MATLAB Script and Results of Example 6.1（continued）

From the results, we can see that the system has two closed-loop feature roots with positive real parts and the system is unstable. In addition, the functions of *eig(G)*, *pole(G)*, and *roots(den)* are the same in this example. They are used to find the closed-loop characteristic roots of the system, and optional according to actual need.

Example 6.2　The closed-loop transfer function of the system is

$$\Phi(s)=\frac{3s^3+16s^2+41s+28}{s^6+14s^5+110s^4+528s^3+1494s^2+2117s+112}$$

Judge the stability of the system and whether the system is the minimum phase system.

The MATLAB script is shown in Figure 6.1.3. The output results show the zero pole and gain of the system. The conclusion is drawn that the system is stable and is the minimum phase system. The distribution of zero-pole output is shown in Figure 6.1.2.

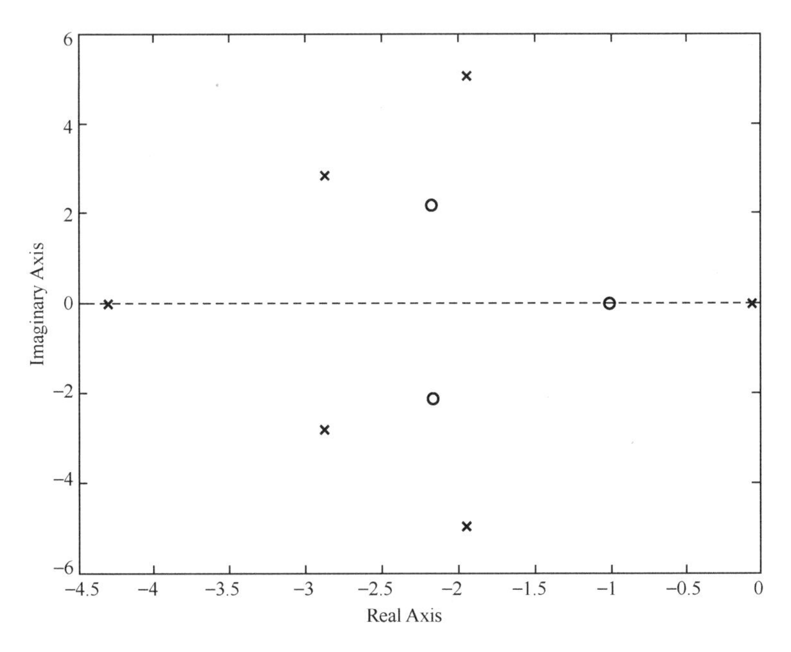

Figure 6.1.2 Distribution of Zero-Pole Output of Example 6.2

```
% MATLAB script used for Example 6.2
clc;
clear;
close all;
num=[3 16 41 28];
den=[1 14 110 528 1494 2117 112];
[z,p,k]=tf2zp(num,den)
% Check the real part of the zero point; find the subscript of the zero point with positive real part
ind_z=find(real(z)>0);
% Find the number of zeros that with positive real part
m=length(ind_z);
% Check the real part of the pole; find the subscript of the pole with positive real part
ind_p=find(real(p)>0);
% Find the number of poles that with positive real part
n=length(ind_p);
% Judging stability
if(n>0)
    disp('the system is unstable')
    disp('the unstable pole is:')
    disp(p(ind_p))
else
    disp('the system is stable')
    if (m>0) % Determine whether the minimum phase system
        disp('the system is a nonminimal phase one')
    else
    disp('the system is a minimal phase one')
    end
end
% Plot zero-pole diagram
```

Figure 6.1.3 MATLAB Script and Results of Example 6.2

```
pzmap(p,z)
```

```
% Results of Example 6.2
z =


  -2.1667 + 2.1538i
  -2.1667 - 2.1538i
  -1.0000
p =


  -1.9474 + 5.0282i
  -1.9474 - 5.0282i
  -4.2998
  -2.8752 + 2.8324i
  -2.8752 - 2.8324i
  -0.0550



k =


     3


the system is stable
the system is a minimal phase one
```

Figure 6.1.3 MATLAB Script and Results of Example 6.2（continued）

Example 6.3 The model of a system is

$$\dot{x}=\begin{bmatrix}1 & 2 & -1 & 2\\ 2 & 6 & 3 & 0\\ 4 & 7 & -8 & -5\\ 7 & 2 & 1 & 6\end{bmatrix}x+\begin{bmatrix}-1\\ 0\\ 0\\ 1\end{bmatrix}u$$

$$\dot{y}=[-2 \quad 5 \quad 6 \quad 1]x+7u$$

Judge the stability of the system and whether the system is the minimum phase system.

The MATLAB script is shown in Figure 6.1.5. The output results show the zero pole and gain of the system. The conclusion is drawn that the system is unstable. The distribution of zero-pole output is shown in Figure 6.1.4.

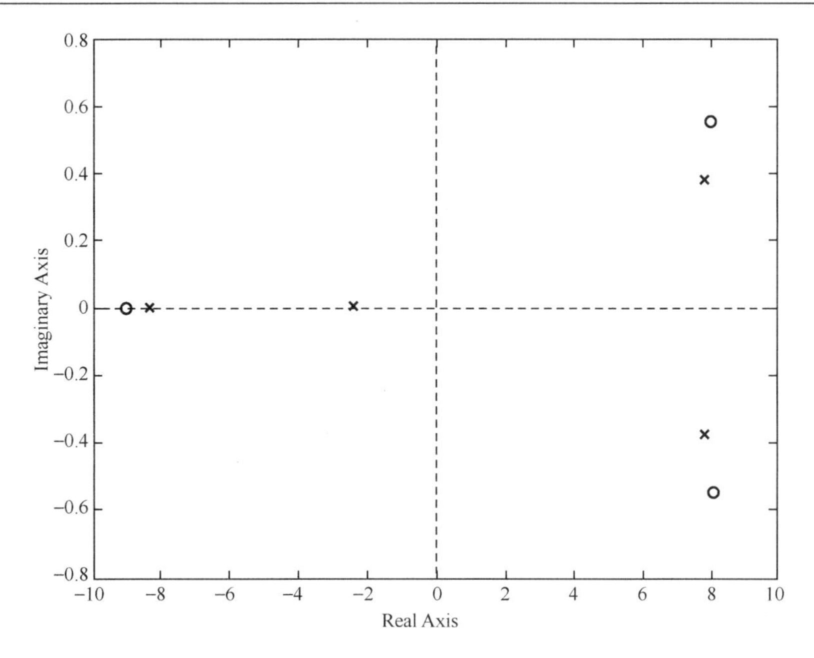

Figure 6.1.4 Distribution of Zero Pole Output of Example 6.3

```
% MATLAB script used for Example 6.3
clc
clear
close all
a=[1 2 -1 2;2 6 3 0;4 7 -8 -5;7 2 1 6];
b=[-1 0 0 1]';
c=[-2 5 6 1];d=7;
[z,p,k]=ss2zp(a,b,c,d)
ind_z=find(real(z)>0);
m=length(ind_z);
ind_p=find(real(p)>0);
n=length(ind_p);
if(n>0)
    disp('the system is unstable')
    disp('the unstable pole is:')
disp(p(ind_p))
else
    disp('the system is stable')
    if(m>0)
        disp('the system is a nonminimal phase one')
    else
        disp('the system is a minimal phase one')
    end
end
pzmap(p,z)
```

Figure 6.1.5 MATLAB Script and Results of Example 6.3

6.2 Routh's Stability Criterion

Routh's stability criterion is to judge whether there exist unstable roots in a characteristic equation instead of finding solution to the equation. This stability criterion can only be applied to polynomials with a finite number of terms. When it is applied to a control system, stability can be judged directly from the coefficients of the characteristic equation.

Routh's stability criterion is a way to determine how many positive real parts there are in the roots of a characteristic equation (that is, how many roots lie in the right-half of s plane).

Consider the following characteristic equation

$$a_0 s^n + a_1 s^{n-1} + \cdots + a_{n-1} s + a_n = 0 \tag{6.2}$$

where the coefficients are real numbers ($a_n \neq 0$).

According to Routh's Stability Criterion, the number of roots in Equation 6.2 to be located with positive real parts in Equation 6.2 is equal to the times of symbol changes in the first column of the array. It should be noted that it is unnecessary to know the exact values of the terms in the first column; We only need to know the signs. The necessary and sufficient condition for all roots in Equation 6.2 to be located in the left-half of s plane is that all the coefficients are positive and all terms in the first column of the Routh array have positive signs.

There is a third-order characteristic equation

$$a_0 s^3 + a_1 s^2 + a_2 s + a_3 = 0$$

where all the coefficients are positive numbers. Apply Routh's Stability Criterion to the above equation and Routh array becomes

$$\begin{array}{lll} s^3 & a_0 & a_2 \\ s^2 & a_1 & a_3 \\ s^1 & \dfrac{a_1 a_2 - a_0 a_3}{a_1} & \\ s^0 & a_3 & \end{array}$$

Clearly, the condition that all roots have negative real parts is given by

$$a_1 a_2 > a_0 a_3$$

Example 6.4 A system has a characteristic equation

$$3s^4 + 10s^3 + 5s^2 + s + 2 = 0$$

Judge the stability of the system by means of Routh's Stability criterion.

Solution: Establishing the Routh array, we have

$$\begin{array}{llll} s^4 & 3 & 5 & 2 \\ s^3 & 10 & 1 & 0 \\ s^2 & \dfrac{47}{10} & 2 & \\ s^1 & -\dfrac{153}{47} & & \\ s^0 & 2 & & \end{array}$$

The signs in the first column changes twice, which indicates that the characteristic equation of the system has two roots in the right-half s plane and the system is thus unstable.

With MATLAB, the roots of the characteristic equation can be calculated by means of '*roots*' function and the results obtained by the Routh's Stabiaty Criterion can be verified, as shown in Figure 6.2.1.

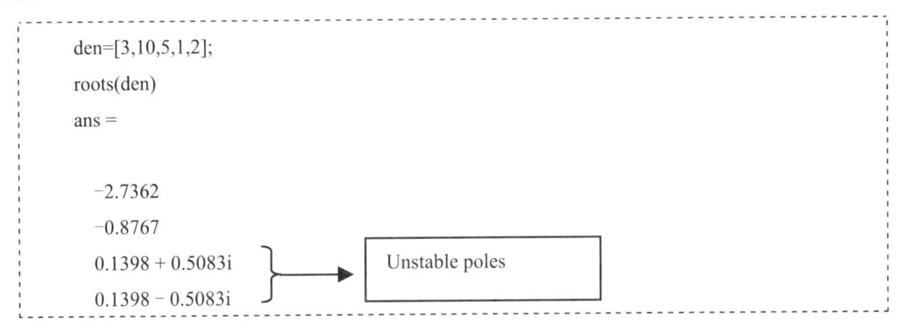

```
den=[3,10,5,1,2];
roots(den)
ans =

  -2.7362
  -0.8767
  0.1398 + 0.5083i
  0.1398 - 0.5083i
```


Figure 6.2.1 Calculate the Roots of the Characteristic Equation

6.3 Human-Controlled Bicycle

This section takes a human-controlled bicycle as an example to illustrate the controller design of implementing the system stability. In daily life, bicycles are widely used and bring convenience. In riding bicycles, it is necessary to maintain a certain posture, speed and stable operation.

In this system, the controlled object is a bicycle, while the controller is the human brain. The human's eyes and ears function as sensors to get the audio and visual signals, which are transmitted to the brain, and form a control decision. The human torso and limbs carry out the decision, adjusting the pose and speed and maintaining stability. (see Figure 6.3.1.)

Figure 6.3.1 Human-Controlled Bicycle

In this chapter, we only discuss the controller parameter design. The controlled object modeling and controller selection are not discussed here. Besides, Routh's Stability Criterion is used to complete the design of the undetermined parameters of the system.

6.3.1 Control Requirements and Parameter Analysis

Consider the human-controlled bicycle shown in Figure 6.3.1. The bicycle will move in a straight line at constant speed V. Let φ denote the angle between the symmetry plane and the vertical plane. When the bicycle is running steadily, the desired angle $\varphi_d(s)$ is

$$\varphi_d(s) = 0 \tag{6.3}$$

1. Control Objective

Control the bicycle in the vertical position, and maintain the prescribed position in the presence of disturbances.

2. Control Variable

The angle of the bicycle deviate from the vertical position.

3. Design Specifications

The closed-loop system must be stable.

This chapter focuses only on the stability of the system rather than its transient response characteristics, so the control specifications are only related to stability.

The bicycle model is given by

$$G(s) = \frac{\Omega_2 s + \Omega_3}{s^2 - \Omega_1} \tag{6.4}$$

where $\Omega_1 = g/h$, $g = 9.806\text{m/sec}^2$, and h is the height of the barycenter of bicycle above the ground (see Figure 6.3.1). $\Omega_2 = VL/(hw)$, $\Omega_3 = V^2/(hw)$ (see Figure 6.3.1). The forward speed of the bicycle is denoted by V, and the wheelbase (the distance between the wheel centers) is by w. The length, L, is the horizontal distance between the front-wheel axle and the barycenter of bicycle.

The bicycle is unstable with poles at $s = \pm\sqrt{\Omega_1}$. With human working as a controller, it can be expressed as

$$G_c(s) = \frac{K_1 s + K_2}{\tau s + 1} \tag{6.5}$$

The controller parameters τ, K_1 and K_2 jointly adjust the system's response.

The front wheel rotation on the vertical plane is ignored in the transfer functions modeling . Also, assume that the forward speed is constant, the feedbacks are needed to increase so as to regulate the actual forward speed . Physical parameters of the system are given in Table 6.2.

Table 6.2 Physical Parameters

Parameters	Values	Units
τ	0.5	sec
Ω_1	12.56	1/sec^2
Ω_2	1.86	1/sec^2
Ω_3	4.66	1/sec

(continued)

Parameters	Values	Units
h	0.78	m
V	2.0	m/sec
L	0.8	m
w	1.1	m
K_1	0.1	
K_2	1	

Formation of the feedback system is shown in Figure 6.3.2.

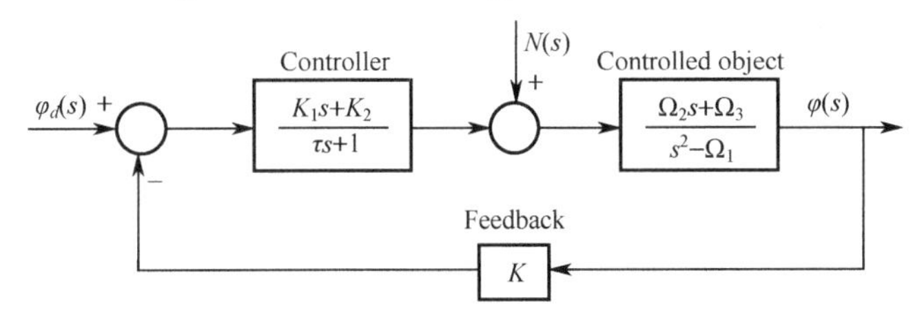

Figure 6.3.2 Block Diagram Human-Controlled Bicycle System

From Figure 6.3.2, the controlled object is expressed in terms of the inherent parameters h, L, w and V of the physical system. Controller is expressed by the parameters τ, K_1, K_2. In this case, assuming that τ, K_1 and K_2 are given, the parameters to be adjusted are the parameter K in the feedback.

4. Key undetermined Parameters

Feedback gains K is a key undetermined parameter.

It should be noted, that such parameters are not constantly in the forward path; in fact, they may exist in any subsystem in the block diagram.

Next, the stability of the system is designed and the Routh's stability criterion is applied to analyze the stability of the closed-loop system. Then what values of K lead to closed-loop stability?

6.3.2 Stability Analysis

The closed-loop transfer function from $\varphi_d(s)$ to $\varphi(s)$ is

$$T(s)=\frac{(K_1 s+K_2)(\Omega_2 s+\Omega_3)}{\Delta(s)} \tag{6.6}$$

where

$$\Delta(s)=\tau s^3+(1+K_1\Omega_2 K)s^2+(K_1\Omega_3 K+K_2\Omega_2 K-\tau\Omega_1)s+K_2\Omega_3 K-\Omega_1 \tag{6.7}$$

The characteristic equation is

$$\Delta(s)=0 \tag{6.8}$$

So, the question to be answered is that, what values of K we may have when all the roots in Equation 6.8 are located in the left-half s plane.

Set the following Routh array,

$$\begin{array}{l|ll} s^3 & \tau & K_1\Omega_3K + K_2\Omega_2K - \tau\,\Omega_1 \\ s^2 & 1+K_1\Omega_2K & K_2\Omega_3K - \Omega_1 \\ s^1 & \alpha & \\ s^0 & K_2\Omega_3K - \Omega_1 & \end{array} \tag{6.9}$$

where

$$\alpha = \frac{(1+K_1\Omega_2K)(K_1\Omega_3K + K_2\Omega_2K - \tau\Omega_1) - \tau(K_2\Omega_3K - \Omega_1)}{1+K_1\Omega_2K} \tag{6.10}$$

To ensure the stability of system, the 1st column of the Routh array must satisfy

$$\begin{cases} \tau > 0 \\ K > -\dfrac{1}{K_1\Omega_2} \\ \alpha > 0 \\ K > \dfrac{\Omega_1}{K_2\Omega_3} \end{cases} \tag{6.11}$$

Obviously, $\tau > 0$, because $K_1 > 0$, $\Omega_2 > 0$, the 2nd inequality can also be satisfied.

Plug the numbers in Table 6.2 into the 4th inequality. We can obtain $K > 2.695$. Therefore, when $K > 2.695$, the system is stable.

In the human-controlled bicycle system, the control objective of is to keep the bicycle upright, i.e. $\varphi_d(s) = 0$ is desired. Therefore, regardless of the situation when $\varphi_d(s) \neq 0$ caused by input or by disturbance, the corresponding output should reach or approach the state when $\varphi_d(s) = 0$. Therefore, it is necessary to choose a suitable K value to make it achieve the control objective.

Let's select K value by MATLAB simulation.

First, take $K = 2.7$, and the output response of the system under the unit step input is shown in Figure 6.3.3. Clearly, the system is stable, but the oscillation a mplitude and the steady-state error are excessive, which does not meet the requirment for stability control over bicycle.

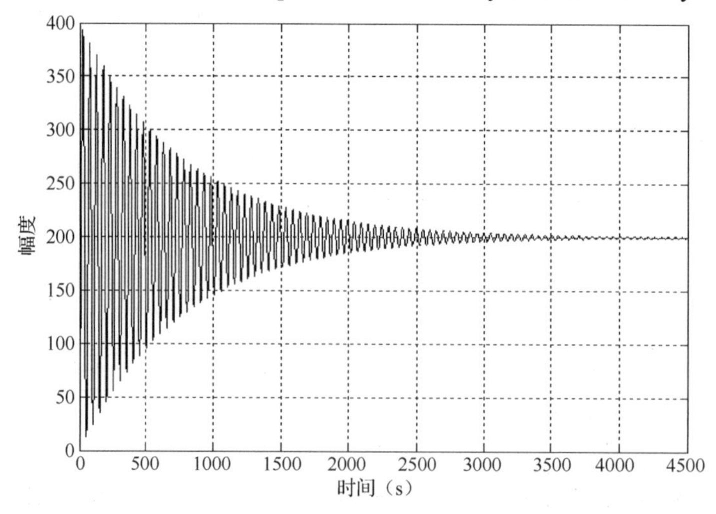

Figure 6.3.3 Unit Step Response of Human-Controlled Bicycle System($K = 2.7$)

Besides, take $K = 6$, and the output response of the system at a unit step input is shown in Figure 6.3.4. It can be seen that as the system can be stable and the amplitude of oscillation declines dramatically, but the steady-state error is about 1/3 of the amplitude of the input signal, which remains to be excessive. It is difficult for the bicycle to keep running at a large inclination, thus failing to achieve the desired control objective.

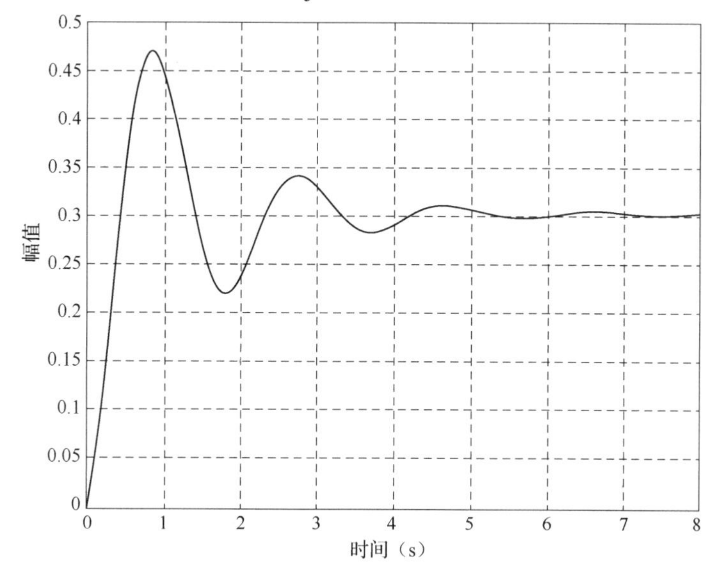

Figure 6.3.4 Unit Step Response of Human-Controlled Bicycle System($K = 6$)

Last, take $K = 26$, the output response of the system under the unit step input is shown in Figure 6.3.5. Clearly, the system can be stable, and the oscillation amplitude and the steady-state error are small. The adjustment time is about 1.2 seconds. The output response is in line with the requirements for stability of a human-controlled bicycle, i.e. achieving the desired control objective.

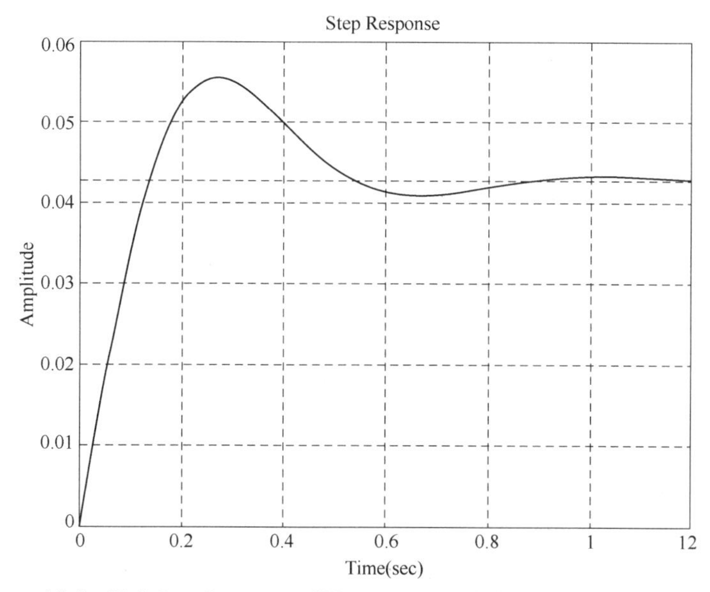

Figure 6.3.5 Unit Step Response of Human-Controlled Bicycle System($K = 26$)

MATLAB script is shown in Figure 6.3.6.

```
%MATLAB script used for Figure 6.3.3, Figure 6.3.4 and Figure 6.3.5
clc;
clear;
close all;
h=0.78;v=2;L=0.8;w=1.1;g=9.8;
Omiga1=g/h;Omiga2=v*L/(h*w);Omiga3=v^2/(h*w);
tau=0.5;k1=0.1;k2=1;
k=2.7;
%k=6;
%k=26;
nc=[k1 k2];dc=[tau 1];
nb=[Omiga2 Omiga3];db=[1 0 -Omiga1];
nf=[0 k];df=[0 1];
[na,da]=series(nc,dc,nb,db);
[num,den]=feedback(na,da,nf,df);
Print sys(num,den)
figure(1);
step(num,den),grid
```

Figure 6.3.6 MATLAB Script

6.3.3 Disturbance Response Analysis

Consider the bicycle itself, i.e. without controllers and feedback, the block diagram of the system from the disturbance $N(s)$ to the output $\varphi(s)$ is shown in Figure 6.3.7.

$$N(s) \rightarrow \boxed{\frac{\Omega_2 s+\Omega_3}{s^2-\Omega_1}} \rightarrow \varphi(s)$$

Figure 6.3.7 Bicycle System Structure Under Disturbance Input

The transfer function from the disturbance $N(s)$ to the output $\varphi(s)$ is

$$T_N(s) = \frac{\varphi(s)}{N(s)} = \frac{\Omega_2 s + \Omega_3}{s^2 - \Omega_1} \tag{6.12}$$

The characteristic equation is

$$q(s) = s^2 - \Omega_1 = 0 \tag{6.13}$$

So the system poles are

$$\begin{aligned} s_1 &= -\sqrt{\Omega_1} \\ s_2 &= +\sqrt{\Omega_1} \end{aligned} \tag{6.14}$$

We can see that there is a pole in the right-half s plane and the bicycle is thus unstable. Without controllers and feedback, any external disturbance will result in the falling of bicycle. Therefore, it is necessary to control the bicycle. With the feedback and controller, the closed-loop transfer function from the disturbance $N(s)$ to the output $\varphi(s)$ is

$$T_N(s)=\frac{(\tau s+1)(\Omega_2 s+\Omega_3)}{\Delta(s)} \tag{6.15}$$

$\Delta(s)$ is shown in Equation 6.7.

The response to a step disturbance $N(s)=1/s$ is shown in Figure 6.3.8, in which the response is stable. The oscilbition amplitude and steady-state error of the system are small. Finally, when $\varphi=0.0425$ (rad), the slight inclination can meet the requirements for vertical running of the bicycle. MATLAB script is shown in Figure 6.3.9.

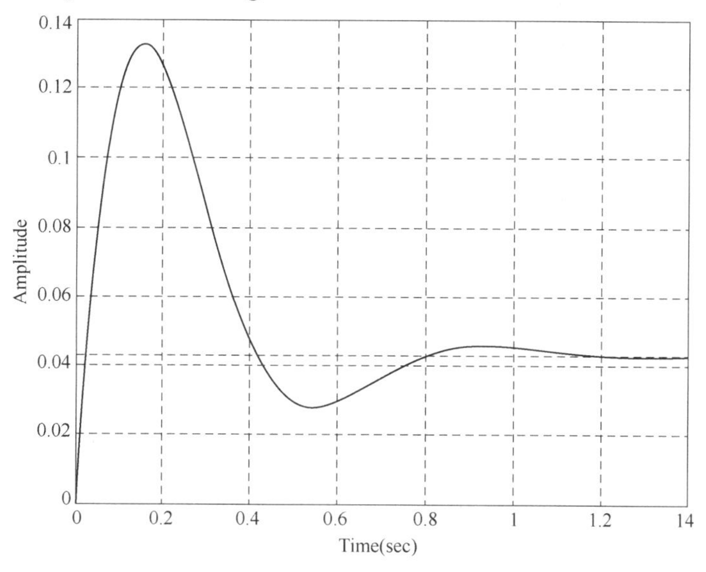

Figure 6.3.8 Step Disturbance Response($K=26$)

```
%%MATLAB script used for Figure 6.3.8
clc;
clear;
char close all;
h=0.78;v=2;L=0.8;w=1.1;g=9.8;
Omiga1=g/h;Omiga2=v*L/(h*w);Omiga3=v^2/(h*w);
tau=0.5;k1=0.1;k2=1;
k=26;
nc=[k1 k2];dc=[tau 1];
nb=[Omiga2 Omiga3];db=[1 0 -Omiga1];
nf=[0 k];df=[0 1];
[na,da]=series(nf,df,nc,dc);
[num,den]=feedback(nb,db,na,da);
printsys(num,den)
figure(2);
step(num,den,'k'),grid
```

Figure 6.3.9 MATLAB Script

6.3.4 Analysis of Speed Effect

As we all know, when riding a bicycle, the slower the speed, the more difficult it is to control the stability. This can be verified by changing the value of speed V.

When the feedback gain $K = 26$ is taken, the forward speed can be continuously changed. The roots of the corresponding characteristic equation are shown in Figure 6.3.10. When the bicycle is at a low speed, for example, $V = 0.6\text{m/s}$, the roots characteristic equation are $r_1 = -2.5391 + 3.6222\text{i}$, $r_2 = -2.5391 - 3.6222\text{i}$, $r_3 = 0.1692$. Since there is a root in the right-half s plane, the system is unstable. As speed increases, when $V = 0.7\text{m/s}$, the roots of characteristic equation are $r_1 = -2.5990 + 4.0709\text{i}$, $r_2 = -2.5990 - 4.0709\text{i}$, $r_3 = -0.1959$, all the roots are in the left-half s plane and the system is stable. As the speed continues to increase, the roots of the characteristic equation remains in the left-half s plane and the system is stable. MATLAB script is shown in Figure 6.3.11.

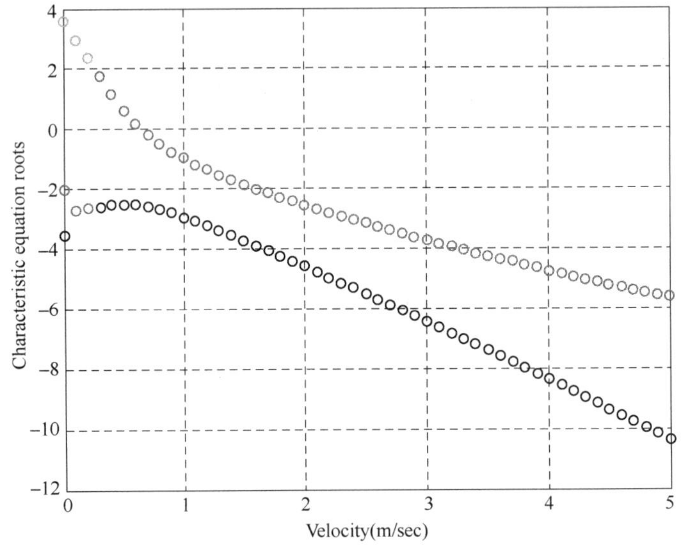

Figure 6.3.10 Effect of Velocity on Characteristic Equation Roots

```
% MATLAB Script used for Figure 6.3.10
clc;
clear;
close all;
h=0.78;L=0.8;w=1.1;g=9.8;
v=[0:0.1:5];
Omiga1=g/h;
Omiga2=v*L/(h*w);
Omiga3=v.^2/(h*w);
tau=0.5;k1=0.1;k2=1;
k=26;
nc=[k1 k2];dc=[tau 1];
nf=[0 k];df=[0 1];
for i=1:length(v)
nb=[Omiga2(i) Omiga3(i)];db=[1 0 -Omiga1];
[na,da]=series(nc,dc,nb,db);
[num,den]=feedback(na,da,nf,df);
```

Figure 6.3.11 MATLAB Script

```
        r(:,i)=roots(den);
    end
    plot(v,r,'o'),grid
    xlabel('Velocity(m/sec)');ylabel('Characteristic equation roots');
```

Figure 6.3.11 MATLAB Script（continued）

Summary

1. Definition of the concept of stability and the Routh's stability criterion are introduced.

2. Stability of the human-controlled bicycle system is analyzed and the undetermined parameters are decided by means of the the Routh's stability criterion.

3. Stable control of human-controlled bicycle system is achieved through disturbance response analysis and speed response analysis.

生词注解

bounded-input and bounded-output　有界输入和有界输出
polynomial　n. 多项式
　adj. 多项式的
eigenvalues　n. 特征值，特征根
oscillation　n. 振荡，振动
criterion　n. 评判标准
amplitude　n. 振幅
undetermined parameter　未定参数，待调参数

第7章 系统性能的实例分析

本章重点介绍飞机飞行期间的波动控制实例。首先简单介绍实例中需要了解的性能指标的相关知识，以及实例用到的相关 MATLAB 函数。然后根据实际的性能指标要求，具体分析飞机飞行期间的波动控制系统的动态性能及其稳态误差。动态性能主要从输入暂态响应与扰动响应两个角度进行仿真分析。最后，通过对三个 PID 控制器的比较，确定了合适的控制器。

7.1　典型的输入信号与时域的性能指标

控制系统性能的评价分为动态性能指标和稳态性能指标两类。为了求解系统的时间响应，必须了解输入信号的解析表达式。但是，在一般情况下，控制系统的输入信号具有随机性而无法预先确定，因此，需要选择几个典型输入信号。

7.1.1　典型的输入信号

为了便于对系统进行分析、设计和比较，根据实际系统常遇到的输入信号形式，在数学描述上加以理想化的一些基本输入函数，称为典型输入信号。

控制系统中常用的典型输入信号有单位阶跃、单位斜坡（速度）函数、单位加速度（抛物线）函数、单位脉冲函数和正弦函数。本节只介绍本节用到的两种输入信号——单位阶跃与单位脉冲函数，如表 7.1 所示。

表 7.1　时域分析法中的典型输入信号

名　称	$r(t)$	时域图形	$R(s)$	例
单位脉冲函数	$\delta(t)=\begin{cases}\infty, & t=0\\ 0, & t\neq 0\end{cases}$ $\int\delta(t)\mathrm{d}t=1$	$\delta(t)$ O t	1	撞击作用后坐力电脉冲
单位阶跃函数	$1(t)=\begin{cases}1, & t\geqslant 0\\ 0, & t<0\end{cases}$	$1(t)$ 1 O t	$\frac{1}{s}$	开关输入

7.1.2　时域的性能指标

1．动态性能

系统动态性能是以系统阶跃响应为基础来衡量的。一般认为阶跃输入对系统而言是比较严峻的工作状态，若系统在阶跃函数作用下的动态性能满足要求，那么系统在其他形式的输入作用下，其动态响应也应是令人满意的。

动态性能指标通常有如下几项。

上升时间 t_r： 阶跃响应从终值的 10%上升到终值的 90%所需的时间；对有振荡的系统，也可以定义为从 0 到第一次达到终值所需的时间。

峰值时间 t_p： 阶跃响应越过终值 $h(\infty)$达到第一个峰值所需的时间。

调节时间 t_s： 阶跃响应到达并保持在终值 $h(\infty)\pm5\%$误差带内所需的最短时间；有时也用终值的±2%误差带来定义调节时间。除非特别说明，本书所说的调节时间均以终值的±5 误差带定义。

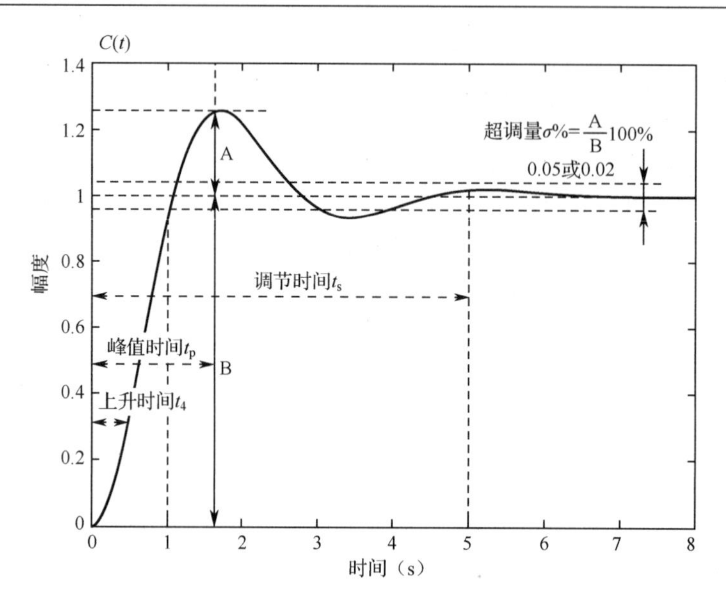

图 7.1.1 系统的典型阶跃响应及动态性能指标

超调量σ%：峰值 $h(t_p)$超出终值 $h(\infty)$的百分比，即

$$\sigma\% = \frac{h(t_p)-h(\infty)}{h(\infty)}\times 100\% \tag{7.1}$$

在上述动态性能指标中，工程上最常用的是调节时间 t_s（反映过渡过程的长短，描述“快”），超调量σ%（反映过渡过程的波动程度，描述“匀”）以及峰值时间 t_p，它们也是本书重点讨论的动态性能指标。

2. 稳态性能

稳态误差是时间趋于无穷时系统实际输出与理想输出之间的误差，是系统控制精度或抗干扰能力的一种度量。稳态误差有不同定义，通常在典型输入下进行测定或计算。

应当指出，系统性能指标的确定应根据实际情况而有所侧重。例如，歼击机要求机动灵活，响应迅速，允许有适当的超调；对于一些启动之后便需要长期运行生产过程则往往更强调稳态精度。

7.1.3 仿真分析函数

本章在系统性能指标仿真分析时用到的 MATLAB 中的函数，主要有 step、impulse 和 dcgain。这些函数的具体使用格式及其说明，详见表 7.2。

表 7.2 仿真分析函数介绍

函　数	说　明
step(sys)	获取动态系统的阶跃响应图
[y,t] = step(sys)	获取动态系统的阶跃响应数据
impulse(sys)	获取动态系统的脉冲响应图
[y,t] = impulse(sys)	获取动态系统的脉冲响应数据
k = dcgain(sys)	连续系统的 DC 值，是在频率 $s = 0$ 时的传递函数值

7.2　稳态误差

稳态误差是指瞬态响应消失后，系统持续响应的误差。

考虑图 7.2.1 所示的系统，其闭环传递函数为

$$\frac{C(s)}{R(s)}=\frac{G(s)}{1+G(s)} \tag{7.2}$$

误差信号 $e(t)$与输入信号 $r(t)$之间的传递函数为

$$\frac{E(S)}{R(S)}=1-\frac{C(S)}{R(S)}=\frac{1}{1+G(s)} \tag{7.3}$$

其中，误差 $e(t)$是输入信号和输出信号之差。

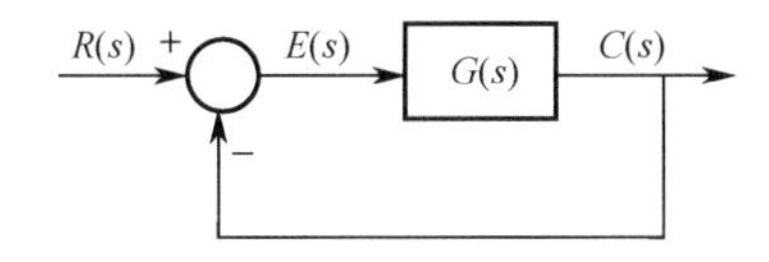

图 7.2.1　控制系统

终值定理为求稳定系统的稳态误差提供了一种简便方法。因为 $E(s)$为

$$E(s)=\frac{1}{1+G(s)}R(s) \tag{7.4}$$

所以稳态误差为

$$e_{ss}=\lim_{t\to\infty}e(t)=\lim_{s\to 0}sE(s)=\lim_{s\to 0}\frac{sR(s)}{1+G(s)} \tag{7.5}$$

例 7.1　设有控制系统，如图 7.2.2 所示。

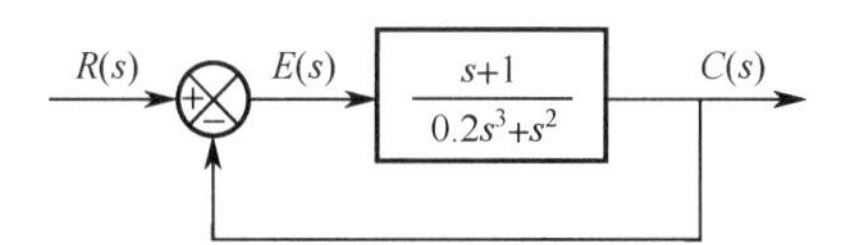

图 7.2.2　例 7.1 的控制系统

当输入为 $r(t)=t^2$， $t>0$ 时，确定系统的稳态误差 e_{ss} 。

解： 首先判定系统的稳定性，图 7.2.3 和图 7.2.4 是例 7.1 的程序及其运行结果。

```
%例 7.1 的程序
num=[1,1];
den=[0.2,1,0,0];
[numc,denc]=cloop(num,den);
tf(numc,denc)
roots(denc)
```

图 7.2.3　例 7.1 的程序

```
%例 7.1 的程序结果
ans =

          s + 1
```

图 7.2.4　程序运行结果

```
    ---------------------
    0.2 s^3 + s^2 + s + 1

Continuous-time transfer function.


ans =

  -4.0739 + 0.0000i
  -0.4630 + 1.0064i
  -0.4630 - 1.0064i
```

图 7.2.4 程序运行结果（续）

得到系统的三个闭环极点，全在 s 左半平面，因此系统是稳定的。系统误差为

$$E(s)=\frac{R(s)}{1+G(s)}=\frac{0.2s^3+s^2}{0.2s^3+s^2+s+1}R(s) \tag{7.6}$$

当 $r(t)=t^2$ 时，

$$R(s)=\frac{2}{s^3} \tag{7.7}$$

根据终值定理，我们可以得到稳态误差为

$$e_{ss}=\lim_{s\to 0}sE(s)=\lim_{s\to 0}\frac{(0.2s+1)s^2}{0.2s^3+s^2+s+1}\cdot\frac{2}{s^3}\cdot s=2 \tag{7.8}$$

7.3 飞机飞行期间的波动控制

飞机是指由具有一个或多个发动机的动力装置产生前进的推力或拉力，由机身的固定机翼产生升力，在大气层内飞行的重于空气的航空器。飞机是最常见的一种固定翼航空器，按照其使用的发动机类型，可被分为喷气飞机和螺旋桨飞机。

在一个现代化的飞行操作中，保证飞机平稳飞行是飞行员的责任。许多重要参数，如飞行姿态、高度、速度、侧向航迹和仰角控制等飞行波动控制都是由飞行员控制在可接受的范围内。当然，为了确保飞机飞行的安全，飞机平稳的飞行必须维持在整个航行过程。这是一个对过程全面控制的人机交互实例。显然，飞机的平稳飞行是最终目的。因此，我们的控制目标是设计一个调节飞机飞行时的波动控制的自动化系统。

7.3.1 FV 控制系统分析

我们将飞行波动率（Flight Volatility，FV）作为飞机平稳飞行最为可靠的表征指标。飞行员通过监控 FV 的水平来确定飞机飞行时的各个参数。基于飞行员的飞行经验以及操作规程，我们决定将飞行波动率作为受控变量。

根据控制系统设计角度，控制目标可以用更具体的术语来说明。

控制目标：调节飞行波动率到任何期望的设定点，且在有干扰时稳定在规定的设定点。

根据所述控制目标，我们确定控制变量。

控制变量：飞行波动率（FV）。

因为我们想设计一个确保飞机航行平稳的系统，所以有必要建立实际的设计规范。一般来说，在满足控制规范的前提下，控制系统应具有最小复杂性。最小复杂性，换言之为提高系统的可靠性且降低成本。

随着 FV 的设定点改变，闭环系统应该快速、平滑地响应，且无超调。闭环系统应该使外界干扰的影响减到最小。重要的干扰类型有：气流干扰，主要是对流，即气流上下运动的情况，主要是飞机在起飞和着陆过程中需要穿过对流层，使飞行波动变大。最后，因为我们想将同一个控制系统应用于许多不同的飞行情况，毕竟我们不能（因为实际理由）对每种情况设有不同的模型，我们必须设计一个对飞行情况参数变化不敏感的闭环系统，即对不同的飞行情况都满足的指标。

基于经验，控制指标可明确地表述为

- 设计指标 1：当 FV 的期望输出是 10 倍单位阶跃的幅值变化时，系统的调节时间小于 15 秒。
- 设计指标 2：当 FV 的期望输出是 10 倍单位阶跃的幅值变化时，系统的超调量小于 15%。
- 设计指标 3：系统在 FV 输入为阶跃信号时，系统不存在稳态误差。
- 设计指标 4：系统在飞行扰动输入为阶跃信号（幅值<50）时，系统不存在稳态误差，且最大响应范围在 FV 的期望输出的±5%之内。
- 设计指标 5：系统对不同飞行情况的参数变化的灵密度最低。

在图 7.3.1 所示的系统结构中，将控制器、传感器和飞机作为主要系统元素。系统的输入 $R(s)$为 FV 的期望值，输出 $C(s)$为实际的 FV 的变化量。控制器利用预期的 FV 和传感器测量得到的 FV 之差，将其作为控制信号，来调整控制器输入到飞机的稳定运行值。

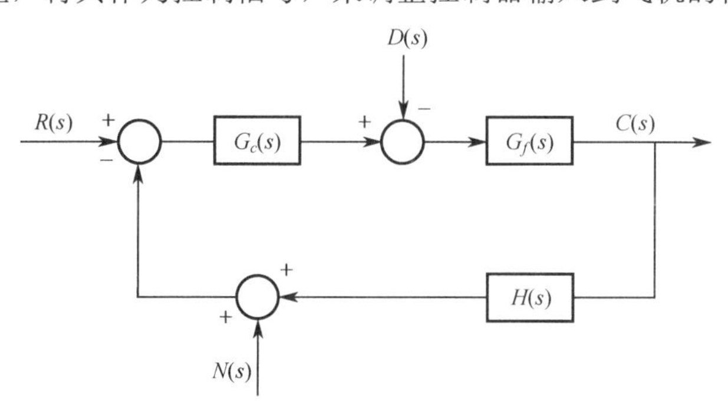

图 7.3.1　飞机的系统结构

飞机飞行的建模过程很复杂，主要原因是很难预测飞机飞行时遇到气流的所有情况。即使建立了类似的模型，通常也是一种多输入多输出的非线性时变模型。这类模型不能直接应用于我们所要建立的单输入单输出的线性时不变模型中。

系统的模型并不是唯一的，对同一个系统可以产生相应于不同层次的多种模型。模型的建立主要取决于系统的控制目标，不同的控制目标，可以得到复杂程度不同的系统模型。针对飞机飞行波动的控制系统，仅仅实现飞机飞行时的波动控制，其他均不考虑。满足这一个控制目标，建立飞机飞行系统的模型。如果限定 FV 只在其设定值附近微小波动，就可以认为在 FV 设定值附近的小范围内，飞机呈现出线性时不变的特性。这种分析方法非常适用于 FV 维持在给定值附近的要求，因此直接给出飞机的传递函数模型，如式（7.9）所示。

$$G_f(s)=\frac{1}{s(s+f)^2} \tag{7.9}$$

其中，$f=2$，时间t的单位为秒。参数f与飞机飞行情况有关，不同飞行情况的参数f不同。

假定传感器处于理想工作状态，其传递函数为

$$H(s)=1 \tag{7.10}$$

在这个应用中，控制器是采用比例-积分-微分（PID）控制器，其传递函数为

$$G_c(s)=K_\mathrm{P}+sK_\mathrm{D}+\frac{K_I}{s}=\frac{K_\mathrm{D}s^2+K_\mathrm{P}s+K_\mathrm{I}}{s} \tag{7.11}$$

于是，得到一个单位负反馈系统，该系统的闭环传递函数为

$$\phi(s)=\frac{G_c(s)G_f(s)}{1+G_c(s)G_f(s)H(s)} \tag{7.12}$$

7.3.2 稳态误差分析

系统如图 7.3.1 所示，令$D(s)=0$与$N(s)=0$的稳态误差为

$$E(s)=R(s)-C(s)=\frac{1}{1+G_c(s)G_f(s)H(s)}R(s) \tag{7.13}$$

或者

$$E(s)=\frac{s^4+2fs^3+f^2s^2}{s^4+2fs^3+(f^2+K_\mathrm{D})s^2+K_\mathrm{P}s+K_I}R(s) \tag{7.14}$$

由终值定理，可以得到系统的稳态误差为

$$\lim_{s\to 0}sE(s)=\lim_{s\to 0}\frac{R_0(s^4+2fs^3+f^2s^2)}{s^4+2fs^3+(f^2+K_\mathrm{D})s^2+K_\mathrm{P}s+K_\mathrm{I}}=0 \tag{7.15}$$

其中，$R(s)=\dfrac{R_0}{s}$是幅值为R_0的阶跃输入信号。因此

$$\lim_{t\to\infty}e(t)=0 \tag{7.16}$$

对于 PID 控制器，只要K_P、K_D和K_I的值都不为零，在阶跃输入条件下，系统的稳态误差总是为零。因为$G_C(s)G_f(s)H(s)$是II型系统，如式（7.17）所示，只有在加速度输入情况下才存在误差。因此，设计指标 3 得到满足。

$$G_c(s)G_f(s)H(s)=\frac{K_\mathrm{D}s^2+K_\mathrm{P}s+K_\mathrm{I}}{s^2(s+f)^2} \tag{7.17}$$

当考虑到阶跃干扰信号输入时，则令$R(s)=0$，$N(s)=0$。在阶跃干扰输入时，我们期望输出信号 $C(s)$的稳态值为 0。系统干扰信号 $D(s)$与输出信号 $C(s)$之间的传递函数为

$$\frac{C(s)}{D(s)}=\frac{-G_f(s)}{1+G_c(s)G_f(s)} \tag{7.18}$$

$$C(s)=\frac{-s}{s^4+2fs^3+(f^2+K_\mathrm{D})s^2+K_\mathrm{P}s+K_\mathrm{I}}D(s) \tag{7.19}$$

当

$$D(s)=\frac{D_o}{s} \tag{7.20}$$

可以得到

$$\lim_{s\to 0} sC(s)=\lim_{s\to 0}\frac{-D_0 s}{s^4+2fs^3+(f^2+K_D)s^2+K_P s+K_I}=0 \tag{7.21}$$

即有

$$\lim_{t\to\infty} C(t)=0 \tag{7.22}$$

由此可见，当扰动信号是幅值为 D_0 的阶跃信号时，系统的稳态输出像期望的一样为零。

7.3.3 暂态响应

假设我们想通过改变 10 倍的单位阶跃输入幅值，来看输出 FV。相应的输入为

$$R(s)=\frac{R_0}{s}=\frac{10}{s} \tag{7.23}$$

控制器是比例-积分-微分（PID）控制器，其传递函数如式（7.11）所示。其中 K_P，K_I 和 K_D 为控制器增益，需要根据设计指标的要求来确定。

本节首先从选择比例 P 控制器开始，然后选择比例-积分 PI 控制器，最后选择比例-积分-微分 PID 控制器。此外，分析了系统在三种控制器下的暂态响应特性，以及各自状态下的性能指标。最后，以最佳满足性能指标作为条件，配置 PID 控制器的参数，从三种 PID 控制器中选择一个参数配置最佳 PID 控制器。

1. P 控制器

首先，我们加入比例作用的 P 控制器，此时微分和积分作用为零。使得控制器传递函数为

$$G_C(s)=K_P \tag{7.24}$$

假设飞行波动率的输入变化为 10 倍的单位阶跃幅值，即输入为 $R(s)=\frac{R_0}{s}=\frac{10}{s}$，比例控制器系数值分别选为 $K_P=[1,3,8]$ 三个值，控制器 $G_c(s)$ 的其他参量都设置为 $K_D=0$， $K_I=0$ 。得到系统在三个不同比例系数时的阶跃响应曲线，如图 7.3.2 所示。

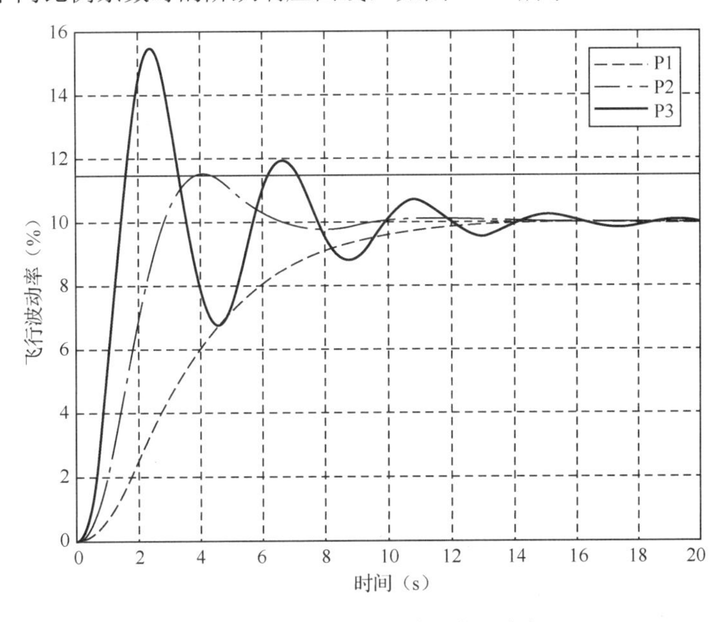

图 7.3.2 P 控制器的系统阶跃响应

由图 7.3.2 可看出，当采用 P 比例控制器时，随着比例系数 K_P 值的逐渐增加，系统响应振幅也随之增加。系统从开始在 $K_\mathrm{P}=1$ 时，没有超调，到 $K_\mathrm{P}=3$ 和 $K_\mathrm{P}=8$ 时，超调依次增加到 15.25%和 54.36%，这说明，比例系数 K_P 值愈大，系统振荡幅度愈大。在 $K_\mathrm{P}=1$ 时，系统的调节时间为 11.9s，在 $K_\mathrm{P}=3$ 和 $K_\mathrm{P}=8$ 时，调节时间依次为 8.3s 和 15.4s。随着比例系数 K_P 值增大，振荡幅度会增大，同时系统的调节时间也增加了。程序和结果见图 7.3.3。

```
%图 7.3.2 的 MATLAB 程序
clc
clear;
close all
tf=20;
t=[0:0.1:tf];
R0=10;
f=2;
den=conv([1,0,0],conv([1,f],[1,f]));
KP=[1,3,8];KD=0;KI=0;                          %采用 P 时赋值
%KP=[1,3,8];KD=0;KI=[2,1,4];                   %采用 PI 时赋值
%KP=[8,7,5];KD=[11,4,7];KI=[1,1,2];            %采用 PID 时赋值
for ii=1:3;
   num=[KD,KP(ii) KI];                         %采用 P 时赋值
   %num=[KD,KP(ii) KI(ii)];                    %采用 PI 时赋值
   %num=[KD(ii),KP(ii) KI(ii)];                %采用 PID 时赋值
   [numc,denc]=cloop(num,den);
   y=step(R0*numc,denc,t);
   finalvalue=dcgain(R0*numc,denc);
   [yss,n]=max(y);
   TP(ii)=t(n);
   [PO(ii),TS(ii)]=poandts(t,y,yss,finalvalue);
   yo(:,ii)=y;
 end
plot(t,yo(:,1),'b--',t,yo(:,2),'k-.',t,yo(:,3),'m-',[0 tf],[11.5 11.5])
legend('P1','P2','P3', -1)                     %采用 P 时
legend('PI1','PI2','PI3', -1)                  %采用 PI 时
legend('PID1','PID2','PID3', -1)               %采用 PID 时
grid
ylabel('Percent decrease in Flight Volatility(%)')
xlabel('Time(seconds)')
TP
PO
TS
%上图程序中用到的函数文件
function[PO,TS]=poandts(t,y,yss,finalvalue)
PO=100*(yss-finalvalue)/finalvalue;
k=length(t);
while (y(k)>0.98*finalvalue)&(y(k)<1.02*finalvalue)
```

图 7.3.3 MATLAB 程序及其结果

```
 k=k-1;
end
TS=t(k);
```

```
% MATLAB 程序输出到命令窗口的结果
TP =

   20.0000    4.0000    2.4000


PO =

   -0.0912   15.2474   54.3551


TS =

   11.9000    8.3000   15.4000
```

图 7.3.3　MATLAB 程序及其结果（续）

2. PI 控制器

采用 PI 控制器，控制器 $G_C(s)$ 的微分系数都设置为 $K_D = 0$，比例和积分分别设置了三组值，$K_p = [1,3,8]$ 三个值；对应的积分器系数的值为 $K_I = [2,1,4]$。三个参量分别按照上面的值进行设置，构成三个 PI 控制器。PI1 的比例和积分系数值分别为 1 和 2，PI2 的比例和积分系数值分别为 3 和 1，PI3 的比例和积分系数值分别为 8 和 4。

采用图 7.3.3 中程序，得到系统阶跃响应曲线，如图 7.3.4 所示。三个 PI 控制器的输出不同。PI1 控制器的输出是不稳定的，K_p 的值为 1，是三个数据中最小的，K_I 的值为 2。PI2 控制器的输出是稳定的，最终趋于稳态值 10，无稳态误差。PI3 控制器的输出是等幅振荡，处于临界状态。

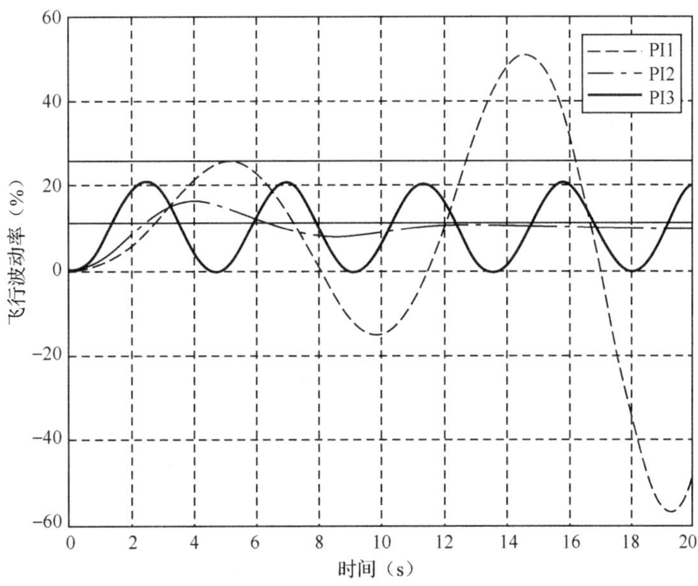

图 7.3.4　PI 控制器的系统阶跃响应

3. PID 控制器

PID 控制器的三个参数的选取基本上是一个三维空间的搜索问题。由于在搜索空间中的三个点对应于 PID 控制器的三个参数值，因此，在参数空间中选取不同的点即可确定 PID 控制器。比如，对阶跃输入的不同阶跃响应，可以用试探的方法在搜索空间中移动点，以确定一个 PID 控制器。

采用 PID 控制器，通过多次仿真，搜索 PID 三个系数值，得到接近性能指标要求的参数值。三组数值为 $K_{\mathrm{p}}=[8,7,5]$，$K_{\mathrm{D}}=[11,4,7]$，$K_{\mathrm{I}}=[1,1,2]$，分别对应构成的三个 PID 控制器。PID1 的比例、微分和积分的系数为 8、11 和 1，PID2 的比例、微分和积分的系数为 7、4 和 1，PID3 的比例、微分和积分的系数为 5、7 和 2。

采用图 7.3.3 中的程序，得到系统阶跃响应曲线，如图 7.3.5 所示。三个 PID 控制器的输出不同。从图 7.3.5 和图 7.3.6 可看出，PID1 控制器展现了最佳的动态性能，超调是 7.52%，调节时间是 12.1s，都满足性能指标。PID2 控制器在动态性能上是接近指标要求的，超调是 18.22%，比要求的 15%略大，不满足超调的指标要求。调节时间是 10.6s，满足性能指标。PID3 控制器在动态性能上，超调是 14.22%，调节时间是 13.5s，都满足性能指标。比较 PID1、PID2 与 PID3 三个控制器，在相同的阶跃信号输入 $R(s)$=10/s 情况下，PID1 和 PID3 两个控制器都是满足了指标 1 和指标 2，系统的超调和调节时间均满足要求。

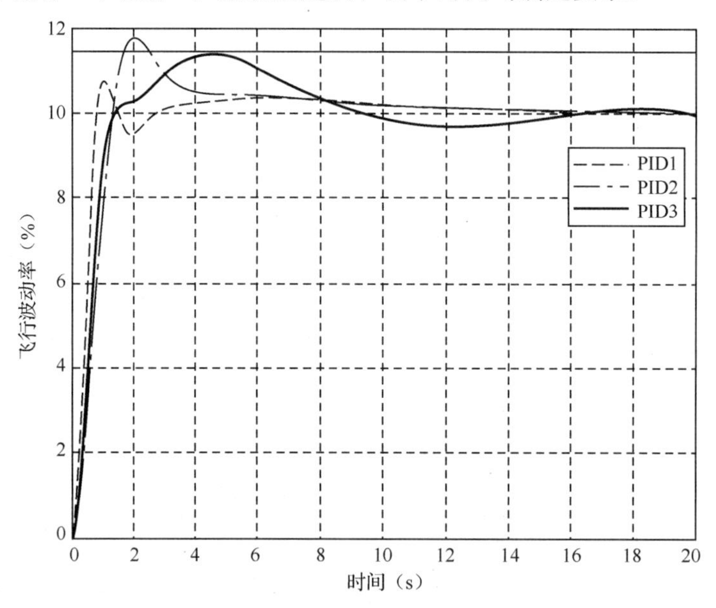

图 7.3.5 PID 控制器的系统阶跃响应

下面，再看一下扰动加入时系统性能是否能够满足指标 4。

为了方便比较，把响应的超调与调节时间放到一个表格中，如表 7.3 所示。

```
% MATLAB 程序输出到命令窗口的结果
TP =

    1.0000    2.0000    4.4000
```

图 7.3.6 MATLAB 命令窗口的结果

```
PO =

    7.5164    18.2151    14.2186


TS =

    12.1000    10.6000    13.5000
```

图 7.3.6　MATLAB 命令窗口的结果（续）

表 7.3　PID 控制器增益和系统性能

PID	KP	KD	KI	超调量（%）	调节时间（s）
1	8	11	1	7.52	12.1
2	7	4	1	18.22	10.6
3	5	7	2	14.22	13.5

7.3.4　扰动响应

从前面的分析可以知道，干扰信号 $D(s)$与系统输出 $C(s)$之间的传递函数为

$$\begin{aligned} C(s) &= \frac{-G_f(s)}{1+G_c(s)G_f(s)}D(s) \\ &= \frac{-s}{s^4+2ps^3+(p^2+K_{\mathrm{D}})s^2+K_{\mathrm{P}}s+K_{\mathrm{I}}}D(s) \end{aligned} \tag{7.25}$$

为了验证是否满足指标设计 4，在扰动信号为

$$D(s)=\frac{D_0}{s}=\frac{50}{s} \tag{7.26}$$

时，计算干扰阶跃响应。

扰动信号的最大幅值为 50。因为阶跃干扰的幅值越小，其输出响应的最大值越小。当确定系统指标 4 是否满足要求时，只需要考察幅值最大的干扰信号输入。

依然选取上节中的三个 PID 控制器。三个不同控制器下的系统对阶跃干扰信号的输出响应如图 7.3.7 所示。可以看出，三个控制器的稳态误差均为 0，满足了设计指标 4 中系统不存在误差要求；控制器 PID3 的最大响应在 FV 设定值的±5%的范围内，因此满足设计指标 4 的要求，而控制器 PID1 和 PID2 则都超出了设计指标 4 的要求。MATLAB 程序如图 7.3.8 所示。

综上所述，在给定的三个 PID 控制器中，只有 PID3 满足了所有的设计指标要求，同时也满足系统对于飞机参数变化的灵密度要求。因此选择 PID3 作为飞行控制系统的控制器。

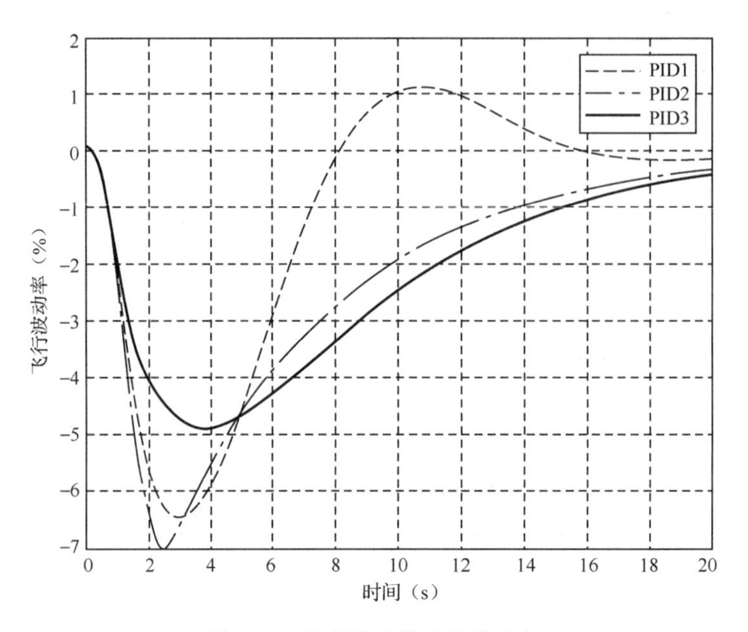

图 7.3.7　飞行波动扰动阶跃响应

```
%图 7.3.7 的运行程序
clc
clear;
close all
tf=20;
t=[0:0.1:tf];
Do=50
f=2;
num=[1,0];
KP=[8,7,5];KD=[11,4,7];KI=[1,1,2];
dc=[1,0];
for ii=1:3;
   den=[1,2*f,f^2+KD(ii),KP(ii),KI(ii)];
   y=step(-Do*num,den,t);
   yo(:,ii)=y;
 end
plot(t,yo(:,1),'b-',t,yo(:,2),'k-.',t,yo(:,3),'m--')
legend('PID1','PID2','PID3',−1)
grid
ylabel('Percent decrease in Flight Volatility(%)')
xlabel('Time(seconds)')
```

图 7.3.8　MATLAB 程序

本章小结

（1）介绍了时域的系统性能指标和典型的输入信号的 MATLAB 函数的使用；

（2）分析了稳态误差，通过实例介绍了采用 MATLAB 方法求取系统的稳态误差；

（3）针对飞机飞行的波动控制系统，分析了阶跃输入下和扰动输入下的稳态误差；

（4）重点研究了飞机飞行的波动控制问题。在 P、PI 和 PID 三种控制器作用下，分别分析了系统在阶跃输入和扰动输入两种情况下的不同响应。在 PID 控制器中，发现有一个 PID 控制器满足所有的设计规范要求。

Chapter 7 Case Analysis of System Performance

This chapter focuses on examples of fluctuation control during aircraft flight. First, the basic knowledge of performance specifications and the MATLAB functions used in the examples are briefly introduced. Besides, based on the requirements for the actual performance, the dynamic performance and steady-state error of the fluctuation control system during flight are specified in detail. To conduct the simulation analysis of the dynamic performance is mainly from two perspectives of input transient response and disturbance response. Finally, the appropriate controller is determined by comparison with three PID controllers.

7.1 Typical Input Signals and the Time-Domain Specifications

The performances of a control system are divided into dynamic performance and steady-state performance. In order to solve the time response of system, we must understand the analytical expression of the input signal. However, under usual circumstances, the input signal of the control system is random and cannot be determined in advance. Therefore, we need to select several typical input signals.

7.1.1 Typical Input Signals

According to the input signal form commonly used in actual systems, typical input signals are some idealized basic input functions in mathematical description for analysis, design and comparison.

These typical input signals include unit step, unit ramp (speed) function, unit acceleration (parabola) function, unit pulse function, and sine function. As shown in Table 7.1, we only describe two input signals used in this section, the unit step and the unit pulse function.

Table 7.1 Typical Input Signals in Time Domain Analysis

Name	$r(t)$	Time Domain Graphics	$R(s)$	Example
Unit pulse function	$\delta(t)=\begin{cases}\infty, & t=0\\ 0, & t\neq 0\end{cases}$ $\int \delta(t)\mathrm{d}t=1$	$\delta(t)$, O, t	1	recoil electric pulse
Unit step function	$1(t)=\begin{cases}1, & t\geqslant 0\\ 0, & t<0\end{cases}$	$1(t)$, 1, O, t	$\frac{1}{s}$	Switch input

7.1.2 The Time-Domain Specifications

1. Dynamic Performance

Dynamic performance is measured on the basis of system step response. Generally, the step input is a severe working condition for the system. If the dynamic performance of under the step function satisfies the requirements of the system, its dynamic response under other forms of input is also be satisfactory.

Usually, there are the following items of specifications.

Rise Time t_r: The time required for the step response to increase from 10% of the final value to 90% of the final value. For systems with oscillations, it can also be defined as the time required to reach the final value from 0 for the first time.

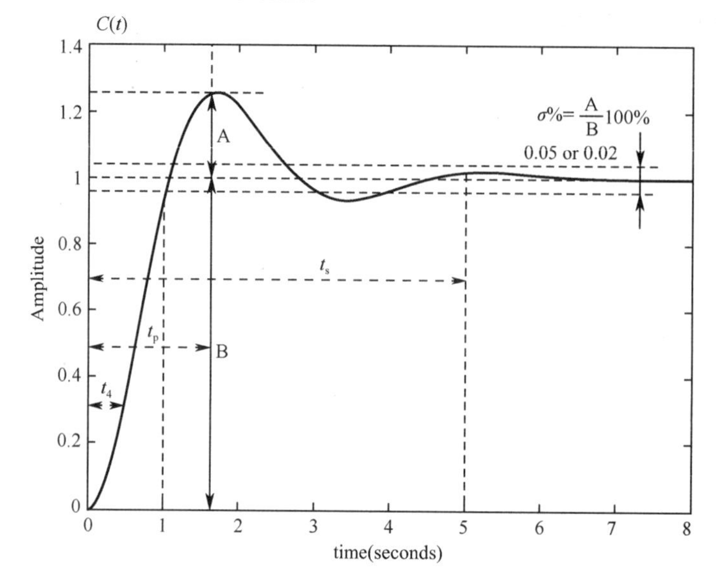

Figure 7.1.1 Typical Step Response and Dynamic Performance Specifications of the System

Peak time t_p: The time required for the step response to cross the final value $h(\infty)$ and reach the first peak time.

Settling time t_s: The minimum time required for the step response to reach and remain within $h(\infty)\pm5\%$ error band of the final value ; the adjustment time is also sometimes defined within ±2% error band of the final value. Unless otherwise specified, the settling time mentioned later in this book is defined within the ±2% error band of the final value.

Overshoot σ%: The percentage of peak $h(t_p)$ beyond the final value $h(\infty)$,

$$\sigma\% = \frac{h(t_p) - h(\infty)}{h(\infty)} \times 100\% \tag{7.1}$$

Among the above-mentioned dynamic performance specifications, the most commonly used in engineering are the settling time t_s (reflecting the length of the transition process and describing "fast"), the overshoot σ% (reflecting the degree of fluctuation of the transition process and describing "uniform") and peak time t_p, all being central specifications discussed in this book.

2. Steady-State Performance

Steady-state error is the error between the actual output and the ideal output of the system when time is approaching infinity. It is a measure of system control accuracy or anti-jamming capability. Steady-state errors are defined differently and are usually measured or calculated at typical inputs.

It should be noted that system performance should be determined according to the actual situation. For example, fighter jets require flexibility and quick response, allowing for proper overshoot; for long-term production processes after startup, steady-state accuracy is more emphasized.

7.1.3 Simulation Analysis Functions

In this chapter, the functions used in MATLAB in the simulation analysis of system performance mainly include: *step, impulse,* and *dcgain*. Their specific formats and instructions are shown in Table 7.2.

Table 7.2 Introduction to Simulation Analysis Functions

Function	Instructions
step(sys)	Get the step response plot of dynamic system
[y,t] = step(sys)	Get the step response data of dynamic system
impulse(sys)	Get the impulse response plot of dynamic system
[y,t] = impulse(sys)	Get the impulse response data of dynamic system
k = dcgain(sys)	DC gain of LTI system, at the frequency s=0.

7.2 Steady-State Error

The steady-state error is the error of the continuous response of the system after the decay of transient response.

Considering the system shown in Figure 7.1.1, the closed-loop transfer function is

$$\frac{C(s)}{R(s)}=\frac{G(s)}{1+G(s)} \tag{7.2}$$

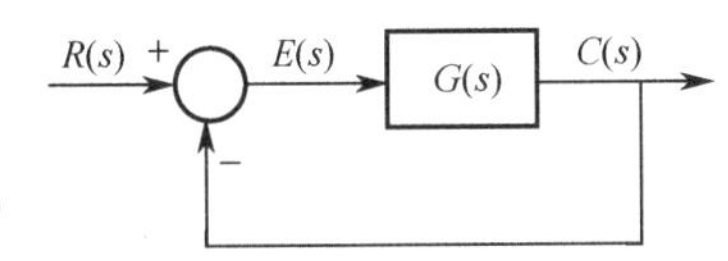

Figure 7.2.1 Control System

The transfer function between the error signal $e(t)$ and the input signal $r(t)$ is

$$\frac{E(S)}{R(S)}=1-\frac{C(S)}{R(S)}=\frac{1}{1+G(s)} \tag{7.3}$$

where the error $e(t)$ is the difference between the input signal and the output signal.

The final-value theorem provides a convenient way to find the steady-state performance of a stable system. Since

$$E(s)=\frac{1}{1+G(s)}R(s) \tag{7.4}$$

the steady-state error is

$$e_{ss}=\lim_{t\to\infty}e(t)=\lim_{s\to 0}sE(s)=\lim_{s\to 0}\frac{sR(s)}{G(s)} \tag{7.5}$$

Example 7.1 Given a control system in Figure 7.2.2.

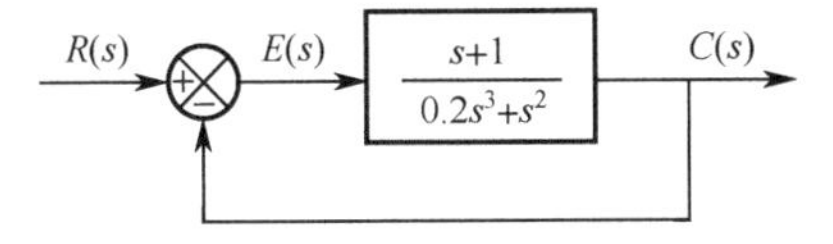

Figure 7.2.2 Control System of Example 7.1

Find the steady-state error e_{ss} when the input is $r(t)=t^2$, $t>0$.

Solution: First of all, judge the stability of the system. The MATLAB script and the results of Example 7.1 is shown in Figure 7.2.3 and Figure 7.2.4 respectively.

```
% The MATLAB script used for Example 7.1
num=[1,1];
den=[0.2,1,0,0];
[numc,denc]=cloop(num,den);
tf(numc,denc)
roots(denc)
```

Figure 7.2.3 The Program of Example 7.1

```
% The result of Example 7.1
ans =

              s + 1
  ----------------------
  0.2 s^3 + s^2 + s + 1

Continuous-time transfer function.


ans =

  -4.0739 + 0.0000i
  -0.4630 + 1.0064i
  -0.4630 - 1.0064i
```

Figure 7.2.4 MATLAB Results

The three closed-loop poles of the system are obtained, all in the left-half s plane, so the system is stable. The system error is

$$E(s)=\frac{R(s)}{1+G(s)}=\frac{0.2s^3+s^2}{0.2s^3+s^2+s+1}R(s) \tag{7.6}$$

When $r(t)=t^2$

$$R(s)=\frac{2}{s^3} \tag{7.7}$$

According to the final-value theorem, we determine that the steady-state error is

$$e_{ss}=\lim_{s\to 0} sE(s)=\lim_{s\to 0}\frac{(0.2s+1)s^2}{0.2s^3+s^2+s+1}\cdot\frac{2}{s^3}\cdot s=2 \tag{7.8}$$

7.3 Fluctuation Control During Airplane Flight

An airplane refers to a heavier-than-air aircraft that generates propulsion or pulling force by a power plant having one or more engines and generates lift by the fixed wing of the fuselage and flies within the atmosphere. Airplanes are the most common type of fixed-wing aircraft. According to the type of engine, airplanes can be divided into jet aircraft and propeller aircraft.

In a modern flight operation, it is the pilot's responsibility to ensure a smooth flight of the aircraft. Many important parameters, such as flight attitude, altitude, speed, lateral track and elevation angle, are all controlled by pilots within acceptable limits. Of course, in order to ensure the safety of the flight, the smooth flight of the aircraft must be maintained during the whole process. This is an example of human-computer interaction with full control of a process. Obviously, the smooth flight of the aircraft is the ultimate goal. Therefore, our control objective is to develop an automated system that regulates the fluctuation of airplane during flight.

7.3.1 Analysis of FV Control Systems

Flight Volatility (FV) serves as the most reliable indicator of a smooth flight. The pilot determines the parameters of the flight by monitoring the level of the FV. Based on pilot's flight experience and operating procedures, we decide to use FV as a controlled variable.

From the perspective of system design, the control goal can be stated in more concrete terms.

Control Objective: Adjust FV to any desired set-point and stabilize at a specified set-point in the presence of disturbances.

Based on the stated control objective, we identify the control variable.

Control Variable: Flight Volatility (FV).

Since we want to develop a system that ensures the smooth operation of aircraft, it is necessary to establish a practical control specifications for system design. In general, the control system should have minimal complexity while satisfying the control specifications. In other words, the minimum complexity is to increase system reliability and reduce costs.

As the FV set-point changes, the closed-loop system should respond quickly and smoothly without overshooting. The closed-loop systems should minimize the effects of external disturbance. Important disturbance types include airflow disturbance, mainly convection, i.e. where the vertical movement of the air flow, mainly during takeoff and landing, the aircraft needs to pass through the troposphere and the flight fluctuation becomes large. Finally, since we want to apply the same control system to many different flight situations, and we can not (for practical reasons) develop different models for each situation, we have to design a closed-loop system insensitive to changes in flight parameters, which, in other words, meets the specifications for different flight situations.

From experience, we can explicitly state the control specifications as follows.

DS1: Settling time less than 15 seconds when the desired output of FV is 10 times the amplitude variation of the unit step.

DS2: Overshoot less than 15% when the desired output of FV is 10 times the amplitude variation of the unit step.

DS3: Zero steady-state error when the output of FV is step signals.

DS4: Zero steady-state error when the flight disturbance input is a step signal (amplitude <50), and the maximum response range is within $\pm$ 5% of the desired output of FV.

DS5: Minimum sensitivity to parameter changes in different flight situations.

In the system structure, as shown in Figure 7.3.1, controllers, sensors, and aircraft serve as the main system elements.The input $R(s)$ of the system is the expected value of FV, and the output $C(s)$ is the variation of the actual FV. The controller uses the difference between the expected FV and the FV measured by the sensor as a control signal to adjust the value of the stable operation of the controller input to the aircraft.

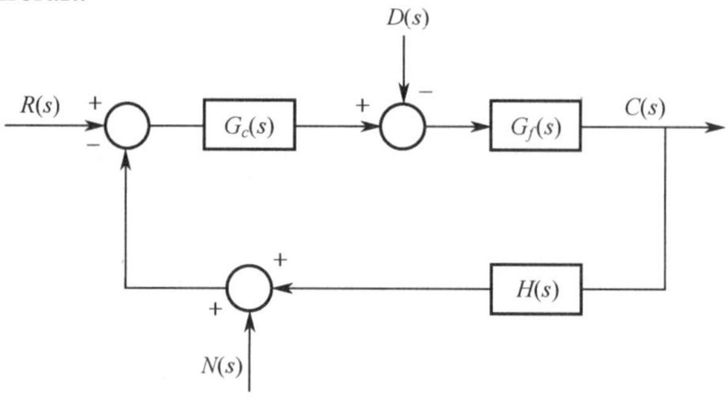

Figure 7.3.1 Aircraft Flight Control System Configuration

The modeling process of aircraft flight is complicated. The main reason is that it is difficult to predict all the air flow conditions when the aircraft is flying. Even if a similar model is established, it is usually a nonlinear time-varying model with multiple inputs and multiple outputs. It cannot be applied directly to the linear time-invariant model with single input and single output that we want to build.

The model of the system is not unique, and multiple models corresponding to different levels can be generated for the same system. The determination of the model mainly depends on the system's control objectives, which help to obtain system models of different complexities. The control system for aircraft flight fluctuations can only implement the fluctuation control during flight, regardless of other factors. If the control objective is satisfied , we can establish a model of the aircraft flight system. If the FV is only limited to fluctuate slightly around its set value, we can assume that the aircraft exhibits a linear time-invariant characteristic within a small range around the FV setting. This method of analysis is very suitable for the requirement that FV be maintained near a given value. Therefore, we directly give the transfer function model of aircraft, as shown in Equation 7.9.

$$G_f(s) = \frac{1}{s(s+f)^2} \tag{7.9}$$

where $f = 2$ and time t is measured in seconds. The parameter f is related to the flight condition of the aircraft. The parameters f of different flight conditions also vary.

Assume that the sensor is in a perfect working condition, its transfer function is

$$H(s) = 1 \tag{7.10}$$

In this application, the controller uses a Proportional-Integral-Derivative (PID) controller with a transfer function of

$$G_c(s) = K_P + sK_D + \frac{K_I}{s} = \frac{K_D s^2 + K_P s + K_I}{s} \tag{7.11}$$

Therefore, we have a unit negative feedback system, whose closed-loop transfer function is

$$\phi(s)=\frac{G_c(s)G_f(s)}{1+G_c(s)G_f(s)H(s)} \tag{7.12}$$

7.3.2 Analysis of Steady-State Error

The steady-state error of systems (shown in Figure 7.3.1 with $D(s) = 0$ and $N(s) = 0$) is

$$E(s)=R(s)-C(s)=\frac{1}{1+G_c(s)G_f(s)H(s)}R(s) \tag{7.13}$$

or

$$E(s)=\frac{s^4+2fs^3+f^2s^2}{s^4+2fs^3+(f^2+K_{\rm D})s^2+K_{\rm P}s+K_{\rm I}}R(s) \tag{7.14}$$

From the final-value theorem, we determine that the steady-state error is

$$\lim_{s\to 0} sE(s)=\lim_{s\to 0}\frac{R_0(s^4+2fs^3+f^2s^2)}{s^4+2fs^3+(f^2+K_{\rm D})s^2+K_{\rm P}s+K_{\rm I}}=0 \tag{7.15}$$

where $R(s)=\frac{R_0}{s}$ is a step input with magnitude R_0. Therefore,

$$\lim_{t\to\infty} e(t)=0 \tag{7.16}$$

For a PID controller, we expect a zero steady-state error (to a step input) for any nonzero values of K1, K2, and K3, since $G_c(s)G_o(s)G_f(s)$ is Type II. As shown in Equation 7.17, there is an error only when the acceleration is input.Thus DS3 is satisfied.

$$G_c(s)G_f(s)H(s)=\frac{K_{\rm D}s^2+K_{\rm P}s+K_{\rm I}}{s^2(s+f)^2} \tag{7.17}$$

When taking the effect of a step disturbance input into consideration, we let $R(s) = 0$ and $N(s) = 0$. We desire the steady-state value of output signal $C(s)$ to be zero with a step disturbance. The transfer function from the disturbance $D(s)$ to the output $C(s)$ is

$$\frac{C(s)}{D(s)}=\frac{-G_f(s)}{1+G_c(s)G_f(s)} \tag{7.18}$$

$$C(s)=\frac{-s}{s^4+2fs^3+(f^2+K_{\rm D})s^2+K_{\rm P}s+K_{\rm I}}D(s) \tag{7.19}$$

When

$$D(s)=\frac{D_o}{s} \tag{7.20}$$

We obtain

$$\lim_{s\to 0} sC(s)=\lim_{s\to 0}\frac{-D_0 s}{s^4+2fs^3+(f^2+K_{\rm D})s^2+K_{\rm P}s+K_{\rm I}}=0 \tag{7.21}$$

Therefore,

$$\lim_{t\to\infty} C(t)=0 \tag{7.22}$$

So, when the disturbance signal is a step signal with an amplitude of D0, the steady-state output of the system is zero as expected.

7.3.3 Transient Response

Assume that we study the FV by changing the amplitude input the 10 times of the unit step, the relevant input is

$$R(s)=\frac{R_0}{s}=\frac{10}{s} \tag{7.23}$$

The controller is a Proportional-Integral-Derivative (PID) controller whose transfer function is shown in Equation 7.11, where K_P, K_I, and K_D are the controller gains, determined by the requirements of the design indicators.

This section covers system control from selecting the proportional controller P, to selecting the Proportional-Integral controller PI, and finally to selecting the PID controller. Besides, the transient response characteristics of the system under the three controllers, and the performance under their respective conditions are analyzed. Finally, on the condition that the parameters of the PID controller are configured to meet the requirement for best performance, one parameter is selected from the three PID controllers to configure the optimal PID controller.

1. P Controller

Firstly, add a proportional controller P, with the differential and integral actions being zero, and we get the controller transfer function as

$$G_c(s)=K_P \tag{7.24}$$

Assuming that the input change of FV is 10 times the amplitude of the unit step, i.e. the input is $R(s)=\frac{R_0}{s}=\frac{10}{s}$, the proportional coefficient values are respectively selected as $K_P=[1,3,8]$ and the other parameters of controller $G_c(s)$ are set to $K_D=0$, $K_I=0$. We get the step response curve in three different proportional coefficient, as shown in Figure 7.3.2.

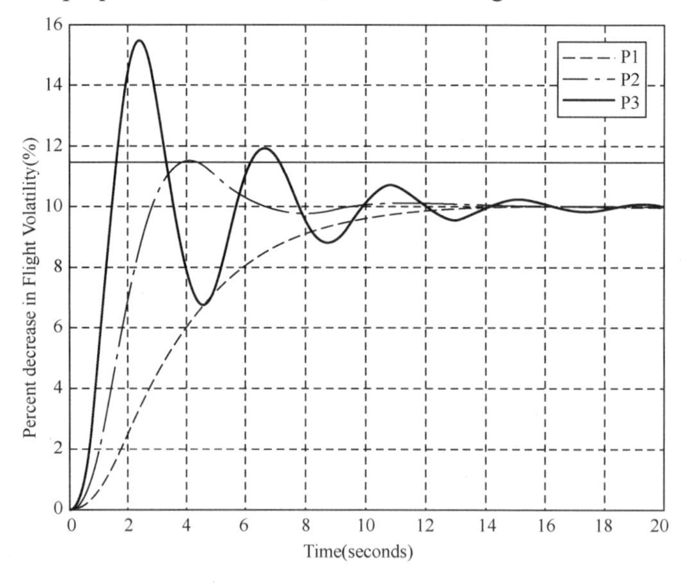

Figure 7.3.2 The Step Response of P Controller

As can be seen from Figure 7.3.2, with the P controller, the amplitude of system response amplifies as the proportional coefficient value K_p increases. From the start, there was no overshoot at $K_p = 1$, while at $K_p = 3$ and $K_p = 8$, the overshoot increased to 15.25% and 54.36%. It indicates that the larger the proportional coefficient K_p, the greater the oscillation amplitude of the system. At $K_p = 1$, the adjustment time of the system is 11.9 seconds, while at $K_p = 3$ and $K_p = 8$, the adjustment time is 8.3 seconds and 15.4 seconds respectively. As the proportional coefficient K_p rises, the oscillation amplitude increases and the adjusting time of the system also goes up. The program and results are shown in Figure 7.3.3.

```
% The MATLAB script used for Figure 7.3.2
clc
clear;
close all
tf=20;
t=[0:0.1:tf];
R0=10;
f=2;
den=conv([1,0,0],conv([1,f],[1,f]));
KP=[1,3,8];KD=0;KI=0;                    %when P
% KP=[1,3,8];KD=0;KI=[2,1,4];             % when PI
% KP=[8,7,5];KD=[11,4,7];KI=[1,1,2];      % when PID
for ii=1:3;
   num=[KD,KP(ii) KI];                    %when P
   %num=[KD,KP(ii) KI(ii)];               % when PI
   %num=[KD(ii),KP(ii) KI(ii)];           % when PID
   [numc,denc]=cloop(num,den);
   y=step(R0*numc,denc,t);
   finalvalue=dcgain(R0*numc,denc);
    [yss,n]=max(y);
   TP(ii)=t(n);
   [PO(ii),TS(ii)]=poandts(t,y,yss,finalvalue);
   yo(:,ii)=y;
  end
plot(t,yo(:,1),'b--',t,yo(:,2),'k-.',t,yo(:,3),'m-',[0 tf],[11.5 11.5])
legend('P1','P2','P3', -1)                % when P
%legend('PI1','PI2','PI3', -1)            % when PI
%legend('PID1','PID2','PID3', -1)         % when PID
grid
ylabel('Percent decrease in Flight Volatility(%)')
xlabel('Time(seconds)')
TP
PO
TS
```

Figure 7.3.3 MATLAB Script and Results

```
% MATLAB function of above program.
function[PO,TS]=poandts(t,y,yss,finalvalue)
PO=100*(yss-finalvalue)/finalvalue;
k=length(t);
while (y(k)>0.98*finalvalue)&(y(k)<1.02*finalvalue)
 k=k-1;
end
TS=t(k);
```

```
% Results of Command window
TP =

   20.0000    4.0000    2.4000


PO =

   -0.0912   15.2474   54.3551


TS =

   11.9000    8.3000   15.4000
```

Figure 7.3.3 MATLAB Script and Results（continued）

2. PI Controller

With PI controller, the differential coefficient of the controller $G_c(s)$ is set to $K_D = 0$. The proportional coefficient and the integral coefficient are given three sets of values respectively. With the three proportional coefficient $K_p = [1,\ 3,\ 8]$, and the three corresponding integral coefficient $K_I = [2,\ 1,\ 4]$, the three sets of parameters are set accordingly to make three PI controllers. Therefore, the proportional coefficient and integral coefficient of PI1 are 1 and 2, those of PI2 are 1 and 3, those of PI3 are 4 and 8 respectively.

With the script in Figure 7.3.3, the step response curve of the system is obtained, as shown in Figure 7.3.4. The three PI controllers display different outputs: PI1 shows an unstable output, with $K_p = 1$, the smallest value of the group, and $K_I = 2$; PI2 demonstrates a stable output, eventually reaching 10, a steady-state value, with zero steady-state error; PI3 indicates a constant amplitude oscillation in a critical state.

3. PID Controller

The selection of three parameters of the PID controller is basically a search problem in three-dimensional space. Since three points in the search space correspond to the three parameters of PID controller, we determine the PID controller by selecting different points in the parameter space. For example, for different step responses to a step input, heuristic methods can be used to move points in the search space to determine a PID controller.

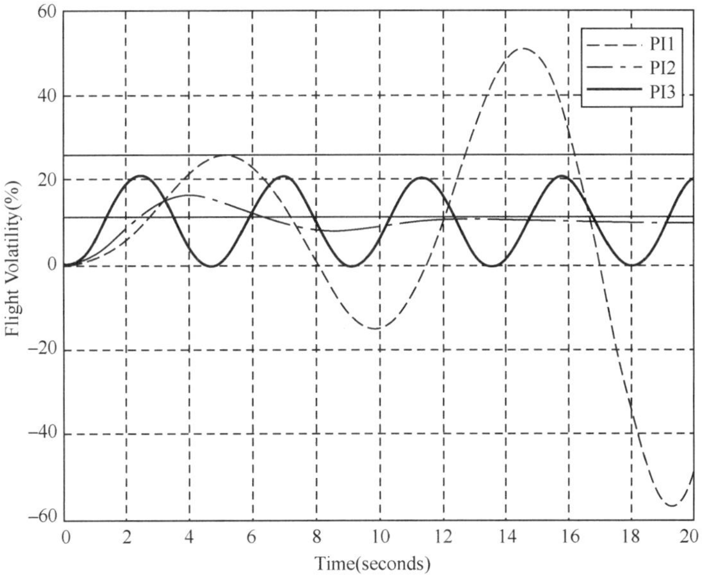

Figure 7.3.4 Step Response of PI Controllers

With PID controller, three PID coefficients close to the required performance can be searched through multiple simulations, i.e. $K_{\mathrm{p}}=[8,7,5]$, $K_{\mathrm{D}}=[11,4,7]$, and $K_{\mathrm{I}}=[1,1,2]$, corresponding to three PID controllers respectively. Among them, the proportional, differential, and integral coefficients of PID1 match 8, 11, and 1respectively; those of PID2 are 7, 4, and 1 respectively, while those of PID3 equal to 5, 7, and 2 respectively.

From the script in Figure 7.3.3, the system step response curve is obtained, as shown in Figure 7.3.5. The three PID controllers vary in their outputs. From Figure 7.3.5 and Figure 7.3.6, PID1 controller demonstrates the optimum dynamic performance, with its overshoot 7.52%, its adjustment time 12.1 seconds, which satisfies the requirement for system performance. PID2 controller is near the requirement for the dynamic performance. Its overshoot reaches 18.22%, slightly beyond the required 15%, which does not meet the requirements of the overshoot. The adjustment time is 10.6 seconds, within the required limits of performance. For the dynamic performance of PID3 controllers, the overshoot is 14.22%, the settling time is 13.5 seconds, both within the scope of the requirement. Compare the three controllers, and we find that in the same step signal input, when $R(s)$=10/s, both PID1 and PID3 satisfy DS1 and DS2, and the overshoot and adjustment time are also satisfactory.

Next, let's explore whether the system performance can meet the requirement of DS4 when the disturbance is added.

We put the overshoot of system response and the adjustment time in Table 7.3 for better comparison.

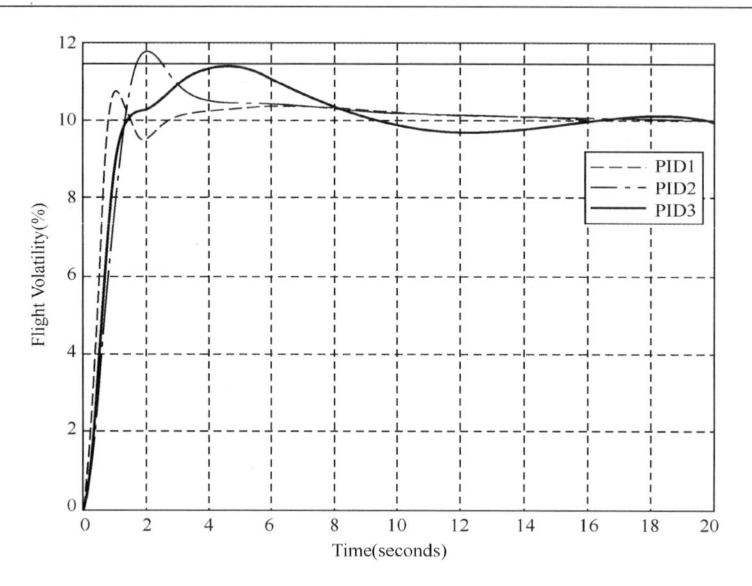

Figure 7.3.5 Step Response of PID Controllers

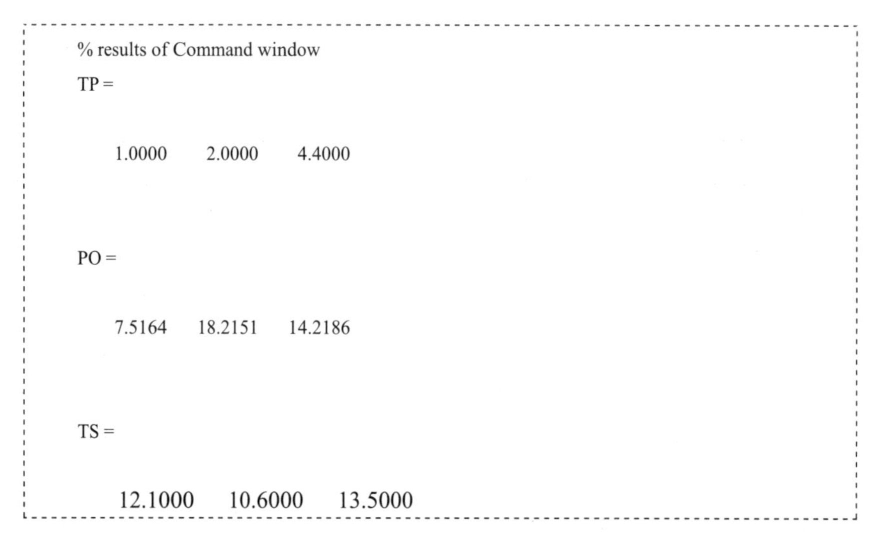

```
% results of Command window
TP =

    1.0000    2.0000    4.4000


PO =

    7.5164    18.2151    14.2186


TS =

    12.1000    10.6000    13.5000
```

Figure 7.3.6 Results of MATLAB Command Window

Table 7.3 Gains of PID Controller and Results of System Performance

PID	KP	KD	KI	Overshoot(%)	Settling Time(second)
1	8	11	1	7.52	12.1
2	7	4	1	18.22	10.6
3	5	7	2	14.22	13.5

7.3.4 Disturbance Response

From the above analysis, we know that the transfer function from the disturbance input $D(s)$ to the output $C(s)$ is

$$\begin{aligned} C(s) &= \frac{-G_f(s)}{1+G_c(s)G_f(s)}D(s) \\ &= \frac{-s}{s^4+2ps^3+(p^2+K_D)s^2+K_Ps+K_I}D(s) \end{aligned} \tag{7.25}$$

To investigate if the system performance can meet DS4, we compute the disturbance step response when the disturbance signal

$$D(s)=\frac{D_0}{s}=\frac{50}{s} \tag{7.26}$$

The maximum amplitude of the disturbance signal is 50. Because the smaller the amplitude of the step disturbance is, the smaller the maximum value of the output response is. When we determine whether the system performance can meet DS4, we only need to examine the largest amplitude of the disturbance input signal.

We still select the three PID controllers in the above section. The output responses of step disturbance signals to the systems controlled by the three PID controllers are shown in Figure 7.3.7. We can see that the steady-state errors of the three controllers are all zero, which satisfies the requirement of zero error in DS4; PID3 meets the requirement of DS4, with its maximum response less than ±5% of the FV set-point, while PID1 and PID2 go beyond the limit. MATLAB program is shown in Figure 7.3.8.

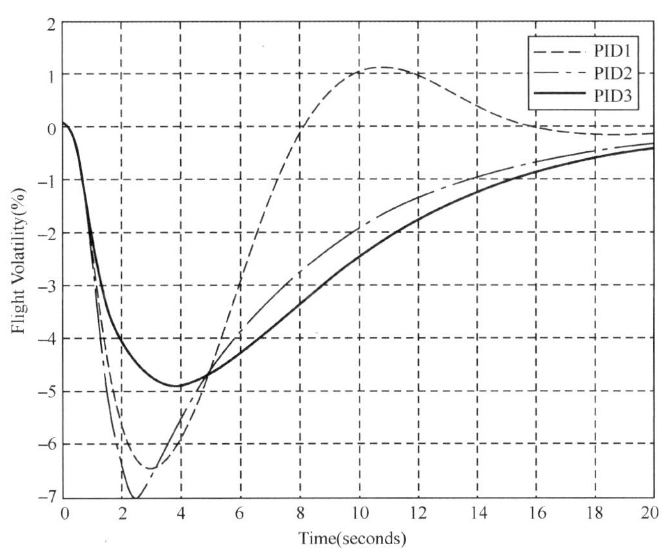

Figure 7.3.7 Disturbance Step Response of FV

```
% Figure 7.3.7 Operation MATLAB script
clc
clear;
close all
tf=20;
t=[0:0.1:tf];
```

Figure 7.3.8 MATLAB Script

```
Do=50
f=2;
num=[1,0];
KP=[8,7,5];KD=[11,4,7];KI=[1,1,2];
dc=[1,0];
for ii=1:3;
   den=[1,2*f,f^2+KD(ii),KP(ii),KI(ii)];
   y=step(-Do*num,den,t);
   yo(:,ii)=y;
 end
plot(t,yo(:,1),'b-',t,yo(:,2),'k-.',t,yo(:,3),'m--')
legend('PID1','PID2','PID3', -1)
grid
ylabel('Percent decrease in Flight Volatility(%)')
xlabel('Time(seconds)')
```

Figure 7.3.8 MATLAB Script（continued）

In summary, we would select PID3 as the controller of the aircraft from the three choices. It meets all requirement of the design specifications while providing a reasonable insensitivity to changes in the parameter.

Summary

1. The system performance of time domain and how to use MATLAB functions of typical input signals are introduced.

2. Steady-state error is analyzed and how to use MATLAB to find the steady-state error of control system is specifically illustrated.

3. The steady-states errors under the step input and under the disturbance input are analyzed concerning the fluctuations control system of aircrafts.

4. The fluctuation control of aircraft flight are explored furtherly. Under the influence of the three P-PI-PID controllers, the system response to the step input and to the disturbance input are both analyzed. Finally, one PID controller is found to meet all the requirements of the design specifications.

生词注解

performance	n. 性能，绩效
evaluation	n. 评价
circumstances	n. 情况
dynamic performance	n. 动态性能指标
severe	adj. 严峻的
specifications	n. 规格
demonstrate	v. 证明
amplitude	n. 振幅
insensitivity	n. 感觉迟钝

第 8 章 主导极点法的 PID 控制器设计实例

在前面第 6 章和第 7 章内容的基础上，本章重点介绍系统控制器的设计。首先介绍 PID 控制器、比例与积分的控制作用、比例与微分的控制作用，以及比例积分微分的控制作用。然后详细论述利用主导极点，配置 PID 控制器参数的方法与具体步骤。最后，以下棋机器人为实例，采用主导极点的方法，实现了对下棋机器人的 PID 控制器设计，使得 PID 控制器满足系统的设计要求。

8.1 PID 控制器的概念

工业控制中广泛采用的是 PID 控制器，具有三个因子，其传递函数为

$$G_c(s)=K_\text{P}+\frac{K_\text{I}}{s}+K_\text{D}s \tag{8.1}$$

其时域输出方程为

$$u(t)=K_\text{P}e(t)+K_\text{I}\int e(t)\text{d}t+K_\text{D}\frac{\text{d}e(t)}{\text{d}t} \tag{8.2}$$

三个因子的控制器就是 PID 控制器，它包括一个比例项、一个积分项和一个微分项，并各自由 K_P，K_I 和 K_D 来表示。典型的 PID 控制器结构如图 8.1.1 所示。

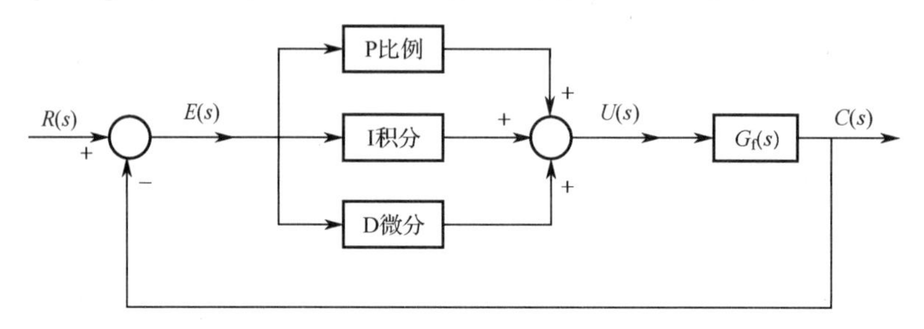

图 8.1.1　PID 控制器结构

如果令 $K_\text{D}=K_\text{I}=0$，PID 控制器就变成了比例控制器

$$G_c(s)=K_\text{P} \tag{8.3}$$

如果令 $K_\text{P}=K_\text{D}=0$，PID 控制器就变成了积分控制器

$$G_c(s)=\frac{K_\text{I}}{s} \tag{8.4}$$

可以把比例控制器和积分控制器放在一起构成一个比例积分控制器。

$$G_c(s)=K_\text{P}+\frac{K_\text{I}}{s}=\frac{K_\text{P}s+K_\text{I}}{s} \tag{8.5}$$

当 $K_\text{P}=K_\text{I}=0$ 时，得到微分控制器

$$G_c(s)=K_\text{D}s \tag{8.6}$$

通常将微分控制器与比例/积分控制器相连接。比例微分控制器为

$$G_c(s)=K_\text{P}+K_\text{D}s \tag{8.7}$$

8.2 比例与积分控制的作用

如果一个控制对象的传递函数中不存在积分器 $1/s$，则当对其进行比例控制时，阶跃输入信号的响应将存在稳态误差，或称为偏差。这个偏差随着比例系数的加大而加大，比例控制的特点就是有差控制。

如果在此控制器中包含积分控制作用，则可以消除这种偏差。因此，积分控制的特点为无差控制。

在积分控制中，控制信号（即控制器的输出信号）在任何瞬间都等于该瞬间之前误差信号曲线之下的面积。积分控制的物理意义是：只要偏差不为零，积分环节就将其不断地积累，并按照负反馈原理产生持续的控制作用，直到使偏差恢复至零，从而有效地提高系统的稳态

性能。

如图 8.2.1 所示，用一个小幅正弦波表示偏差信号，图（a）表示偏差信号通过一个积分器之后的波形。当偏差信号在正弦曲线上的半个周期点 A 时，其为零值。与此点 A 相对应的积分器输出曲线上的点 B 却在波峰上。当误差信号一个周期再次过零点 C 时，由于误差信号与时间横轴的面积大小刚刚正负抵消，因此，在积分后的曲线上与点 C 对应的点 D 的值为零。图（b）是误差信号通过一个比例环节之后，只是幅值变化了。当误差信号曲线上的点 A 为零时，通过比例环节之后的曲线上与点 A 相对应的点 B 也会为零。误差信号通过积分器后，在误差曲线为零的点，相应的通过积分器的曲线在此点可能是零，也可能不是零，但是对于通过比例环节的曲线在此点必然会为零。

注意，在消除偏差（即稳态误差）的同时，积分控制作用也会导致振幅缓慢衰减，甚至使振幅不断增加的振荡响应，这两种情况通常都不是我们所希望的。

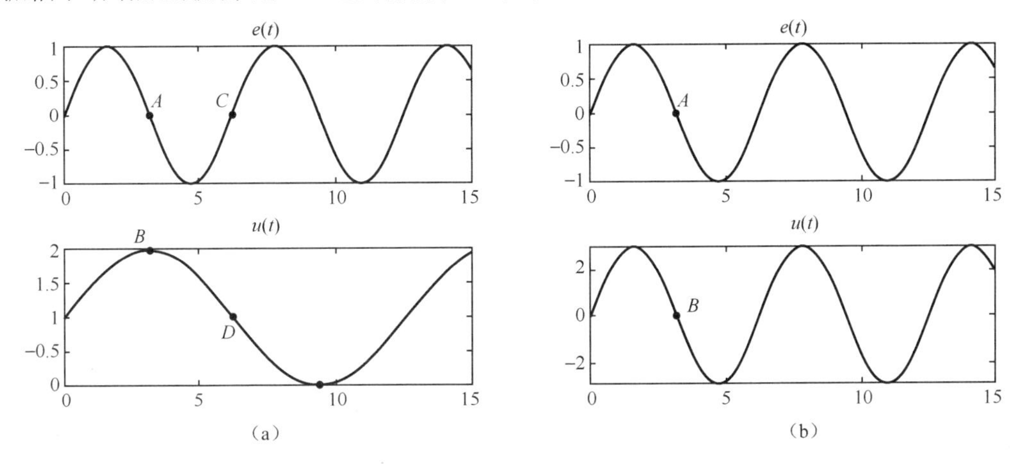

图 8.2.1　比例控制与积分控制：（a）当作用误差信号为零时，非零控制信号的 $e(t)$ 和 $u(t)$ 曲线图（积分控制）；（b）当作用误差信号为零时，控制信号为零的 $e(t)$ 和 $u(t)$ 曲线图（比例控制）

8.2.1　比例控制下的误差

现在我们来证明，对于单位阶跃输入信号，当没有积分器时，系统的比例控制将造成稳态误差。随后还将证明，当控制器中包含积分控制作用时，可以消除这种误差。

考虑图 8.2.2 所示系统，我们来求系统在单位阶跃响应中的稳态误差。系统的开环传递函数为

$$G(s)=\frac{K}{Ts+1} \tag{8.8}$$

因为

$$\frac{E(s)}{R(s)}=\frac{R(s)-C(s)}{R(s)}=1-\frac{C(s)}{R(s)}=\frac{1}{1+G(s)} \tag{8.9}$$

所以误差 $E(s)$ 为

$$E(s)=\frac{1}{1+G(s)}R(s)=\frac{1}{1+\dfrac{K}{Ts+1}}R(s) \tag{8.10}$$

对于单位阶跃输入信号 $R(s)=1/s$，所以得到

$$E(s)=\frac{Ts+1}{Ts+1+K}\frac{1}{s} \tag{8.11}$$

因此，稳态误差为

$$e_{ss}=\lim_{t\to\infty}e(t)=\lim_{s\to 0}sE(s)=\lim_{s\to 0}\frac{Ts+1}{Ts+1+K}=\frac{1}{K+1} \tag{8.12}$$

这种在前向路径中不带积分器的系统，在阶跃响应中总是存在着稳态误差。这种稳态误差称为偏差。

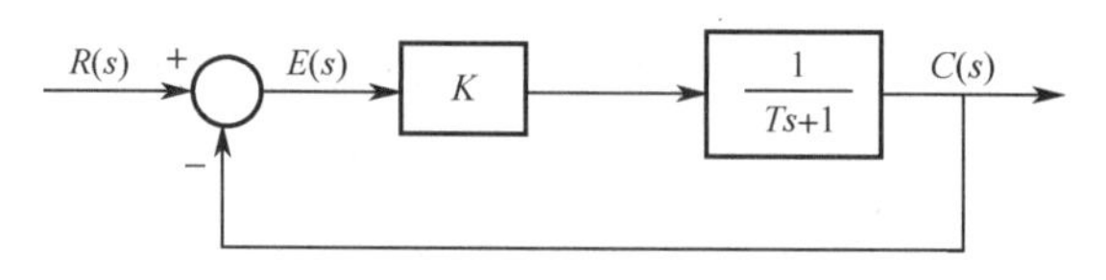

图 8.2.2 比例控制系统

在这个比例控制系统中，选取阶跃输入 $R(s)=1$， $K=10$， $T=0.4$，系统的阶跃输出响应见图 8.2.3，程序和结果见图 8.2.4。

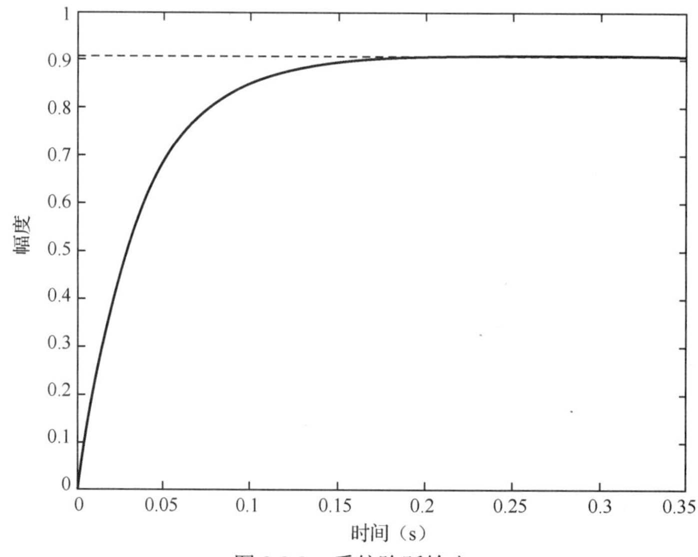

图 8.2.3 系统阶跃输出

```
%图 8.2.3 程序
num=[10];
den=[0.4,1];
[numc,denc]=cloop(num,den);
step(numc,denc)
roots(den)
```

```
%图 8.2.3 程序输出结果
ans =


  -2.5000
```

图 8.2.4 MATLAB 程序及其结果

从图 8.2.3 可以看出，系统的输出是有误差的。稳态输出值为 0.9，稳态误差大约为 0.1，即$1/K+1=1/11$约为 0.1。在实轴上，系统闭环的根为−2.5，系统阻尼比为 1，系统没有振荡，系统的调节时间 0.2s 左右。

8.2.2 消除积分控制中的稳态误差

考虑图 8.2.2 所示系统，系统中的控制器为积分控制器。系统的闭环传递函数为

$$\frac{C(s)}{R(s)}=\frac{K}{s(Ts+1)+K} \tag{8.13}$$

因此

$$\frac{E(s)}{R(s)}=\frac{R(s)-C(s)}{R(s)}=\frac{s(Ts+1)}{s(Ts+1)+K} \tag{8.14}$$

因为系统是稳定的，所以应用终值定理可以求得单位阶跃响应

$$e_{\text{ss}}=\lim_{s\to 0}sE(s)=\lim_{s\to 0}\frac{s^2(Ts+1)}{Ts^2+s+Ks}\frac{1}{s}=0 \tag{8.15}$$

因此，系统的积分控制消除了对阶跃输入响应中的稳态误差。这对于产生偏差的比例控制来说，是一项重要的改进。

在图 8.2.5 积分控制系统中，选取阶跃输入$R(s)=1$，$K=10$，$T=0.4$，系统的阶跃输出响应如图 8.2.6 所示。实现的程序如图 8.2.7 所示。

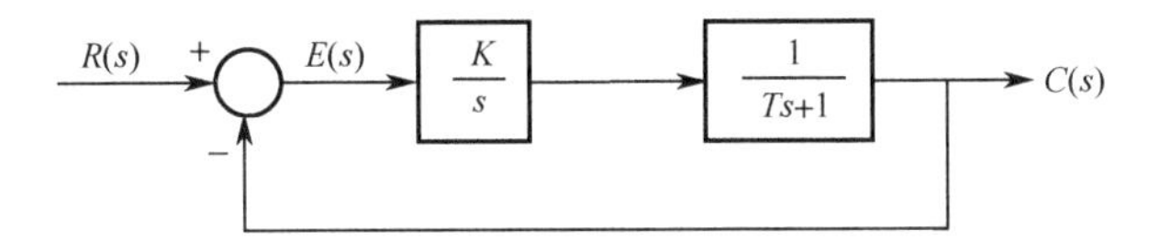

图 8.2.5　积分控制系统

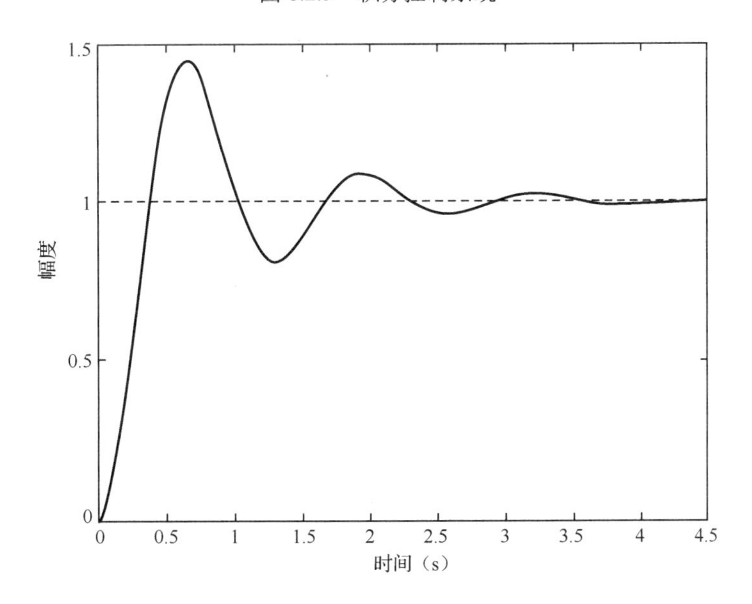

图 8.2.6　积分控制系统阶跃响应

```
%图 8.2.6 程序
num=[10];
den=[0.4,1,0];
[numc,denc]=cloop(num,den);
step(numc,denc)
roots(denc)
```

```
%图 8.2.6 程序结果
ans =


  -1.2500 + 4.8412i
  -1.2500 - 4.8412i
```

图 8.2.7 MATLAB 程序及其结果

从图 8.2.6 可以看出，系统的输出是没有误差的，稳态输出值为 1，稳态误差为 0。系统闭环的根为一对共轭复根，在第二、第三象限，系统阻尼比 $0<\zeta<1$，系统有超调，存在振荡，系统的调节时间在 3.5s 左右。

8.3 比例与微分的控制作用

当把微分控制作用加进比例控制器时，就提供了一种获得高灵敏度控制器的方法。采用微分控制作用的优点是，它能够反映误差信号的变化速度，并在误差的值变得很大之前产生一个有效的修正。因此微分控制可以预测误差，使修正作用提前发生，从而有助于增进系统的稳定性。

虽然微分控制不直接影响稳态误差，但它增加了系统的阻尼，因而允许采用比较大的增益 K 值，这将有助于系统稳态精度的改善。

因为微分控制的工作是基于误差的变化速度的，而不是基于误差本身的，因此这种方法决不能单独应用，它总是与比例控制作用或比例-加-积分控制作用组合在一起应用。

8.3.1 比例控制的不稳定性

在讨论微分控制作用对系统性能的影响之前，先研究惯性负载的比例控制。

控制对象的模型如下

$$G_f(s)=\frac{1}{s^2} \tag{8.16}$$

首先，采用比例反馈来控制

$$G_c(s)=K_{\mathrm{P}} \tag{8.17}$$

带有控制器 $G_c(s)$ 的闭环反馈系统，如图 8.3.1 所示。

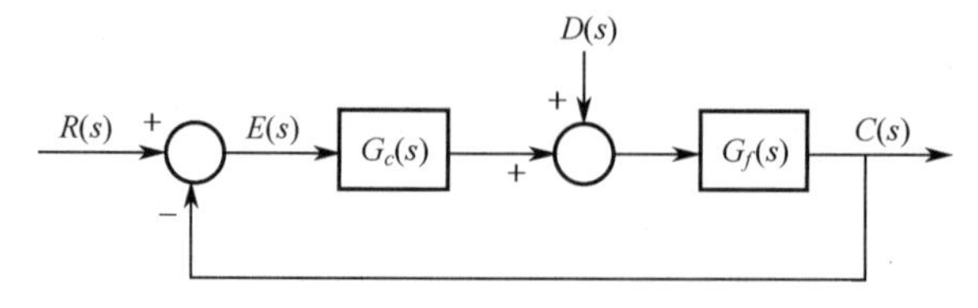

图 8.3.1 控制系统的闭环框图

考虑图 8.3.1 所示的系统，其闭环传递函数为

$$\frac{C(s)}{R(s)}=\frac{G_c G_f}{1+G_c G_f}=\frac{K_\mathrm{P}}{s^2+K_\mathrm{P}} \tag{8.18}$$

特征方程为

$$s^2+K_\mathrm{P}=0 \tag{8.19}$$

采用图 8.3.4 中的程序，可实现闭环系统的零极点分布图的绘制（见图 8.3.2）、系统的阶跃响应及其闭环系统特征根求取。

闭环系统的零极点分布如图 8.3.2 所示。$K_\mathrm{P}=2$ 时，虚轴上有两个极点，系统处于临界状态，且不稳定。若无有效控制，系统在外界干扰下会失衡。

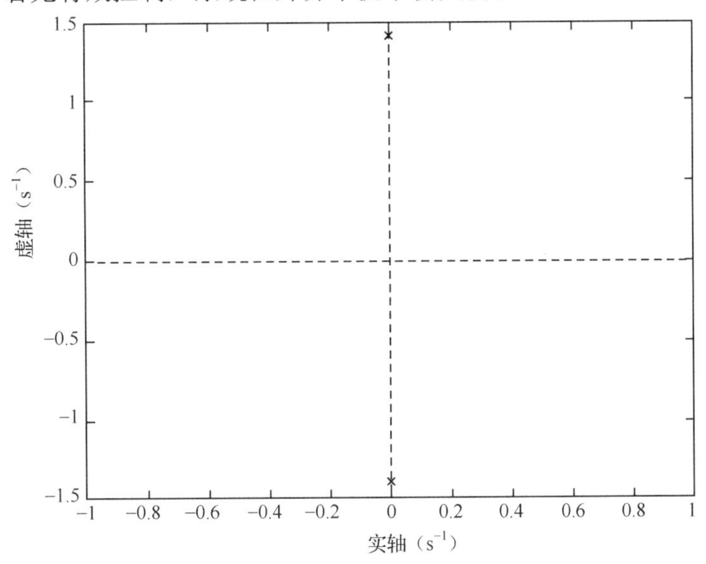

图 8.3.2　闭环系统的零极点分布

系统对单位阶跃输入的响应是一个无限期的持续振荡，如图 8.3.3 所示。控制系统呈现这种响应特性不是我们所希望的。加进微分控制后，使系统稳定。

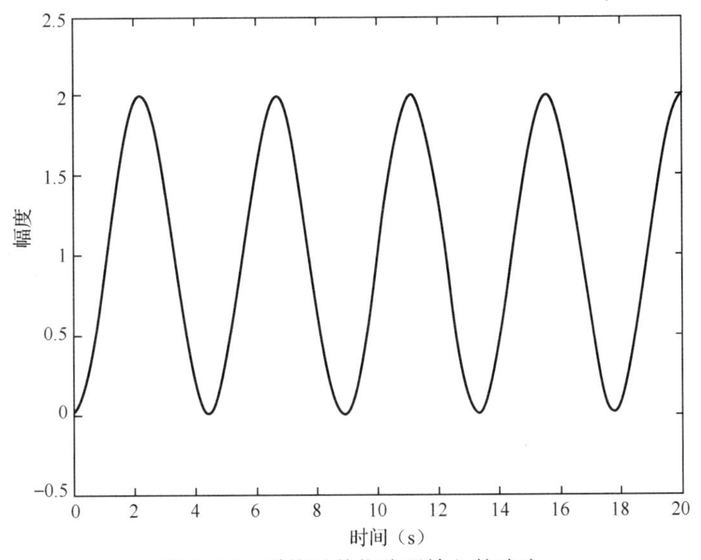

图 8.3.3　系统对单位阶跃输入的响应

```
%图 8.3.2 和图 8.3.3 的程序
num=2;den=[1 0 2];
figure(1),pzmap(num,den)
figure(2),step(num,den)
roots(den)
```

```
%图 8.3.2 的程序结果
ans =

   0.0000 + 1.4142i
   0.0000 - 1.4142i
```

图 8.3.4　MATLAB 程序及其结果

8.3.2　比例微分控制的稳定性

为了使系统稳定，我们需要所有的根轨迹进入 s 左半平面。但不存在使系统的极点都位于 s 左半平面的 K_P 值。很明显，仅通过比例控制不可能使系统稳定。

我们怎样指定 $G_c(s)$ 才能使特征方程的根转向 s 左半平面？因为根轨迹开始于开环系统（$k_P = 0$）的极点，终止于开环系统（$K_P \to \infty$）的零点，因此，使控制器至少有一个零点位于 s 左半平面来吸引根轨迹是有意义的。我们可以考虑控制器

$$G_c(s) = K_P + K_D s = K_D\left(s + \frac{K_P}{K_D}\right) \tag{8.20}$$

方程（8.20）是比例微分（PD）控制器，比值决定了零点的位置。

$$z = K_P / K_D \tag{8.21}$$

稳定性不是设计中的唯一问题，在最终设计中，性能指标将影响 z 的选择。微分控制实际上是超前的，它可以测量瞬时误差速度，提前预测大的超调量，并且在产生过大的超调量之前，产生一个适当的反作用。

考虑图 8.3.1 所示的系统，系统的闭环传递函数为

$$\frac{C(s)}{R(s)} = \frac{G_C G_f}{1 + G_C G_f} \tag{8.22}$$

$$\frac{C(s)}{R(s)} = \frac{K_D s + K_P}{s^2 + K_D s + K_P} \tag{8.23}$$

特征方程为

$$s^2 + K_D s + K_P = 0 \tag{8.24}$$

对于正的 K_D 的值，特征方程有两个负实部的根，系统的阻尼比不为零。因此，微分控制带来了阻尼效应。

闭环系统的零极点分布如图 8.3.5 所示。令 $K_P = 2$，$K_D = 0.8$ 生成零极点图。有两个共轭极点，一个零点在负实轴上，显然系统是稳定的。如图 8.3.6 所示，系统的两条根轨迹也在 s 左半平面，表明系统是稳定的。

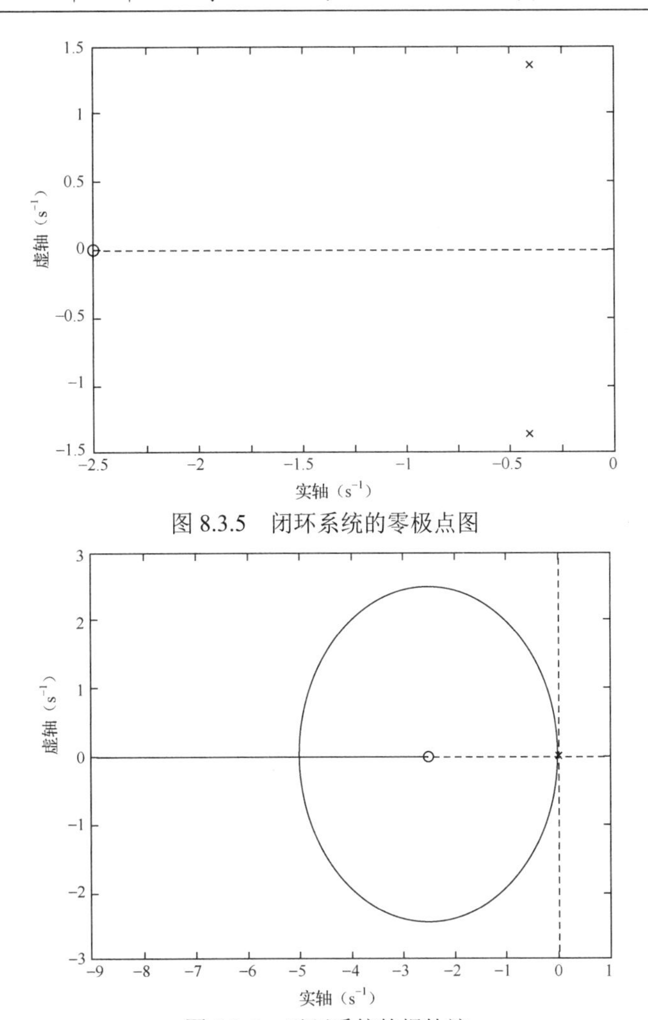

图 8.3.5　闭环系统的零极点图

图 8.3.6　开环系统的根轨迹

采用图 8.3.7 中的程序可实现闭环系统的零极点分布图的绘制、系统的阶跃响应及其闭环系统特征根的求取。

```
% 图 8.3.5、图 8.3.6 和图 8.3.7 的程序
num=[0.8 2];den=[1 0 0];            %开环传递函数 GcGf
[numc,denc]=cloop(num,den);         %闭环传递函数
figure(1),pzmap(numc,denc);         %闭环零极点图
figure(2),rlocus(num,den);          %开环根轨迹
figure(3),step(numc,denc);          %系统阶跃响应
tf(numc,denc)                       %闭环传递函数
```

```
% MATLAB 的 Command Window 输出
ans =

      0.8 s + 2
  ---------------
  s^2 + 0.8 s + 2

Continuous-time transfer function.
```

图 8.3.7　MATLAB 程序及其结果

对单位阶跃输入信号的典型响应曲线 $C(t)$如图 8.3.8 所示。显然，与图 8.3.4 所示的原响应曲线相比，现在的响应曲线有了明显改进。系统由 P 控制器下的振荡响应变成 PD 控制下趋于稳定的响应。系统中有一定超调，稳态误差为零。

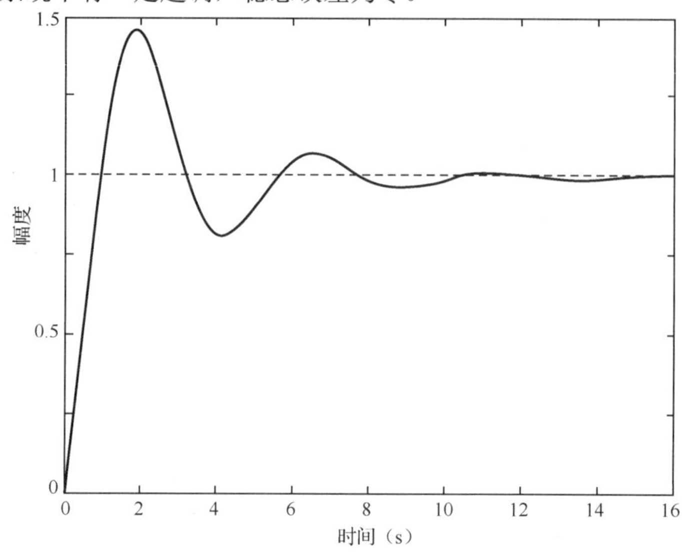

图 8.3.8 单位阶跃输入的响应

8.4 PID 控制器

从 8.2 节 PI 和 8.3 节 PD 的控制来看，单独的 PI 控制使得系统消除了误差，单独的 PD 控制使得系统稳定性能增加，如果综合二者，采用 PID 控制，系统又如何呢？

给定 PID 控制器为

$$G_c(s) = K_P + K_D s + \frac{K_I}{s} = \frac{K_D s^2 + K_P s + K_I}{s} = K_D\left(\frac{s^2 + \frac{K_P}{K_D}s + \frac{K_I}{K_D}}{s}\right) \tag{8.25}$$

上式中积分项的作用是提高了稳态响应。

我们看到，PID 控制器在 $s = 0$ 处有一个极点，在下列情况下有两个零点。

$$s = -\frac{K_P}{2K_D} \pm \frac{1}{2}\sqrt{\left(\frac{K_P}{K_D}\right)^2 - 4\frac{K_I}{K_D}} \tag{8.26}$$

考虑图 8.3.1 所示的系统，系统的闭环传递函数为

$$\frac{C(s)}{R(s)} = \frac{G_c G_f}{1 + G_c G_f} \tag{8.27}$$

$$\frac{C(s)}{R(s)} = \frac{K_D s^2 + K_P s + K_I}{s^3 + K_D s^2 + K_P s + K_I} \tag{8.28}$$

特征方程为

$$s^3 + K_D s^2 + K_P s + K_I = 0 \tag{8.29}$$

采用图 8.4.4 中的程序可实现闭环系统的零极点分布图绘制、系统的阶跃响应以及其闭环系统特征根求取。

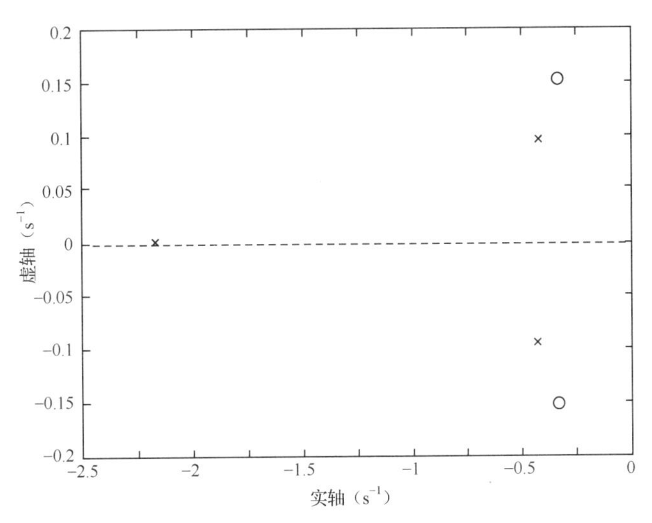

图 8.4.1　闭环系统的零极点分布

闭环系统的零极点分布如图 8.4.1 所示。

令 $K_P = 2$，$K_D = 3$，$K_I = 0.4$ 生成零极点图。系统共有三个极点，有两个共轭极点，另外一个极点在实轴上。实轴上的极点到虚轴的距离为 d_1，两个共轭极点到虚轴的距离为 d_2，d_1 是 d_2 的 5.1 倍，基本可以忽略实轴上的极点对系统的作用，这时的两个共轭极点就是主导极点。因此系统性能由两个共轭复根决定，也就是在 8.5 节将要讲述的主导极点内容。闭环系统的所有根都在 s 左半平面，系统是稳定的。

PID 控制系统的开环根轨迹如图 8.4.2 所示。可以看出，系统三条根轨迹都从原点出发，两根分别在两个共轭零点处终止，另一根沿着实轴的负轴终止于无穷远处。随着 K_D 从 0 变化到 ∞，根轨迹象预期的一样从原点进入到 s 左半平面。有两条根轨迹从原点出发，经过 s 右半平面与虚轴相交，此处为系统稳定的临界点。系统的两条根轨迹在此处进入 s 左半平面，系统便是稳定的。比值 K_P / K_D 和 K_I / K_D 的选择将会受性能指标的影响。

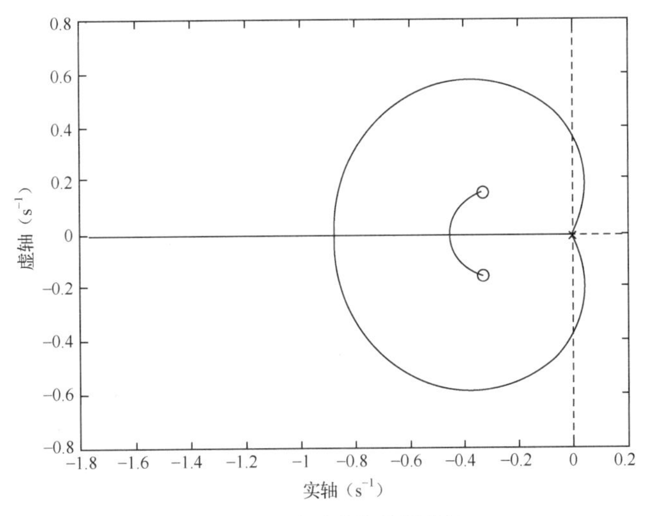

图 8.4.2　开环系统的根轨迹

系统的单位阶跃响应曲线如图 8.4.3 所示。系统有小于 20%的超调，调节时间小于 8s，PID 控制效果不错，其具有稳定的输出、零稳态误差、轻微的超调和快速的响应。

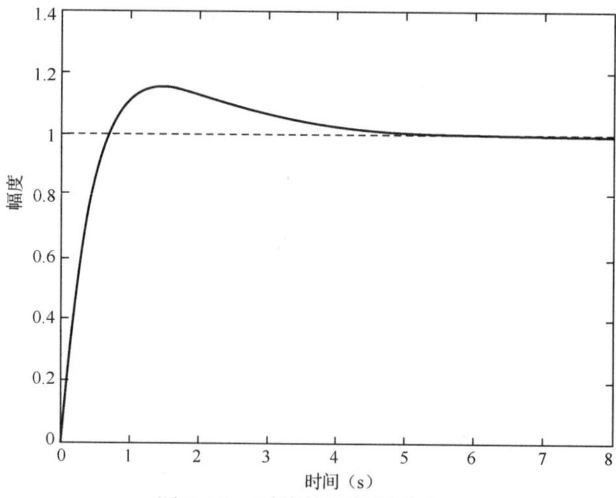

图 8.4.3 系统单位阶跃响应 1

```
%图 8.4.1、图 8.4.2 和图 8.4.3 的程序
num=[3 2 0.4];den=[1 0 0 0];        %开环传递函数 GcGf
[numc,denc]=cloop(num,den);         %闭环传函
figure(1),pzmap(numc,denc);         %闭环零极点图
figure(2),rlocus(num,den);          %开环根轨迹
figure(3),step(numc,denc);          %系统阶跃响应
roots(denc)                         %闭环系统的根
```

```
% MATLAB 的 Command Window 输出

ans =

  -2.1597 + 0.0000i
  -0.4201 + 0.0932i
  -0.4201 - 0.0932i
```

图 8.4.4 MATLAB 程序及其结果

从图 8.4.4 可看出，系统的闭环特征方程的根均在 s 左半平面，系统是稳定的。

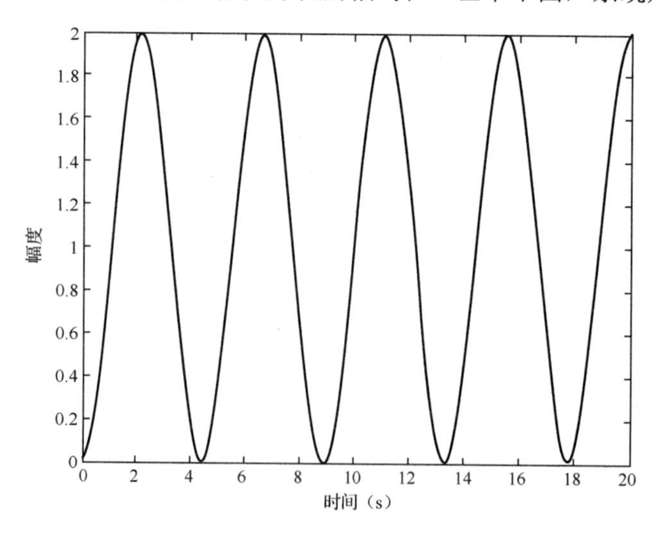

图 8.4.5 系统单位阶跃响应 2

```
% MATLAB 的 Command Window 输出

ans =

    0.0000 + 1.4142i
    0.0000 - 1.4142i
   -0.2000 + 0.0000i
```

图 8.4.6　命令窗口输出

图 8.4.5 是在取 $K_{\mathrm{P}}=2$，$K_{\mathrm{D}}=0.2$，$K_{\mathrm{I}}=0.4$ 时，系统的阶跃响应。系统是临界状态，输出是等幅振荡的曲线，系统受到干扰就会不稳定。从图 8.4.5 可以看出，由于系统的两个根在虚轴上，因此系统处于临界稳定状态。

8.5　闭环主导极点

8.5.1　闭环主导极点概念

闭环极点的相对主导作用，取决于闭环极点的实部的比值，同时也取决于在闭环极点上求得的留数的相对大小。而留数的大小既取决于闭环极点，又取决于闭环零点。

如果实部的比值超过 5，并且在极点附近不存在零点，那么距虚轴最近的闭环极点，将对瞬态响应特性起主导作用，因为这些极点对应于瞬态响应中衰减最慢的项。这些对瞬态响应特性具有主导作用的闭环极点，叫做闭环主导极点。闭环主导极点经常以共轭复数的形式出现。在所有闭环极点中，闭环主导极点是最重要的。

应当指出，常常对高阶系统的增益进行调整，以便能使系统具有一对闭环主导共扼复数极点。稳定系统中这样一对主导极点的存在，将会减小一些非线性因素，如死区、间隙和库仑摩擦，对系统性能的影响。

例 8.1　已知系统的开环传递函数为

$$\phi(s)=\frac{1}{(s+1)(0.01s^2+0.08s+1)} \tag{8.30}$$

试估算系统的性能指标。

解：闭环有三个极点，$s_1=-1$，$s_{2,3}=-4\pm \mathrm{j}9.2$，其中 $s_{2,3}$ 可以忽略不计，s_1 为主导极点。闭环传递函数简化为

$$\phi(s)=\frac{1}{s+1} \tag{8.31}$$

系统的调节时间为 $t_s=3T=3s$ 。

8.5.2　闭环主导极点的设定

控制系统阶跃响应下的性能指标通常用调节时间和超调量来表示。它与闭环的零极点有关，但主要由闭环极点决定。因此, 若要求系统快速性好（即 t_s 较小），则闭环极点应该远离虚轴；若要求平稳性好（即 $\sigma\%$ 小），则闭环极点设置在 s 平面与负实轴成 ±45° 夹角附近。

系统的动态性能与其闭环主导极点之间存在对应关系。一旦根据某种要求提出系统的动

态性能，便可立即在 s 平面上找出所要求的闭环主导极点的位置。在时域内，可以通过系统输出及微分构成的反馈系统，实现系统的极点与主导极点的位置相一致。通过输出反馈，一个或一对闭环极点将处于所要求的主导极点位置，其余各极点则处在其他位置。这样，实现闭环极点在 s 平面上重新分布的目的，从而设定希望的闭环主导极点位置。

例 8.2 已知系统如图 8.5.1 所示，被控的传递函数为

$$G(s)=\frac{1}{s^3+3.5s^2+3.5s+1}=\frac{1}{(s+0.5)(s+1)(s+2)} \tag{8.32}$$

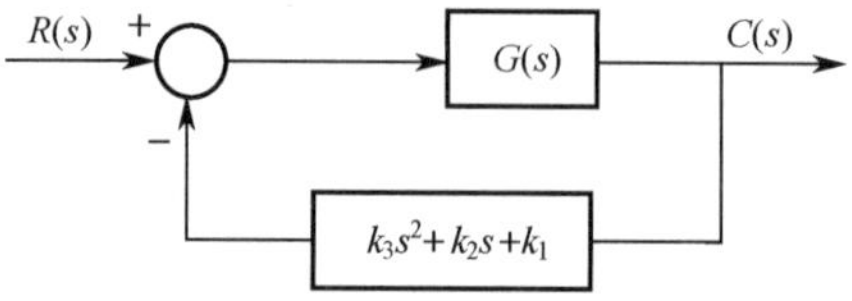

图 8.5.1 输出反馈系统结构图

根据需求，希望闭环性能指标为 $\sigma\%=4.3\%$，$t_s=4s$。试通过输出反馈控制，使其闭环极点位于希望的主导极点位置上。

解：

（1）希望主导极点的计算

由 $\sigma\%=\mathrm{e}^{-\pi\xi/\sqrt{1-\xi^2}}\times100\%$ 和 $t_s=\dfrac{3.5}{\zeta\omega_n}$ 可得两个极点为 $s_{1,2}\approx-0.7\pm0.7\mathrm{j}$，令第三个极点为 $s_3=7\,\mathrm{Re}[s_1]=-4.9$，其中，$s_1$，$s_2$ 为希望的闭环主导极点，s_3 为位于实轴并远离虚轴的极点。

（2）希望的闭环特征式 $D(s)$ 的计算

$$\widehat{D}(s)=(s-s_1)(s-s_2)(s-s_3) \tag{8.33}$$

$$\widehat{D}(s)=s^3+6.3s^2+7.84s+4.8 \tag{8.34}$$

（3）写出输出反馈及其微分反馈的系统闭环特征式

如图 8.5.1 所示，系统的闭环传递函数为

$$\frac{C(s)}{R(s)}=\frac{1}{s^3+(3.5+k_3)s^2+(3.5+k_2)s+(k_1+1)} \tag{8.35}$$

闭环特征式为

$$D(s)=s^3+(3.5+k_3)s^2+(3.5+k_2)s+(k_1+1) \tag{8.36}$$

（4）确定反馈系数 k_1，k_2 和 k_3

令 $\widehat{D}(s)=D(s)$，对应系数相等，得到以下等式

$$k_1+1=4.8\rightarrow k_1=3.8 \tag{8.37}$$

$$3.5+k_2=7.84\rightarrow k_2=4.34 \tag{8.38}$$

$$3.5+k_3=6.3\rightarrow k_3=2.8 \tag{8.39}$$

从图 8.5.2 中看出，系统三个极点的分布情况。一对共轭复根距离虚轴很近，是主导极点，另一个极点在负半实轴上，离虚轴很远，二者距离比是 7 倍。从图 8.5.3 中看出，系统阶跃响应有较小的超调、快速的响应。从图 8.5.4 可知，系统超调大小为 4.2%，调节时间为 5.8s，比较接近希望的闭环性能指标。

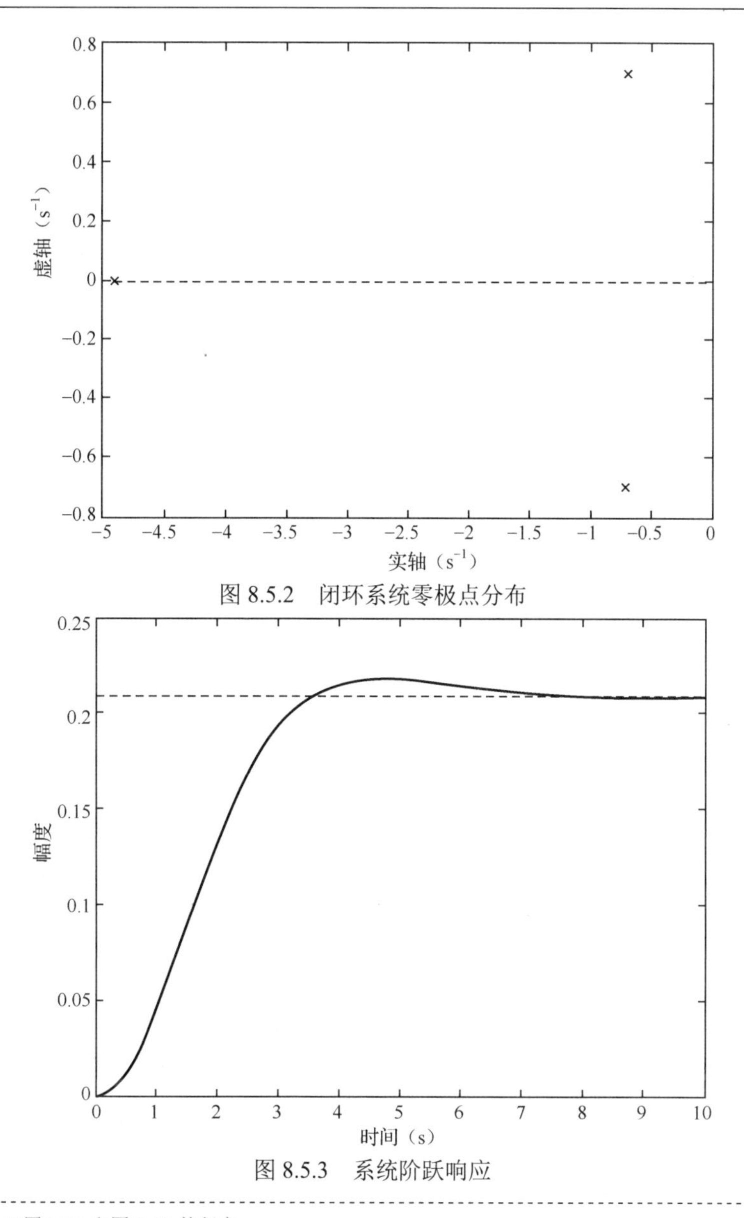

图 8.5.2　闭环系统零极点分布

图 8.5.3　系统阶跃响应

```
%图 8.5.2 和图 8.5.3 的程序
num=[1];den=[1 6.3 7.84 4.8];   %闭环传递函数
t=1:0.1:10;
figure(1),pzmap(num,den)  %闭环传递函数的零极点图
figure(2),step(num,den)  %闭环传递函数的阶跃响应
y=step(num,den);
finalvalue=dcgain(num,den);
[yss,n]=max(y);
[PO,TS]=poandts(t,y,yss,finalvalue);   %调用函数 poandts 计算超调和调节时间
PO
TS
```

图 8.5.4　MATLAB 程序及其结果

```
%上面程序中调用的 function 文件
function[PO,TS]=poandts(t,y,yss,finalvalue)
PO=100*(yss-finalvalue)/finalvalue;   %计算超调
k=length(t);
while (y(k)>0.95*finalvalue)&(y(k)<1.05*finalvalue)%计算调节时间用的是±0.05 误差带
k=k-1;
end
TS=t(k);
```

```
% MATLAB 的 Command Window 输出
PO =

    4.2085


TS =

    5.8000
```

图 8.5.4 MATLAB 程序及其结果（续）

8.6 下棋机器人

这是一个由我国科学家研发的一种单臂下棋机器人，如图 8.6.1 所示，它正在表演下围棋。

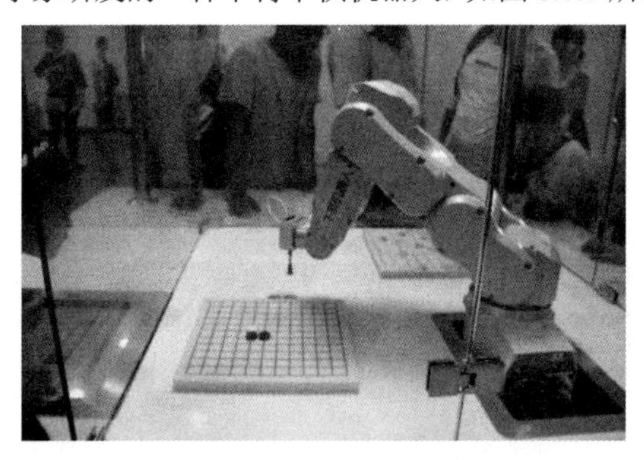

图 8.6.1 单臂下棋机器人

本节只考虑机器人单个棋子控制系统的设计。已知机器人一个棋子的执行器的数学模型，其传递函数为

$$G_f(s)=\frac{1}{s(s^2+10s+37)} \tag{8.40}$$

输入是对执行器发出的一个电压命令信号，输出是机器人下的棋子位置（仅垂直方向）。控制系统的方框图如图 8.6.2 所示。

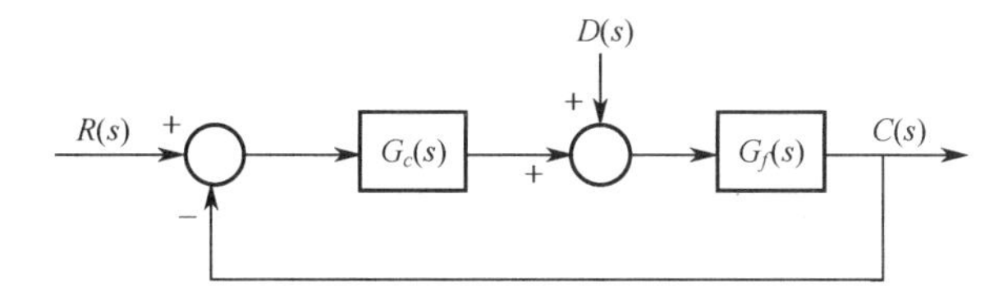

图 8.6.2　单臂下棋机器人控制系统

控制目标：控制棋子的位置以及在有噪声扰动时维持其原有位置。

控制变量：棋子的位置 $C(s)$。

我们想要棋子以最小的超调量以尽可能快的速度移动到指定位置。最初的设计目标是使机器人产生移动，即使移动很慢。换言之，控制系统的起始带宽将很低——约 2Hz，随后，根据需求增加带宽。

控制设计指标：

- 设计指标 1：闭环带宽大于 2Hz。
- 设计指标 2：阶跃输入的超调量小于 15%。
- 设计指标 3：阶跃输入的稳态误差为 0。

设计指标 1 和设计指标 2 的目的是确保可接受的跟踪性能。设计指标 3 事实上在设计中是没问题的：执行器的传递函数是一个 I 型系统，因此阶跃输入时系统稳态误差保证是 0。仅需确定 $G_c(s)G(s)$ 至少仍是 I 型系统。

8.6.1　控制器选择

考虑 PID 控制器

$$G_c(s)=K_{\mathrm{P}}+sK_{\mathrm{D}}+\frac{K_I}{s}=\frac{K_{\mathrm{D}}s^2+K_{\mathrm{P}}s+K_{\mathrm{I}}}{s} \tag{8.41}$$

为了不增加系统积分环节，确定 $G_c(s)G(s)$仍是 I 型系统，把 PID 控制器调整为

$$G_c(s)=\frac{K_{\mathrm{D}}s^2+K_{\mathrm{P}}s+K_{\mathrm{I}}}{s+p} \tag{8.42}$$

上式 PID 控制器中，关键待调参数为 K_{P}， K_{D}， K_{I}， p。

8.6.2　控制器设计

闭环控制系统的响应主要是由主导极点的位置决定的。设计方法是为闭环系统的主导极点设定适当的位置。根据性能指标，利用二阶系统的近似公式来确定这个位置。得到控制器参数，这样闭环系统就有了期望的主导极点，而其他极点的位置对系统响应的作用是可以忽略的。

当闭环频率响应的幅值下降到零值以下 3dB 时，对应的频率 ω_b 称为截止频率。因此

$$\left|\frac{C(\mathrm{j}w)}{R(\mathrm{j}w)}\right|<\left|\frac{C(\mathrm{j}0)}{R(\mathrm{j}0)}\right|-3\mathrm{dB},\omega>\omega_b \tag{8.43}$$

对于 $|C(\mathrm{j}0)/R(\mathrm{j}0)|=0\mathrm{dB}$ 的系统，

$$\left|\frac{C(\mathrm{j}w)}{R(\mathrm{j}w)}\right| < -3\mathrm{dB}, \omega > \omega_b \tag{8.44}$$

闭环系统过滤掉大于截止频率的信号分量，但可以使频率低于截止频率的信号分量通过。

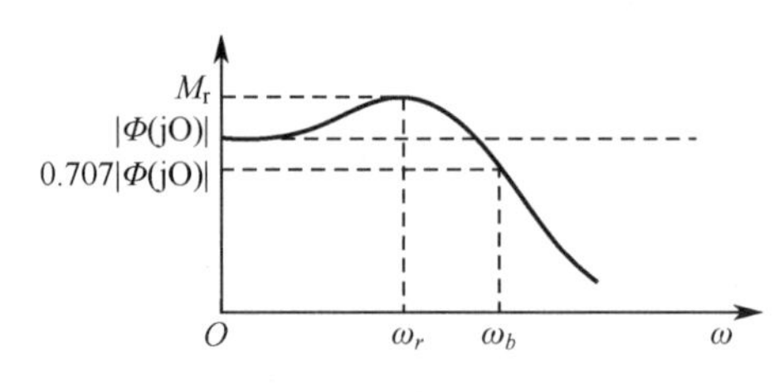

图 8.6.3 系统的带宽

闭环系统的幅值不低于-3dB 时，对应的频率范围 $0 \ll \omega \ll \omega_b$ 称为系统的带宽。带宽表示增益开始从其低频值下降的频率范围。因此，带宽表明系统跟踪正弦输入信号的能力。对于给定的 ω_n，上升时间随着阻尼比 ξ 的增加而延长。另一方面，带宽随着 ξ 的增加而减小。因此，上升时间与带宽之间成反比关系。

带宽的指标取决于下列因素：

（1）对输入信号的再现能力

大的带宽对应小的上升时间，即对应快速响应特性。粗略地说，带宽与响应速度成正比。

（2）对高频噪声必要的滤波特性

为了使系统能够精确地跟踪任意输入信号，系统必须具有大的带宽。但是，从噪声的观点来看，带宽不应当太大。因此，对带宽的要求是矛盾的，好的设计通常需要折中考虑。注意，具有大带宽的系统需要有高性能的元件，因此元件的成本通常随着带宽的增加而增大。

1. 期望二阶系统的主要参量计算

我们期望闭环控制系统的性能是由一对主导极点决定，那么这对主导极点的位置就应该由期望的系统性能指标 1 到指标 3 来确定。期望的二阶系统主要参量有自然振荡频率和阻尼比，如果这两个参量均由设计指标来决定，那么设计的系统就完全满足了性能指标要求，则设计的系统就是被期望系统。

带宽 ω_B 与自然频率 ω_n 的关系近似为

$$\frac{\omega_B}{\omega_n} \approx -1.1961\xi + 1.8508 \ (0.3 \leqslant \xi \leqslant 0.8) \tag{8.45}$$

这个近似关系适用于二阶系统。

按照设计指标 1，我们想要

$$\omega_B = 2\mathrm{Hz} = 12.57\mathrm{rad/s} \tag{8.46}$$

根据超调量指标，可以确定 ξ 的最小值。因为 $P.O. \leqslant 15\%$，根据公式（对二阶系统有效）

$$\sigma\% = \mathrm{e}^{-\pi\xi/\sqrt{1-\xi^2}} \times 100\% \tag{8.47}$$

可以得到

$$\xi = \frac{\ln\dfrac{100}{\sigma}}{\left[\pi^2 + \left(\ln\dfrac{100}{\sigma}\right)^2\right]^{\frac{1}{2}}} \tag{8.48}$$

$$\xi \geqslant 0.52 \tag{8.49}$$

我们将用 $\xi \geqslant 0.52$ 来设计。在边界上选择 $\xi = 0.52$，因为通常 ξ 选择的越大需要的调节时间越长。尽管调节时间不是系统的设计指标，但是，通常在满足所有的设计指标后尝试使系统的响应尽可能快。根据式（8.44）和式（8.45），可以得到

$$\omega_n = \frac{\omega_B}{-1.1961\xi + 1.8508} = 10.19\text{rad / s} \tag{8.50}$$

期望的 2%误差的调节时间约为

$$T_s \approx \frac{4}{\xi\omega_n} = 0.76\,\text{sec} \tag{8.51}$$

分别计算出来的 $\omega_n = 10.19\text{rad / sec}$ 和 $\xi \geqslant 0.52$，两个参量决定了一个二阶系统。

闭环频率特性也可用振幅和相频表示

$$\Phi(\text{j}\omega) = M(\omega)\text{e}^{\text{j}\alpha(\omega)} \tag{8.52}$$

式中，$M(\omega)$ 和 $\alpha(\omega)$ 分别为闭环系统的幅频和相频特性。在系统的阻尼比小于等于 0.707 时，幅频特性会出现谐振峰值 M_r，且这个谐振峰值和阻尼比存在如下关系

$$M_r = \frac{1}{2\xi\sqrt{1-\xi^2}} \qquad (\xi < 0.707) \tag{8.53}$$

谐振频率 ω_r、自然频率 ω_n 和阻尼比 ξ 之间的关系为

$$\omega_r = \omega_n\sqrt{1-\xi^2} \qquad (\xi < 0.707) \tag{8.54}$$

设计系统时，同样可以用上述频域两个参量为二阶系统的设计指标，代入阻尼比 ξ 和 ω_n 计算出来这两个参量值

$$M_r = 1.1299 \qquad \omega_r = 8.7231 \tag{8.55}$$

2. 期望的闭环特征多项式

上节确定了期望的闭环系统，是由主导极点确定的二阶系统。但是被控对象 G_f 是三阶系统，再加上 PID 控制器 G_C 分母是一阶的，那么闭环系统事实就是四阶系统，而不是期望的二阶系统。因此，把实际的四阶系统设计成由两对共轭复根组成，一对就是主导极点，另一对就是非主导极点。期望的四阶闭环特征多项式为

$$\widehat{D}(s) = (\text{s}^2 + 2\xi\omega_n s + \omega_n^2)(s^2 + m_1 s + m_0) \tag{8.56}$$

式中，ξ 和 ω_n 的选择满足设计指标。$s^2 + 2\xi\omega_n s + \omega_n^2 = 0$ 的根是主导极点。相反，我们希望 $s^2 + m_1 s + m_0 = 0$ 的根是非主导极点。主导极点应该位于复平面的垂直线上，与虚轴的距离为 $9\,\omega_n$。令

$$m_1 = 2\alpha\xi\omega_n \tag{8.57}$$

那么 $s^2 + m_1 s + m_0 = 0$ 的根位于复平面上 $s = -\alpha\xi\omega_n$ 的垂直线上。通过选择 $\alpha > 1$，有效地将根移向主导极点的左边。选择的 α 越大，非主导极点位于主导极点左边的距离越远。在本系统中的 α 合理值是 $\alpha = 8$。

可以利用 MATLAB 很容易地尝试不同的 d 值。另外，如果选择

$$m_0 = \alpha^2\xi^2\omega_n^2$$

那么得到两个实根

$$s^2 + m_1 s + m_0 = (s + \alpha\xi\omega_n)^2 = 0 \tag{8.58}$$

尽管选择 $m_0 = \alpha^2\xi^2\omega_n^2$，不是一个必须的选择，但是，是一个合理的选择，因为我们希望非主导极点对系统响应的作用快速且无振荡。

于是，期望的特征多项式是

$$\widehat{D}(s) = (s^2 + 2\xi\omega_n s + \omega_n^2)(s^2 + 2\alpha\xi\omega_n s + \alpha^2\xi^2\omega_n^2) \tag{8.59}$$

则展开为

$$s^4 + [2\xi\omega_n(1+\alpha)]s^3 + [\omega_n^2(1+\alpha\xi^2(\alpha+4)]s^2 + [2\alpha\xi\omega_n^3(1+\xi^2\alpha)]s + \alpha^2\xi^2\omega_n^4 = 0 \tag{8.60}$$

3. PID 控制系统的特征多项式

如图 8.6.2 系统，系统的闭环传递函数为

$$\frac{C(s)}{R(s)} = \frac{G_c G_f}{1+G_c G_f} = \frac{\dfrac{K_D s^2 + K_P s + K_I}{s+p} \times \dfrac{1}{s(s^2+10s+37)}}{1 + \dfrac{K_D s^2 + K_P s + K_I}{s+p} \times \dfrac{1}{s(s^2+10s+37)}} \tag{8.61}$$

$$\frac{C(s)}{R(s)} = \frac{K_D s^2 + K_P s + K_I}{s^4 + (10+p)s^3 + (37+10p+K_D)s^2 + (37p+K_P)s + K_I} \tag{8.62}$$

PID 控制系统的特征多项式为

$$D(s) = s^4 + (10+p)s^3 + (37+10p+K_D)s^2 + (37p+K_P)s + K_I \tag{8.63}$$

4. 确定 PID 控制器系数

令 $\widehat{D}(s) = D(s)$，式（8.60）和式（8.63）对应系数相等，得到以下等式

$$\begin{aligned} &10 + p = 2\xi\omega_n(1+\alpha) \\ &37 + 10p + K_D = \omega_n^2(1+\alpha\xi^2(\alpha+4) \\ &37p + K_P = 2\alpha\xi\omega_n^3(1+\xi^2\alpha) \\ &K_I = \alpha^2\xi^2\omega_n^4 \end{aligned} \tag{8.64}$$

在前面步骤中计算出来的阻尼比 $\xi = 0.52$ 和自然振荡频率 $\omega_n = 10.19\text{rad/s}$，$\alpha = 8$ 分别代入式（8.64）中，得到 PID 控制器的参量为

$$\begin{aligned} &p = 84.82 \\ &K_D = 1882.5 \\ &K_P = 24\,323 \\ &K_I = 184\,410 \end{aligned} \tag{8.65}$$

则 PID 控制器为

$$G_c(s) = \frac{K_D s^2 + K_P s + K_I}{s+p} = 1882.5\frac{s^2 + 12.92s + 97.96}{s + 84.82} \tag{8.66}$$

8.6.3 系统仿真

PID 控制的闭环系统零极点分布如图 8.6.5 所示。

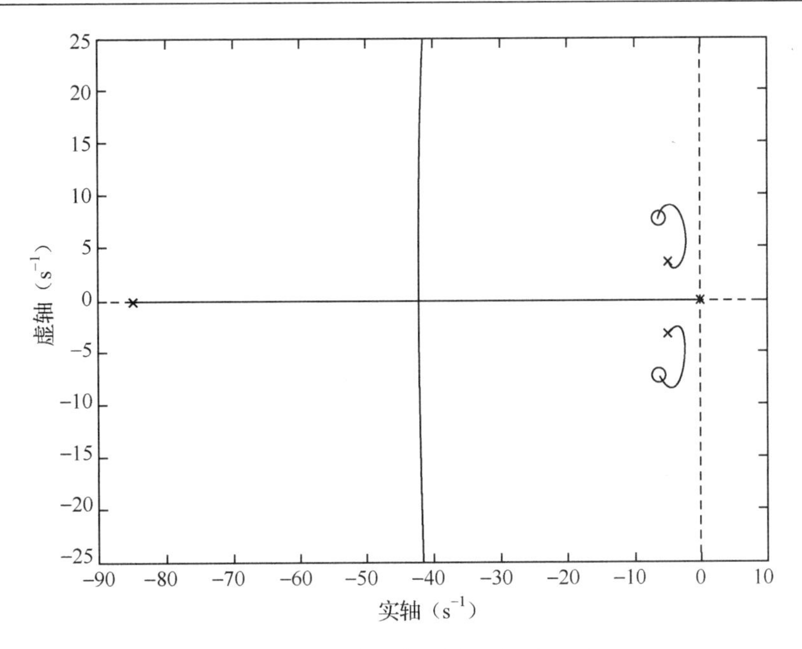

图 8.6.4　PID 控制的开环系统根轨迹图

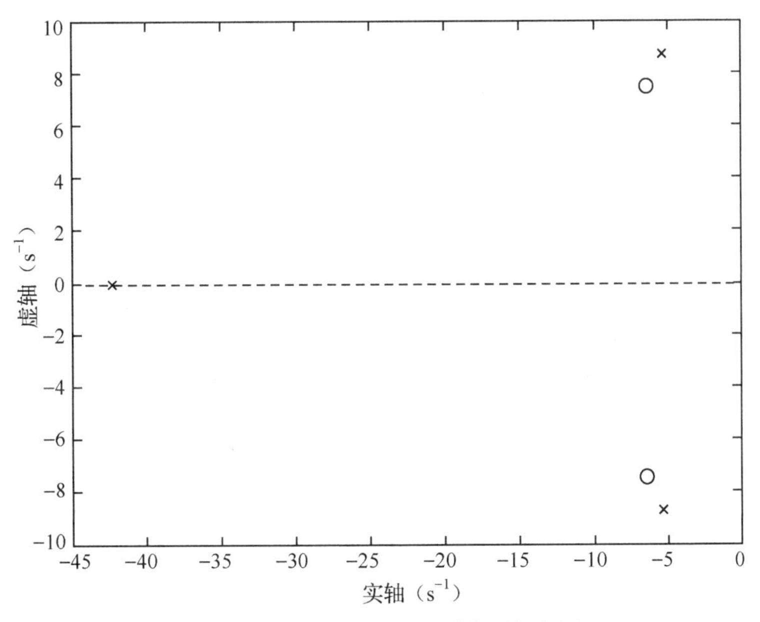

图 8.6.5　PID 控制的闭环系统零极点图

从图 8.6.4 可看出，开环系统的 4 条根轨迹都在 s 左半平面，系统是稳定的。从图 8.6.5 可看出，闭环系统的 4 个极点都在 s 左半平面。一对主导极点，离虚轴很近，距离虚轴距离为 5.2677。另一对非主导极点为重根，在负半轴上，距离虚轴距离为 42.1412，它们距离虚轴的距离比为 8。因此系统的性能主要由离虚轴近的主导极点决定。

从图 8.6.6 可看出，系统的阶跃响应的超调在 20%之内，系统的调节时间不到 1s。

从图 8.6.7 结果可知，系统阶跃输出的超调为 15%，调节时间为 0.57s。系统设计指标：超调小于 15%，期望的调节时间是 0.76s，与期望值相比，都满足设计需求。

时域指标都满足了，下面我们验证频域指标。

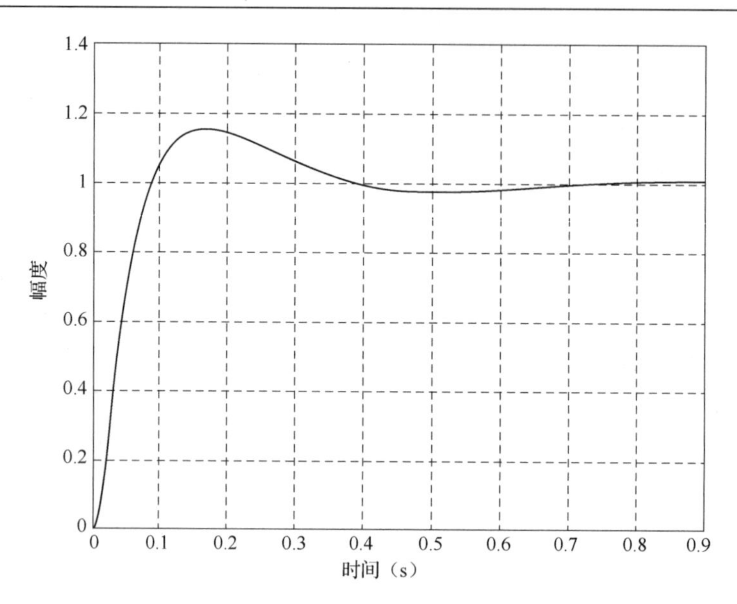

图 8.6.6　PID 控制的系统阶跃响应

```
%图 8.6.4、图 8.6.5 和图 8.6.6 的程序
clc
clear
close all
POI=15;          %输入系统超调量指标 2
wb=2*6.28;       %输入系统闭环带宽指标 1
af=8;            %主导极点与非主导极点距离虚轴的距离比
a=10;b=37;
zeta=log(100/POI)/sqrt(pi^2+(log(100/POI))^2)     %根据指标 2 计算系统阻尼比
wn=wb/(-1.196*zeta+1.8508)                  %根据指标 1 计算系统的自然振动频率
TsI=4/(wn*zeta)                             %计算期望的调节时间
s3=2*zeta*wn*(1+af);                        %公式（8.63）中 s^3 的系数
s2=wn^2*[1+af*zeta^2*(af+4)];               %公式（8.63）中 s^2 的系数
s1=2*af*zeta*wn^3*(1+zeta^2*af);            %公式（8.63）中 s^1 的系数
s0=af^2*zeta^2*wn^4;                        %公式（8.63）中的常数项
nf=[1];df=conv([1,0],[1,a,b]);              %公式（8.40）中数学模型，被控对象模型
p=s3-a;
KP=s1-b*p;
KD=s2-a*p-b;
KI=s0;
nc=[KD,KP,KI];dc=[0,1,p];                   %控制器 Gc
[num,den]=series(nc,dc,nf,df);
[numc,denc]=cloop(num,den);                 %PID 控制器闭环模型
figure(1),rlocus(num,den);
figure(2),pzmap(numc,denc);
figure(3),step(numc,denc);
grid
[y,x,t]=step(numc,denc);
```

图 8.6.7　MATLAB 程序及其结果

```
PO=100*(max(y)-1)                    %计算阶跃响应的超调量
T=find(abs(y-1)>0.02);               %计算阶跃响应的调节时间
Ts=t(T(length(T)))
```

```
% MATLAB 的 Command Window 输出
zeta =

    0.5169
wn =

   10.1903

TsI =

    0.7594


PO =

   15.0127


Ts =

    0.5726


ans =

              1882 s^2 + 2.432e04 s + 1.844e05
  ---------------------------------------------------
  s^4 + 94.82 s^3 + 2768 s^2 + 2.746e04 s + 1.844e05

Continuous-time transfer function.


ans =

  -42.1412 + 0.0000i
  -42.1412 + 0.0000i
  -5.2677 + 8.7231i
  -5.2677 - 8.7231i
```

图 8.6.7 MATLAB 程序及其结果（续）

闭环系统的幅频特性如图 8.6.8 所示。带宽是 $\omega_b = 32.47\text{rad}/\text{s} = 5.64\text{Hz}$，这满足设计指标 1，但是大于设计时的 $\omega_b = 2\text{Hz}$。因此设计结果的带宽比期望值越高，则调节时间比期望值越小。事实上，调节时间是 $T_s = 0.57\text{s}$，而我们期望的 $T_s = 0.76\text{s}$。

峰值是 $M_r = 1.2301$，我们期望的 $M_r = 1.1299$。设计结果的峰值略高，基本也是满足需求的。程序和结果见图 8.6.9。

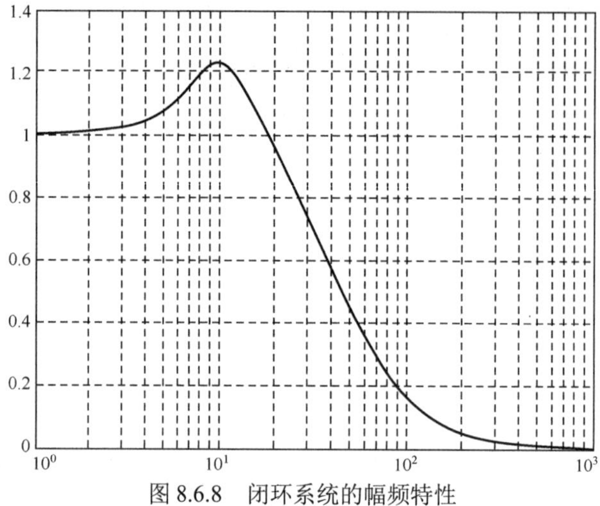

图 8.6.8 闭环系统的幅频特性

```
%图 8.6.8 程序
clc
clear
close all
POI=15;  %输入系统超调量指标 2
wb=2*6.28;  %输入系统闭环带宽指标 1
af=8;  %主导极点与非主导极点距离虚轴的距离比
a=10;b=37;
zeta=log(100/POI)/sqrt(pi^2+(log(100/POI))^2);  %根据指标 2 计算系统阻尼比
wn=wb/(-1.196*zeta+1.8508);  %根据指标 1 计算系统的自然振动频率
TsI=4/(wn*zeta);  %计算期望的调节时间
s3=2*zeta*wn*(1+af);  %公式（8.63）中 s^3 的系数
s2=wn^2*[1+af*zeta^2*(af+4)];  %公式（8.63）中 s^2 的系数
s1=2*af*zeta*wn^3*(1+zeta^2*af);  %公式（8.63）中 s^1 的系数
s0=af^2*zeta^2*wn^4;  %公式（8.63）中常数项
nf=[1];df=conv([1,0],[1,a,b]);  %公式（8.40）中数学模型，被控对象模型
p=s3-a;
KP=s1-b*p;
KD=s2-a*p-b;
KI=s0;
nc=[KD,KP,KI];dc=[0,1,p];  %控制器 Gc
[num,den]=series(nc,dc,nf,df);
[numc,denc]=cloop(num,den);  %PID 控制器闭环模型
[y,x,t]=step(numc,denc);
wrI=wn*sqrt(1-zeta^2)
MrI=1/[2*zeta*sqrt(1-zeta^2)]
w=logspace(-1,2,400);
[mag,phase,w]=bode(numc,denc);

semilogx(w,mag);grid
[y,L]=max(mag);
wb_list=find(mag<0.707);
```

图 8.6.9 MATLAB 程序及其结果

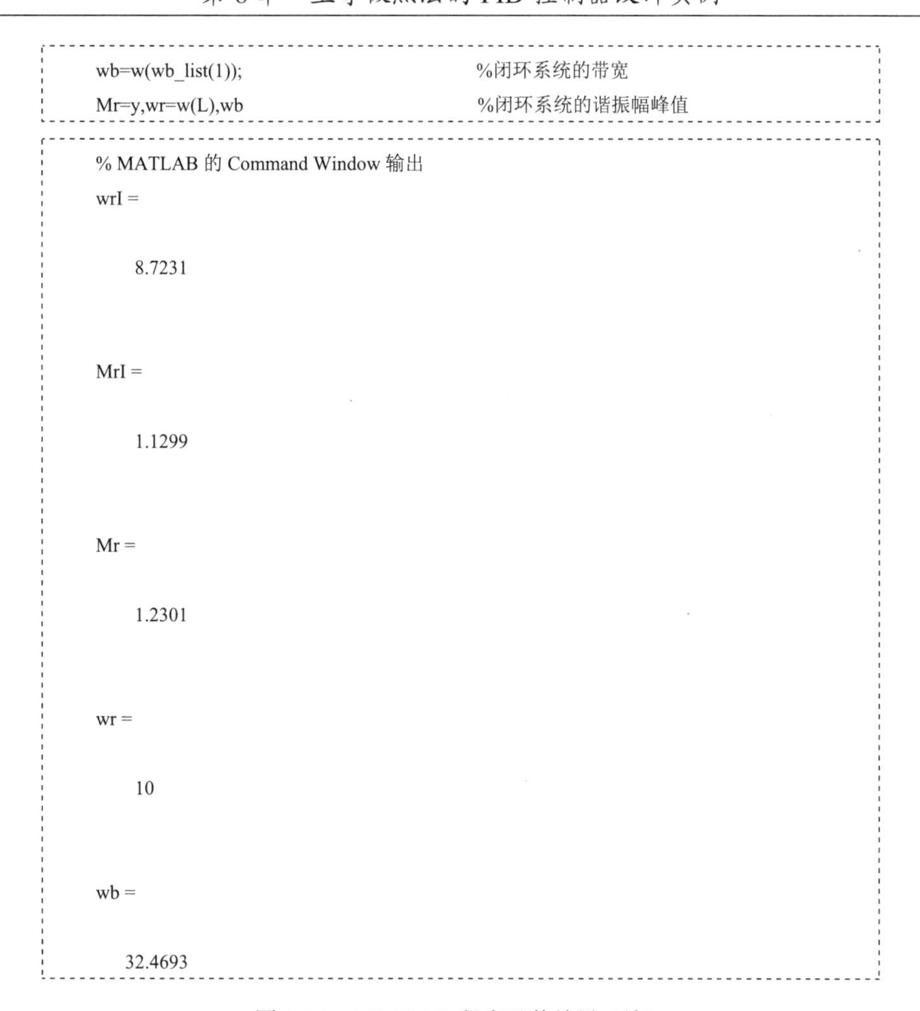

```
wb=w(wb_list(1));                      %闭环系统的带宽
Mr=y,wr=w(L),wb                        %闭环系统的谐振幅峰值
```

```
% MATLAB 的 Command Window 输出
wrI =

    8.7231


MrI =

    1.1299


Mr =

    1.2301


wr =

    10


wb =

   32.4693
```

图 8.6.9　MATLAB 程序及其结果（续）

本章小结

（1）介绍了 PID 控制器概念。

（2）通过比例与积分控制实例的比较、比例与微分控制实例的比较，得到如下结论：比例控制是有差控制；积分控制是无差控制，但是会带来系统的不稳定；微分控制会改善系统的稳定性。

（3）通过 PID 控制实例说明，PID 控制器不仅改善了系统的静态性能，也改善了系统的动态性能。

（4）利用主导极点的概念，为下棋机器人设计了一个满足带宽和超调量指标的控制器。通过简单分析，得到控制器参数的初始设置值，而后证明控制器的设计几乎满足设计要求，但需做一些必要的微调以完全满足设计指标。

Chapter 8

Examples of PID Controller Design by Dominant Poles

Based on chapter 6 and 7, this chapter focuses on the design of system controllers. Firstly, the PID controller, proportional and integral control action, proportional and differential control action, and Proportional-Integral- Derivative control action are introduced. Then the methods and specific steps of configuring the parameters of PID controller are discussed in detail by means of dominant poles. Finally, by taking the chess robot as an example, the PID controller design of the robot is realized with the method of the dominant poles, making the PID controller meet the design requirements of the system.

8.1 Concept of PID Controller

PID controller is widely used in industrial control. With three factors, its transfer function is

$$G_c(s) = K_P + \frac{K_I}{s} + K_D s \tag{8.1}$$

Its output equation in time domain is

$$u(t) = K_P e(t) + K_I \int e(t)\mathrm{d}t + K_D \frac{\mathrm{d}e(t)}{\mathrm{d}t} \tag{8.2}$$

The three-factor controller is called a PID controller because it contains a proportional, an integral, and a derivative term represented by K_P, K_I and K_D respectively. The structure of a typical PID controller is shown in Figure 8.1.1.

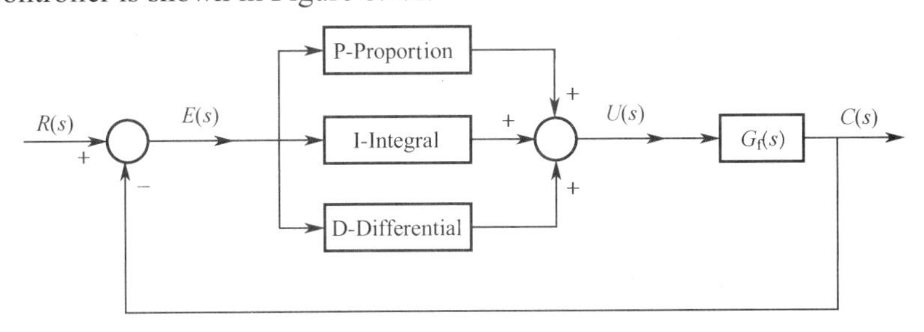

Figure 8.1.1 Structure of PID Controller

If we set $K_D = K_I = 0$, the PID controller is reduced to a proportional controller

$$G_c(s) = K_P \tag{8.3}$$

If we set $K_P = K_D = 0$, the PID controller is reduced to an integral controller

$$G_c(s) = \frac{K_I}{s} \tag{8.4}$$

The proportional controller and integral controller can be put together to form a Proportional-Integral(PI) controller,

$$G_c(s) = K_P + \frac{K_I}{s} = \frac{K_P s + K_I}{s} \tag{8.5}$$

when $K_P = K_I = 0$, we get a derivative controller

$$G_c(s) = K_D s \tag{8.6}$$

Generally we use a derivative controller in conjunction with the proportional and/or integral terms. The Proportional-Derivative (PD) controller is given by

$$G_c(s) = K_P + K_D s \tag{8.7}$$

8.2 Proportional and Integral Control Action

If there is not an integrator 1/s in the transfer function of a control object, then in the process of the proportional control, a steady-state error, or deviation, may occur in response to a step input. And this deviation will go up with the increase of the proportional coefficient. The characteristics of proportional control is deviation control.

Such deviation can be eliminated if the integral control action is included in the controller.

Therefore, integral control is featured by zero deviation control.

In the integral control, the control signal (the output signal of the controller) at any instant is equal to the area under the error signal curve prior to that instant. The physical implication of the integral control is that as long as the deviation is not zero, in the integral element, it will continuously accumulate and generate a continuous control action according to the negative feedback until the deviation is restored to zero, hence improving the steady-state performance of the system.

As shown in Figure 8.2.1, a sine wave is used to represent the deviation signal. Figure (a) shows the waveform after the deviation signal passes through an integrator. When the deviation signal is at the point A, the half cycle point on the sinusoid, its value is zero. The point B, on the integrator output curve corresponding to point A, is at the peak point, and the value is the largest. When the error signal crosses the zero point C again, since the area of the error signal and the horizontal axis of the time is offset positively and negatively, the value of the point D, corresponding to the point C on the integrated curve, is zero. Figure (b) shows that the amplitude changes after the error signal passes through a proportional element. When the point A on the error signal curve is zero, the point B, corresponding to the point A on the curve after passing the proportional element, will also be zero. After the error signal passes through the integrator, the curve through the integrator may or may not be zero where the error signal is zero, but the curve through the proportional element must be zero at this point.

Note that while removing deviation (steady-state error), integral control may also lead to oscillatory response of either decaying slowly or ever increasing in amplitude. Both are usually not what we want.

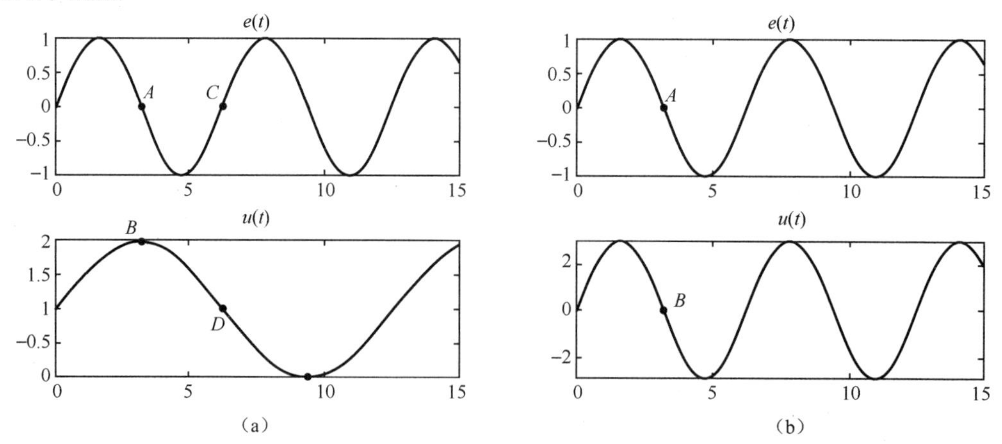

Figure 8.2.1 Proportional Control and Integral Control：（a） Plots of $e(t)$ and $u(t)$ curves showing nonzero control signal when the actuating error signal is zero (integral control); （b） plots of $e(t)$ and $u(t)$ curves showing zero control signal when the actuating error signal is zero (proportional control)

8.2.1 Error Under Proportional Control

We shall show that the proportional control of a system will result in a steady-state error for unit step inputs while without an integrator. We shall then show that such an error can be eliminated if integral control action is included in the controller.

Consider the system shown in Figure 8.2.2. Let us find the steady-state error of the system in the unit step response. Define the open-loop transfer function of the system as

$$G(s)=\frac{K}{Ts+1} \tag{8.8}$$

Since

$$\frac{E(s)}{R(s)}=\frac{R(s)-C(s)}{R(s)}=1-\frac{C(s)}{R(s)}=\frac{1}{1+G(s)} \tag{8.9}$$

the error $E(s)$ is given by

$$E(s)=\frac{1}{1+G(s)}R(s)=\frac{1}{1+\dfrac{K}{Ts+1}}R(s) \tag{8.10}$$

For the unit step input $R(s)=1/s$, we get

$$E(s)=\frac{Ts+1}{Ts+1+K}\frac{1}{s} \tag{8.11}$$

The steady-state error is

$$e_{ss}=\lim_{t\to\infty}e(t)=\lim_{s\to 0}sE(s)=\lim_{s\to 0}\frac{Ts+1}{Ts+1+K}=\frac{1}{K+1} \tag{8.12}$$

Such a system without an integrator in the forward path always has a steady-state error in the step response, which is called deviation.

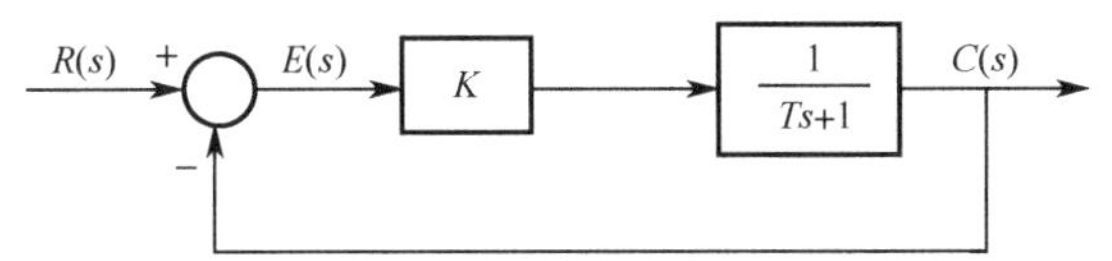

Figure 8.2.2 Proportional Control System

For the proportional control system, with the step input $R(s)=1$, $K=10$, $T=0.4$, the system's step output response is shown in Figure 8.2.3, and the program and results are shown in Figure 8.2.4.

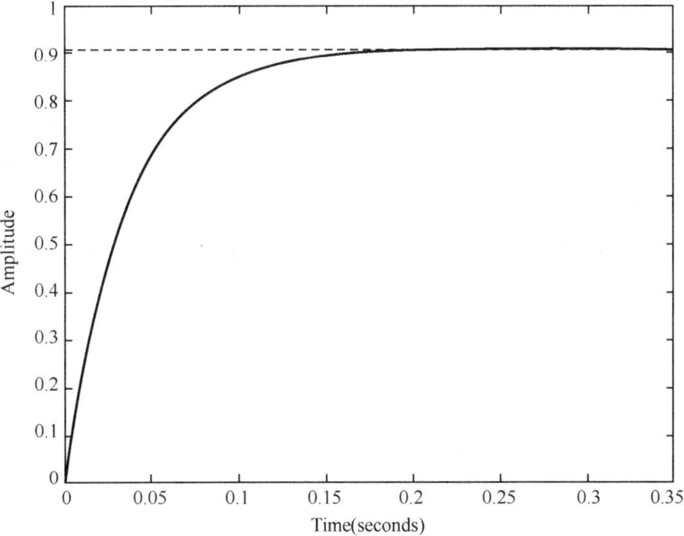

Figure 8.2.3 The Step Response

```
% The MATLAB script used for Figure 8.2.3
num=[10];
den=[0.4,1];
[numc,denc]=cloop (num,den);
step(numc,denc)
roots(den)
```

```
% Figure 8.2.3 Program results
ans =


    -2.5000
```

Figure 8.2.4 MATLAB Program and Its Results

As can be seen from Figure 8.2.3, the output of the system is erroneous. The output of steady-state is 0.9 and the steady-state error is about 0.1, that is, $1 / K + 1 = 1/11$ is about 0.1. On the physical axis, the root of closed-loop is −2.5, and the system damping ratio is 1. The system does not oscillate, and the adjustment time is about 0.2 seconds.

8.2.2 Eliminating Steady-State Error in Integral Control

Consider the system shown in Figure 8.2.2. The controller is an integral controller. The closed-loop transfer function of the system is

$$\frac{C(s)}{R(s)}=\frac{K}{s(Ts+1)+K} \tag{8.13}$$

Hence

$$\frac{E(s)}{R(s)}=\frac{R(s)-C(s)}{R(s)}=\frac{s(Ts+1)}{s(Ts+1)+K} \tag{8.14}$$

Since the system is stable, the steady-state error for the unit-step response can be obtained by applying the final-value theorem,

$$e_{ss}=\lim_{s\to 0} sE(s)=\lim_{s\to 0}\frac{s^2(Ts+1)}{Ts^2+s+Ks}\frac{1}{s}=0 \tag{8.15}$$

Integral control of the system thus eliminates the steady-state error in the response to the step input. This is an important improvement for the proportional control of deviation.

The integral control system is showed in Figure 8.2.5. Let the step input $R(s)=1$, $K=10$, $T=0.4$, and the step output of the system is shown in Figure 8.2.6. The implemented program is shown in Figure 8.2.7.

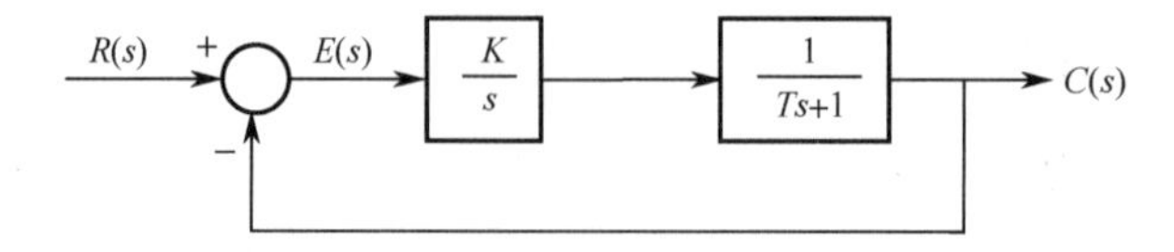

Figure 8.2.5 Integral Control System

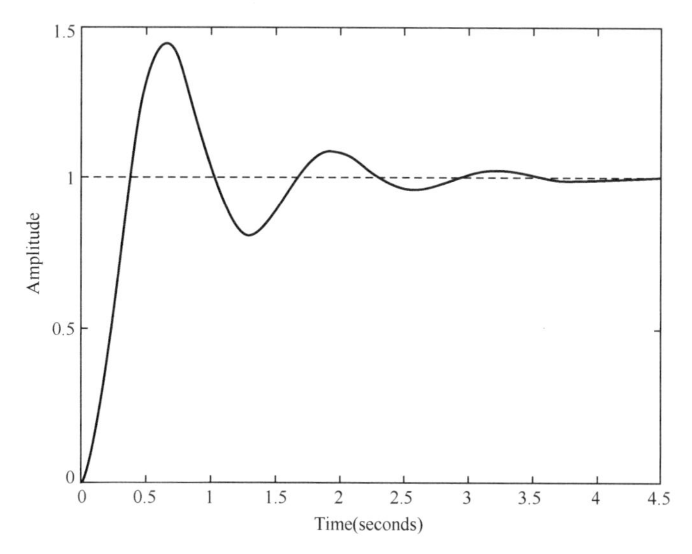

Figure 8.2.6 Step Response of Integral Control Systems

```
% The MATLAB script used for Figure 8.2.6
num=[10];
den=[0.4,1,0];
[numc,denc]=cloop(num,den);
step(numc,denc)
roots(denc)
```

```
%Figure 8.2.6 Program results
ans =

  -1.2500 + 4.8412i
  -1.2500 - 4.8412i
```

Figure 8.2.7 MATLAB Program and Its Results

As can be seen from Figure 8.2.6, the output of the system is error-free, the steady-state output is 1, and the steady-state error is zero. The roots of closed-loop system are a pair of conjugate complex root in the second and third quadrants. The damping ratio of the system is $0<\zeta<1$. There is overshoot and oscillation of the system, whose adjustment time is about 3.5 seconds.

8.3 Proportional and Integral Control Action

Derivative control action, when added to a proportional controller, provides a means of obtaining a controller with high sensitivity. An advantage of using derivative control action is that it responds to the rate of change of the error signals and can produce a significant correction before the magnitude of the error becomes too large. Derivative control thus anticipates the error, initiates an early corrective action, and tends to increase the stability of the system.

Although derivative control does not affect the steady-state error directly, it adds damping to the system and thus permits the use of a larger value of the gain K, which will result in an improvement in the steady-state accuracy.

Because derivative control operates on the rate of change of the error rather than the error itself, this mode can never be used alone. Instead, it is used in combination with proportional or proportional-plus-integral control action.

8.3.1 Instability of Proportional Control

Before we discuss further the effect of derivative control action on system performance, we shall consider the proportional control of an inertia load.

The model of the control object is as follows,

$$G_f(s) = \frac{1}{s^2} \tag{8.16}$$

This can be controlled by proportional feedback .

$$G_c(s) = K_P \tag{8.17}$$

The closed-loop feedback system with a controller, $G_c(s)$, is shown in Figure 8.3.1

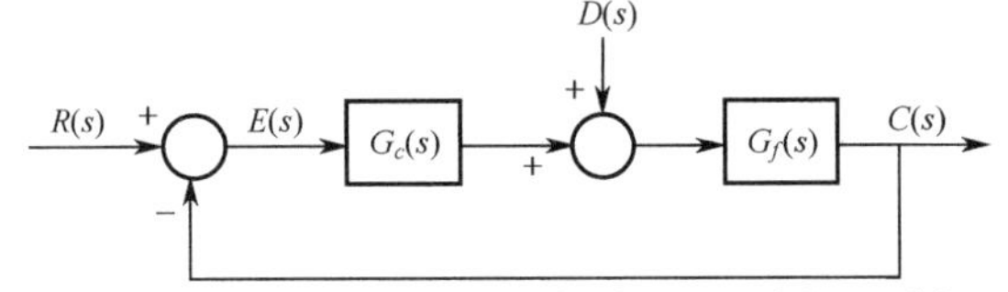

Figure 8.3.1 A Closed-loop Block Diagram of Control System

Consider the system shown in Figure 8.3.1. The closed-loop transfer function is obtained as

$$\frac{C(s)}{R(s)} = \frac{G_c G_f}{1+G_c G_f} = \frac{K_P}{s^2 + K_P} \tag{8.18}$$

Since the characteristic equation is

$$s^2 + K_P = 0 \tag{8.19}$$

With the program of Figure 8.3.4, the diagram of the zero-pole distribution of the closed-loop system (See Figure 8.3.2), the step response of the system, and the characteristic roots of the closed-loop system can be achieved.

The zero-pole distribution of the closed-loop system is shown in Figure 8.3.2. When $K_P = 2$, there are two poles at the imaginary axis. The system is in a critical state and unstable. If without active control, the system will overbalance under external disturbances.

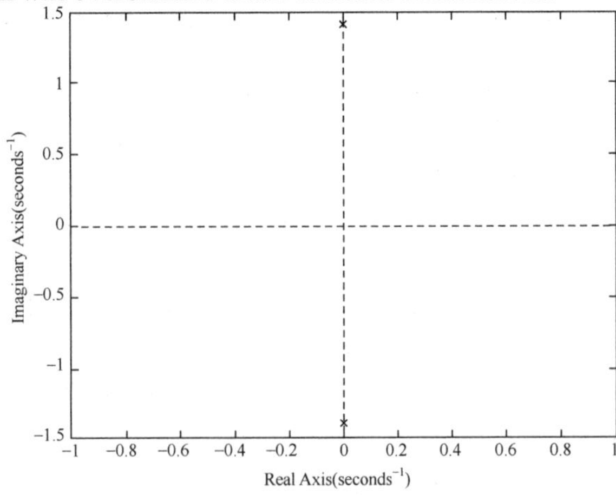

Figure 8.3.2 Zero-Pole Distribution of a Closed-Loop System

The system's response to a unit step input is an indefinitely continuous oscillation, as shown in Figure 8.3.3. Such response of the control system is not desirable. We shall see that the addition of derivative control will stabilize the system.

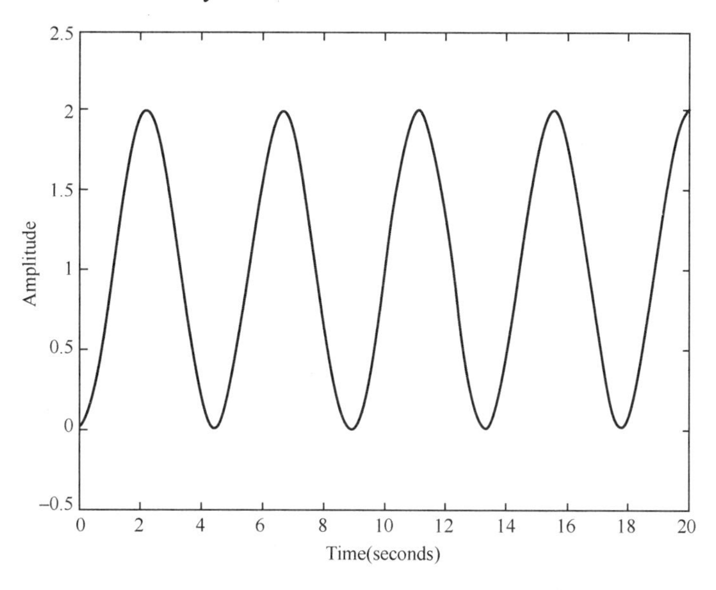

Figure 8.3.3 System Response to Unit Step Input

```
% The MATLAB script used for Figure 8.3.2 and Figure 8.3.3
num=2;den=[1 0 2];
figure(1),pzmap(num,den)
figure(2),step(num,den)
roots(den)
```

```
% Figure 8.3.2 Program results
ans =

   0.0000 + 1.4142i
   0.0000 - 1.4142i
```

Figure 8.3.4 MATLAB Program and Its Results

8.3.2 Stability of Proportional and Differential Control

For stability, we need the loci to enter into the left-half s plane. There is no value of K_P for which the system poles lie in the left-half s plane. Clearly it is not possible to stabilize the system with proportional control alone.

How can we specify $G_c(s)$ such that the roots of the characteristic equation bend into the left-half s plane? Since the root locus begins at the poles of the open-loop system ($K_P = 0$) and ends at the zeros ($K_P \to \infty$), it makes sense to locate at least one zero point of the controller in the left-half s plane to attract the loci. We might consider the controller

$$G_c(s) = K_P + K_D s = K_D\left(s + \frac{K_P}{K_D}\right) \tag{8.20}$$

Equation (8.20) is known as a Proportional-Derivative (PD) controller. The ratio determines the location of the zero,

$$z = K_P / K_D \tag{8.21}$$

Stability is not the only issue in design; the performance specifications will influence the choice of z in the final design. Derivative control is essentially anticipatory, which measures the instantaneous error velocity, and predicts the large overshoot ahead of time and produces an appropriate counteraction before too large an overshoot occurs.

Consider the system shown in Figure 8.3.1. The closed-loop transfer function is

$$\frac{C(s)}{R(s)} = \frac{G_C G_f}{1 + G_C G_f} \tag{8.22}$$

$$\frac{C(s)}{R(s)} = \frac{K_D s + K_P}{s^2 + K_D s + K_P} \tag{8.23}$$

Since the characteristic equation is

$$s^2 + K_D s + K_P = 0 \tag{8.24}$$

here it has two roots with negative real parts for positive values of K_D and the damping ratio of the sytem is not zero. Thus derivative control brings about a damping effect.

The zero-pole distribution of the closed-loop system is shown in Figure 8.3.5. Let $K_P = 2$ and $K_D = 0.8$, a zero-pole diagram with two conjugate poles is generated. One zero is on the negative real axis, and the system is obviously stable. As showed in Figure 8.3.6, the two root loci of the system are also on the left-half of the s plane, indicating that the system is stable.

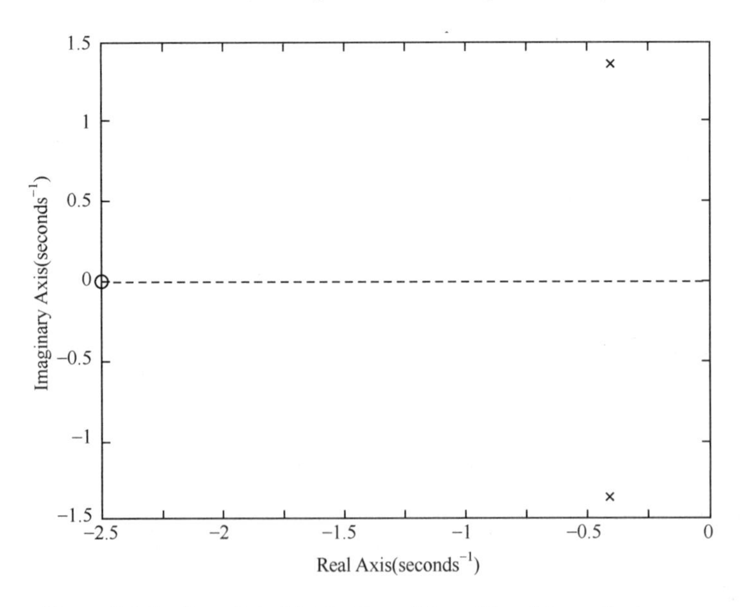

Figure 8.3.5 Zero-Pole Distribution of the a Closed-Loop System

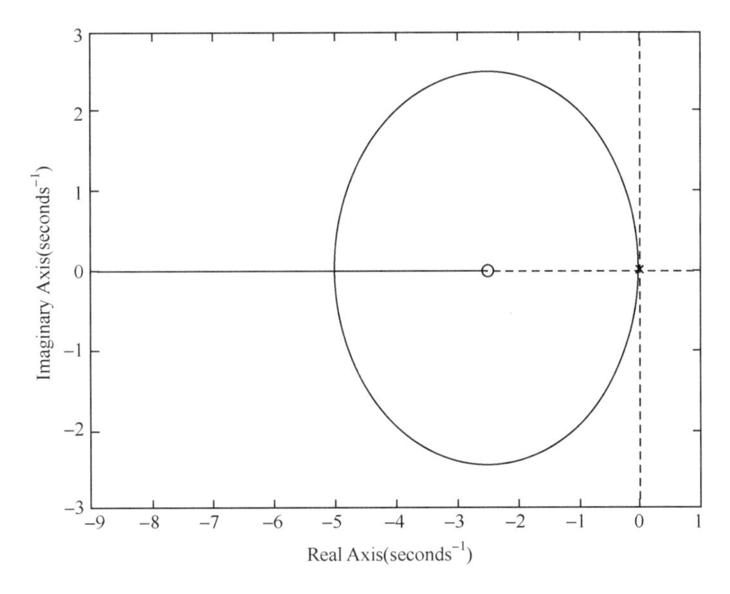

Figure 8.3.6 Root Loci of Open-Loop System

The zero-pole distribution of the closed-loop system, the step response of the system, and the characteristic roots of the closed-loop system can be implemented by applying the procedure in Figure 8.3.7.

```
% The MATLAB script used for Figure 8.3.5, Figure 8.3.6, and Figure 8.3.7
num=[0.8 2];den=[1 0 0];              % Open loop transfer function GcGf
[numc,denc]=cloop(num,den);           %Closed loop transfer function
figure(1),pzmap(numc,denc);           %Closed loop zero point graph
figure(2), rlocus(num,den);           % Open loop root locus
figure(3), step(numc,denc);           %System step response
tf(numc,denc)                         %Closed loop transfer function
```

```
% Result of Command Window
ans =

      0.8 s + 2
  ---------------
  s^2 + 0.8 s + 2

Continuous-time transfer function.
```

Figure 8.3.7 MATLAB Program and Its Results

The typical response curve to the unit step input signal, $C(t)$, is shown in Figure 8.3.8. Obviously, compared to the original response curve shown in Figure 8.3.4, the current response curve has been significantly improved. The system response changes from oscillation under the Pcontroller to a steady-state under the PD Controller There is certain overshoot in the system and the steady-state error is zero.

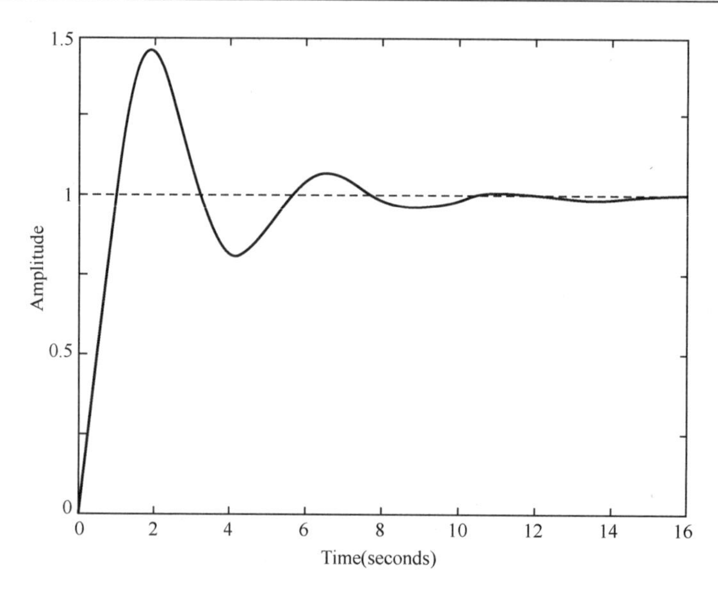

Figure 8.3.8 Step Response

8.4 PID Controller

From the PI controller in Section 8.2 and the PD controller in Section 8.3, single PI control eliminates system errors and single PD control stabilizes the system performance. If they are combined to form a PID controller, will it work wonders?

A PID controller is given by

$$G_c(s) = K_P + K_D s + \frac{K_I}{s} = \frac{K_D s^2 + K_P s + K_I}{s} = K_D \left(\frac{s^2 + \dfrac{K_P}{K_D} s + \dfrac{K_I}{K_D}}{s} \right) \tag{8.25}$$

The action of the integral factor improves the steady-state response.

We see that there is a pole in the PID controller at $s = 0$ and there are two zeros at

$$s = -\frac{K_P}{2K_D} \pm \frac{1}{2} \sqrt{\left(\frac{K_P}{K_D} \right)^2 - 4\frac{K_I}{K_D}} \tag{8.26}$$

Consider the system shown in Figure 8.3.1. The closed-loop transfer function is

$$\frac{C(s)}{R(s)} = \frac{G_c G_f}{1 + G_c G_f} \tag{8.27}$$

$$\frac{C(s)}{R(s)} = \frac{K_D s^2 + K_P s + K_I}{s^3 + K_D s^2 + K_P s + K_I} \tag{8.28}$$

Since the characteristic equation is

$$s^3 + K_D s^2 + K_P s + K_I = 0 \tag{8.29}$$

The zero-pole distribution of the closed-loop system, the step response of the system, and the characteristic root of its closed-loop system can be implemented by applying the procedure in Figure 8.4.4.

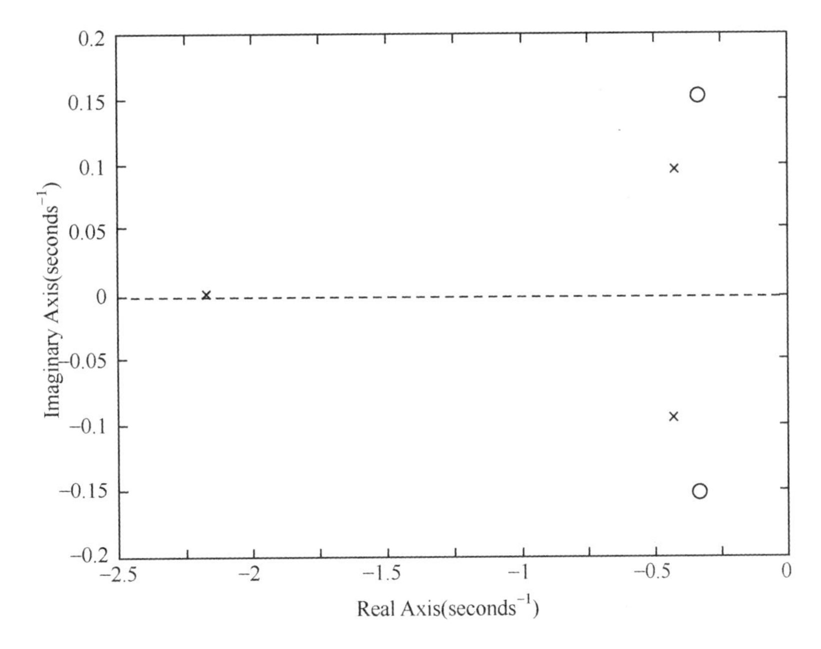

Figure 8.4.1 Zero-Pole Distribution of the System

The zero-pole distribution of the closed-loop system is shown in Figure 8.4.1.

Let $K_P = 2$, $K_D = 3$ and $K_I = 0.4$, and a zero-pole diagram is generated. The system has a total of three poles, two conjugate poles and one on the real axis. The distance from the pole on the real axis to the imaginary axis is d_1, while the distance between the two conjugate poles to the imaginary axis is d_2. d_1 is 5.1 times of d_2. Basically, the effect of the pole on the real axis to the system can be ignored, considering the two conjugate poles are the dominant poles. So the system performance is determined by two conjugate multiple roots, which is to be discussed in Section 8.5. All the roots of the closed-loop system are in the left-half s plane and the system is stable.

The open-loop root loci of the PID control system are shown in Figure 8.4.2. We can see that the three root loci of the system proceed from the origin, with two ends at two conjugate zeros, and one ends at the infinity along the negative axis of the real axis. When K_D changes from 0 to ∞, the root locus enters the left-half s plane from the origin as expected. There are two root trajectories starting from the origin and intersecting with the imaginary axis through the right-half s plane, where is the critical point of the system stability. The two root loci of the system enter the left-half s plane here, and the system is stable. The selection of the ratios K_P / K_D and K_I / K_D will be influenced by the performance specifications.

The unit step response curve of the system is shown in Figure 8.4.3. The system has an overshoot of less than 20% and the adjustment time is less than 8 seconds. The effect of PID controller is pretty good, with a stable output, zero steady-state errors, a mild overshoot and a fast response.

As can be seen from Figure 8.4.4, the roots of the closed-loop characteristic equations of the system are in the left-half s plane, and the system is stable.

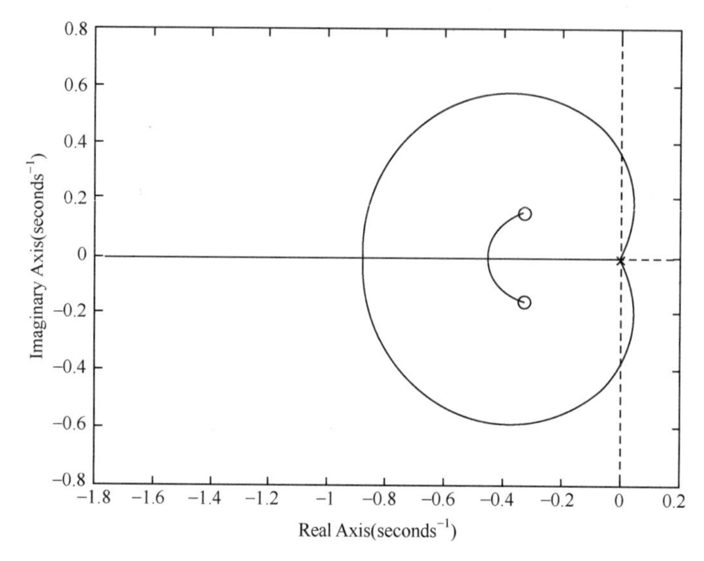

Figure 8.4.2 Root Loci of Open Loop System

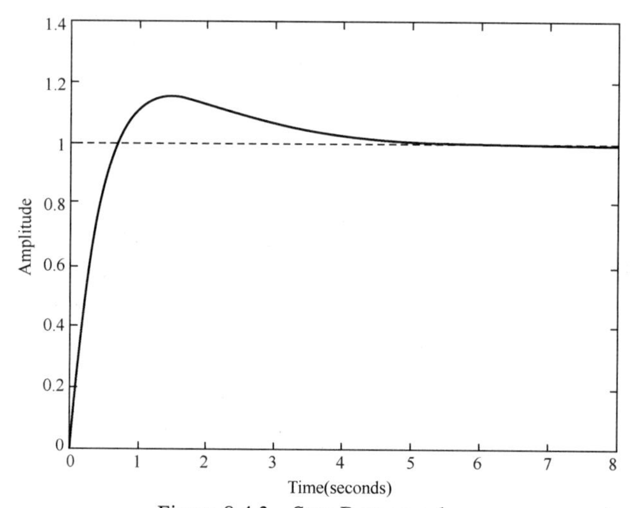

Figure 8.4.3 Step Response 1

```
% The MATLAB script used for Figure 8.4.1,Figure 8.4.2 and Figure 8.4.3
num=[3 2 0.4];den=[1 0 0 0];        % Open loop transfer function GcGf
[numc,denc]=cloop(num,den);         %Closed loop transfer function
figure(1),pzmap(numc,denc);         %Closed loop zero point graph
figure(2),rlocus(num,den);          % Open loop root locus
figure(3),step(numc,denc);          %System step response
roots(denc)                         %The root of a closed loop system
```

```
% Results of Command Window

ans =

  -2.1597 + 0.0000i
  -0.4201 + 0.0932i
  -0.4201 - 0.0932i
```

Figure 8.4.4 MATLAB Program and Its Results

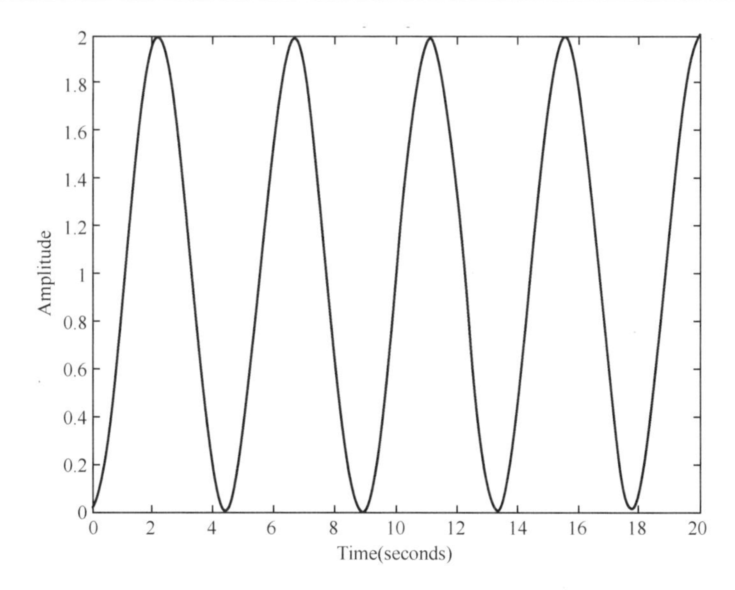

Figure 8.4.5 System Response 2 to Unit Step Input

```
% Results of Command Window

ans =

   0.0000 + 1.4142i
   0.0000 - 1.4142i
  -0.2000 + 0.0000i
```

Figure 8.4.6 Results of Command Window

When $K_{\mathrm{P}} = 2$, $K_{\mathrm{D}} = 0.2$ and $K_{\mathrm{I}} = 0.4$, the step response of system is shown in Figure 8.4.5. The system is in the critical stable-state. The output is a curve of continuous oscillation, and the system is unstable due to interference. From Figure 8.4.5, the two roots of the system are on the imaginary axis, so the system is in the critical stability.

8.5 Dominant Closed-Loop Poles

8.5.1 Basic Concept

The relative dominance of closed-loop poles is determined by the ratio of the real parts of the closed-loop poles, as well as by the relative magnitudes of the residues evaluated at the closed-loop poles. The magnitudes of the residues depend on both the closed-loop poles and the closed-loop zeros.

If the ratios of the real parts exceed 5 and there are no zero nearby, then the closed-loop poles closest to the imaginary axis will dominate in the transient-response behavior because these poles correspond to the slowest decaying term in the transient response. Those closed-loop poles that dominant the transient-response behavior are called dominant closed-loop poles. Quite often, the dominant closed-loop poles occur in the form of a complex-conjugate pair. The dominant closed-loop poles are most important among all closed-loop poles.

Note that the gain of a higher-order system is often adjusted so that there will exist a pair of dominant complex conjugate closed-loop poles. The presence of such poles in a stable system reduces the effects of such nonlinear factors such as dead zone, backlash, and coulomb friction on system performance.

Example 8.1 Given the open-loop transfer function of the system

$$\phi(s)=\frac{1}{(s+1)(0.01s^2+0.08s+1)} \tag{8.30}$$

Find the performance specifications of system.

Solution: Closed-loop has three poles: $s_1=-1$, $s_{2,3}=-4\pm j9.2$, where $s_{2,3}$ can be ignored, s_1 is the dominant pole. The closed-loop transfer function is simplified as

$$\phi(s)=\frac{1}{s+1} \tag{8.31}$$

Then the adjustment time of the system is $t_s=3T=3s$.

8.5.2 Setting of Dominant Closed-Loop Poles

The performance specifications under the step response of the control system are usually represented by the adjustment time and overshoot. It is related to the zero pole of a closed-loop, but it is mainly determined by the closed-loop pole. Therefore, if the system is required to be fast (i.e. t_s is small) , the closed-loop pole should be away from the imaginary axis; If the smoothness is required to be outstanding(i.e. $\sigma\%$ is small), the closed-loop pole is set near the angle of $\pm 45^\circ$ between the s plane and the negative real axis.

There is the correspondence between the dynamic performance of the system and its dominant closed-loop pole. Once the dynamic performance of the system is proposed according to the requirement, the required dominant closed-loop pole can be located on the s plane immediately. In the time domain, the agreement between positions of the system pole and the dominant pole can be realized through system output and the feedback system formed by differentials. Through output feedback, one or a pair of closed-loop poles are located at the position of the required dominant pole, and the rest of the poles are at other positions. This achieves the goal of redistributing the closed-loop poles in the s plane, thereby setting the desired position of dominant closed-loop pole.

Example 8.2 Given the transfer function showed in Figure 8.5.1.

$$G(s)=\frac{1}{s^3+3.5s^2+3.5s+1}=\frac{1}{(s+0.5)(s+1)(s+2)} \tag{8.32}$$

According to requirements, closed-loop performance specifications is $\sigma\%=4.3\%, t_s=4s$. Set the closed-loop pole at the desired position of dominant pole by means of output feedback control.

Solution:

（1）The calculation of the desired dominant pole

Noting that $\sigma\%=\mathrm{e}^{-\pi\xi/\sqrt{1-\xi^2}}\times 100\%$, and $t_s=\dfrac{3.5}{\zeta\omega_n}$, the two poles are

$$s_{1,2}\approx -0.7\pm 0.7\mathrm{j}$$

Let the third pole be

$$s_3 = 7\,\mathrm{Re}[s_1] = -4.9$$

where s_1 and s_2 are the desired closed-loop dominant poles and s_3 is the pole on the real axis and away from the imaginary axis.

(2) The calculation of the desired closed-loop characteristic equation $D(s)$.

$$\widehat{D}(s) = (s - s_1)(s - s_2)(s - s_3) \tag{8.33}$$

$$\widehat{D}(s) = s^3 + 6.3s^2 + 7.84s + 4.8 \tag{8.34}$$

(3) Write a systematic closed-loop characteristic equation of output feedback and its differential feedback.

As shown in Figure 8.5.1, the closed-loop transfer function of the system is

$$\frac{C(s)}{R(s)} = \frac{1}{s^3 + (3.5 + k_3)s^2 + (3.5 + k_2)s + (k_1 + 1)} \tag{8.35}$$

The closed-loop characteristic equation is

$$D(s) = s^3 + (3.5 + k_3)s^2 + (3.5 + k_2)s + (k_1 + 1) \tag{8.36}$$

(4) Determination of feedback coefficient k_1, k_2 and k_3.

Let $\widehat{D}(s) = D(s)$, and the corresponding coefficients are equal. We obtain

$$k_1 + 1 = 4.8 \rightarrow k_1 = 3.8 \tag{8.37}$$

$$3.5 + k_2 = 7.84 \rightarrow k_2 = 4.34 \tag{8.38}$$

$$3.5 + k_3 = 6.3 \rightarrow k_3 = 2.8 \tag{8.39}$$

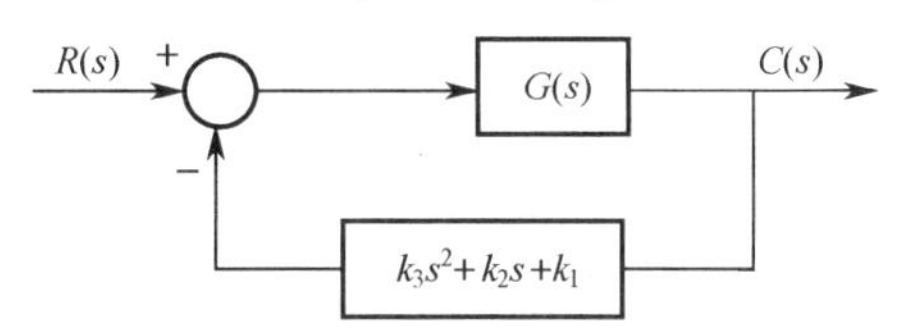

Figure 8.5.1 Structure Diagram of Output Feedback System

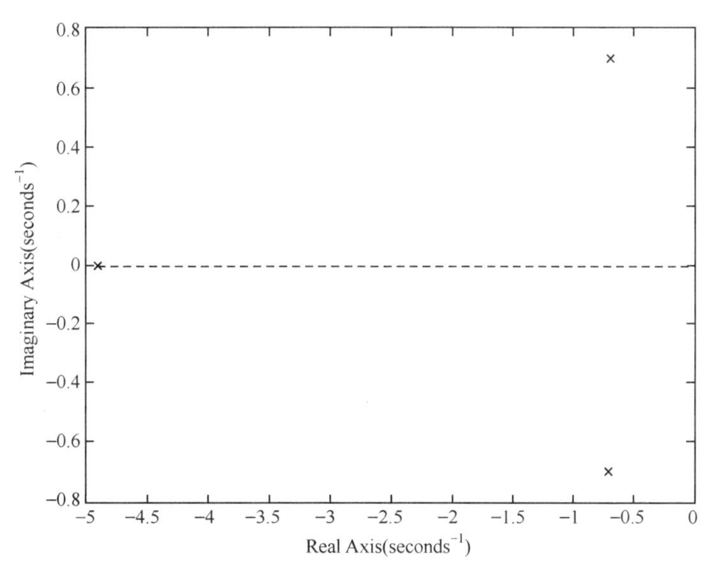

Figure 8.5.2 Zero-Pole Distribution of Closed-Loop Systems

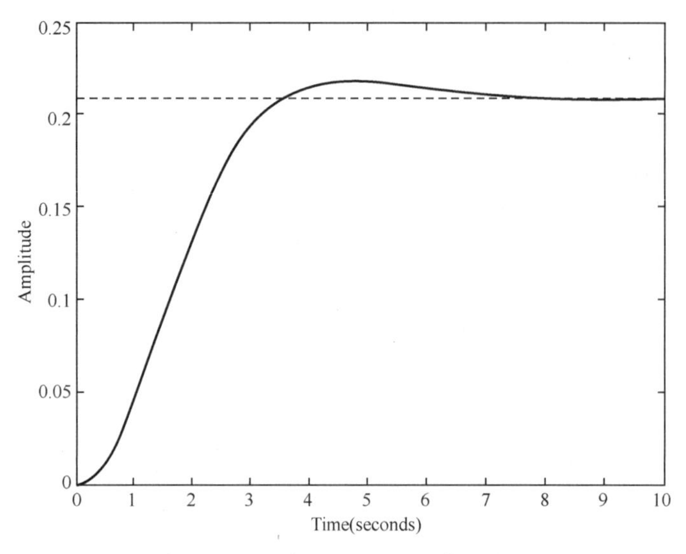

Figure 8.5.3 Step Response of Systems

```
% The MATLAB script used for Figure 8.5.2 and Figure 8.5.3
num=[1];den=[1 6.3 7.84 4.8];          %Closed-loop transfer function
t=1:0.1:10;

figure(1),pzmap(num,den)               %Zero-pole diagram of closed-loop transfer function
figure(2),step(num,den)                % Response of closed-loop transfer function
y=step(num,den);
finalvalue=dcgain(num,den);
[yss,n]=max(y);
[PO,TS]=poandts(t,y,yss,finalvalue); %Call function po and ts to calculate overshoot and
adjustment time
PO
TS
```

```
% Function file used
function[PO,TS]=poandts(t,y,yss,finalvalue)
PO=100*(yss-finalvalue)/finalvalue;                    %Overshoot calculation
k=length(t);
while (y(k)>0.95*finalvalue)&(y(k)<1.05*finalvalue) %   adjustment time calculation
k=k-1;
end
TS=t(k);
```

```
%Results of Command Window
PO =

    4.2085


TS =

    5.8000
```

Figure 8.5.4 MATLAB Program and Its Results

The distribution of the three poles of the system is showed in Figure 8.5.2. A pair of conjugate complex roots, close to the imaginary axis, are the dominant poles, while another pole is on the negative semi-real axis, far from the imaginary axis, with 7 times of distance ratio. Figure 8.5.3 shows that, the step response of the system has small overshoot and fast response. According to Figure 8.5.4, the overshoot is 4.2% and the adjustment time is 5.8 seconds, which is close to the desired closed-loop performance specifications.

8.6 Chess Robot

The following performance is given by a single-armed chess robot developed by scientists in China, as shown in Figure 8.6.1.

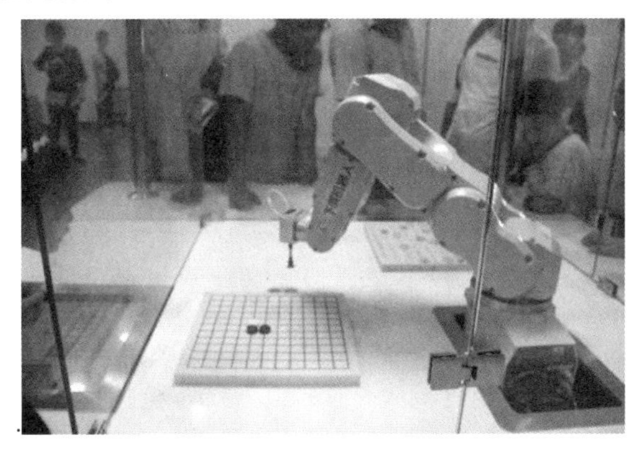

Figure 8.6.1 Single-Armed Chess Robot

In this section we only consider the system design of controlling a single chess piece for the robot. Given the mathematical model of the actuator of a chess piece by robot, the transfer function is

$$G_f(s) = \frac{1}{s(s^2 + 10s + 37)} \tag{8.40}$$

The input is a voltage command to the actuator, and the output is the position of chess piece (only referred to vertical position). A block diagram of the control system is shown in Figure 8.6.2.

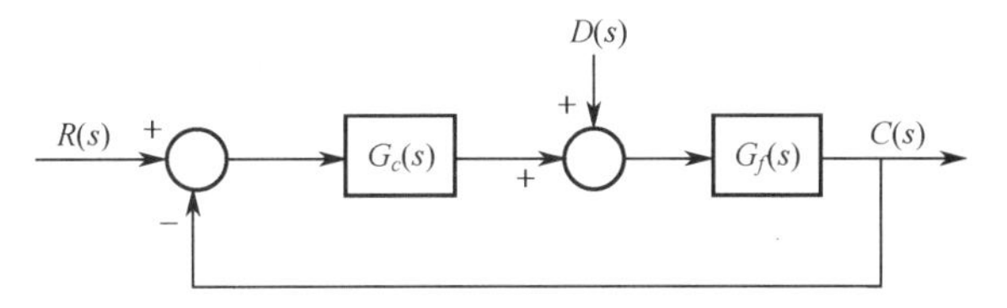

Figure 8.6.2 The Control System of Single-Armed Chess Robot

Control Objective: controlling the position of chess piece and maintaining the original position in the presence of noise disturbance.

Control Variable: the position of chess piece $C(s)$.

We want the chess piece to move to the commanded position with minimal overshoot as fast as possible. The initial goal of design is to make the robot work, even if it works slowly. In other words, the initial bandwidth of control system will be low—around 2Hz. We will later amplify the bandwidth accordingly.

The control design specifications are

DS1: Closed-loop bandwidth greater than 2 Hz.

DS2: Step input overshoot less than 15%.

DS3: Zero steady-state error.

DS1 and DS2 are intended to ensure acceptable tracking performance. DS3 is actually a non-issue in our design: the transfer function of actuator is a Type-I system to guarantee a zero steady-state error in a step input. We simply need to ensure that $G_c(s)G(s)$ remains at least a Type-I system.

8.6.1 Controller Selection

Consider the PID controller

$$G_c(s) = K_P + sK_D + \frac{K_I}{s} = \frac{K_D s^2 + K_P s + K_I}{s} \tag{8.41}$$

In order to avoid the integrator of the system and guarantee that $G_c(s)$ $G(s)$ remains a Type-I system , we adjust the PID controller to

$$G_c(s) = \frac{K_D s^2 + K_P s + K_I}{s + p} \tag{8.42}$$

In the above PID controller, the key parameters to be adjusted are K_P, K_D, K_I, p.

8.6.2 Controller Design

The response of the closed-loop control system is mainly determined by the position of the dominant pole. Our design method is to set the appropriate position for the dominant pole of the closed-loop system. According to the performance specifications, we determine this position by means of the approximate formula of the second-order system. We obtain the controller parameters so that the closed-loop system has the desired dominant poles, and the effect of other pole positions on the system response can be ignored.

The frequency ω_b at which the magnitude of the closed-loop frequency response is reduced to 3dB below zero frequency value is called the *cutoff frequency*. Thus

$$\left|\frac{C(jw)}{R(jw)}\right| < \left|\frac{C(j0)}{R(j0)}\right| - 3\text{dB}, \omega > \omega_b \tag{8.43}$$

For systems in which $|C(j0)/R(j0)| = 0\text{dB}$,

$$\left|\frac{C(jw)}{R(jw)}\right| < -3\text{dB}, \omega > \omega_b \tag{8.44}$$

The closed-loop system filters out the signal components with frequencies greater than the cutoff frequency and passes those signal components with frequencies lower than the cutoff frequency.

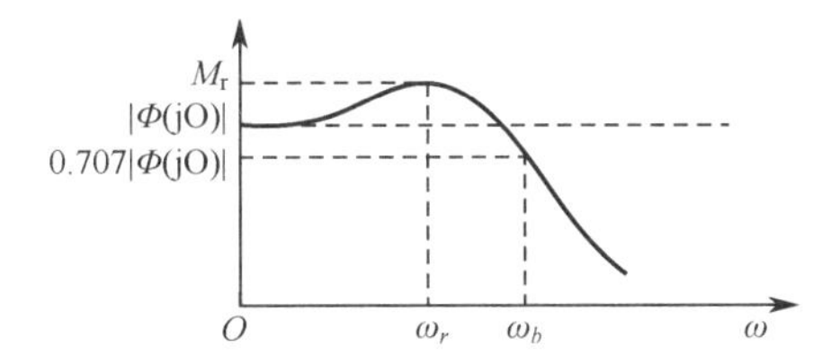

Figure 8.6.3 System Bandwidth

When the amplitude of the closed-loop system is no less than -3dB, the corresponding frequency range $0 \ll \omega \ll \omega_b$ is called the bandwidth of the system, which indicates the frequency where the gain starts to decline from its low-frequency value. Therefore, the bandwidth suggests the ability to track sinusoidal input signals of the system. For a given ω_n, the rise time extends with the increase of damping ratio ξ. On the other hand, the bandwidth decreases with the increase of ξ. Therefore, the rise time and the bandwidth are inversely proportional to each other.

The specification of the bandwidth is determined by the following factors:

(1) The ability to reproduce the input signal

A large bandwidth corresponds to small rise time, or fast response. Roughly speaking, the bandwidth is proportional to the speed of response.

(2) The necessary filtration performance for high frequency noise

In order to track any input signal accurately, the system must possess a high bandwidth. However, considering noise, the bandwidth should not be excessive. Thus, there are conflicting requirements for the bandwidth, and a compromise is usually necessary for a good design. Note that a high-bandwidth system requires components of superior performance, hence the cost of components usually increases with the rise of bandwidth.

1. Calculation of primary parameter in expected second-order systems

It is expect that the performance of the closed-loop control system is determined by a pair of dominant poles. Accordingly, the position of the dominant poles should be determined by the expected system performance specifications DS1 to DS3. The main parameters of the expected second-order system include natural oscillation frequency and damping ratio. If they are both determined by the design, then the designed system can fully satisfy the requirements for performance specifications and thus is the desired system.

The relationship between bandwidth ω_B and natural frequency ω_n is approximately as follows,

$$\frac{\omega_B}{\omega_n} \approx -1.1961\xi + 1.8508 \ (0.3 \leqslant \xi \leqslant 0.8) \tag{8.45}$$

The approximation can apply to second-order systems.

According to DS1, we want

$$\omega_B = 2\text{hz} = 12.57\text{rad} / \sec \tag{8.46}$$

The minimum value of ξ can be determined according to the overshoot specification. Considering $\text{P.O.} \leqslant 15\%$, according to the formula (valid for second-order systems)

$$\sigma\% = \mathrm{e}^{-\pi\xi/\sqrt{1-\xi^2}} \times 100\% \tag{8.47}$$

We obtain

$$\xi = \frac{\ln\frac{100}{\sigma}}{\left[\pi^2 + \left(\ln\frac{100}{\sigma}\right)^2\right]^{\frac{1}{2}}} \tag{8.48}$$

$$\xi \geqslant 0.52 \tag{8.49}$$

We will apply $\xi \geqslant 0.52$ in our design. We select $\xi = 0.52$ on the boundary because the longer ξ is, the longer the adjustment time will be. Although the adjustment time is not a system design specifications, we usually attempt to fasten the system response as much as possible after all the design specifications is satisfied. According to Equations 8.44 and 8.45, we obtain

$$\omega_n = \frac{\omega_B}{-1.1961\xi + 1.8508} = 10.19\text{rad}/\sec \tag{8.50}$$

The expected 2% error adjustment time is

$$T_s \approx \frac{4}{\xi\omega_n} = 0.76\sec \tag{8.51}$$

The two separately calculated parameters, $\omega_n = 10.19\text{rad}/\sec$ and $\xi \geqslant 0.52$, determine a second-order system.

Additionally, the closed-loop frequency characteristics can be expressed as the amplitude and phase frequency.

$$\Phi(\mathrm{j}\omega) = M(\omega)\mathrm{e}^{\mathrm{j}\alpha(\omega)} \tag{8.52}$$

In the formula, $M(\omega)$ and $\alpha(\omega)$ are the amplitude frequency and phase frequency characteristics of a closed-loop system respetively. When the damping ratio of the system is less than or equal to 0.707, a resonance peak Mr, may occur in the amplitude frequency characteristics, and form the relationship with damping ratio as follows,

$$M_r = \frac{1}{2\xi\sqrt{1-\xi^2}} \qquad (\xi < 0.707) \tag{8.53}$$

The relationship between the resonant frequency ω_r, the natural frequency ω_n and the damping ratio ξ is

$$\omega_r = \omega_n\sqrt{1-\xi^2} \qquad (\xi < 0.707) \tag{8.54}$$

In system design, we can use the above two parameters in the frequency domain as the design index of the second-order system, introducing the damping ratio ξ and ω_n to calculate the following parameter values,

$$M_r = 1.1299\text{，}\ \omega_r = 8.7231 \tag{8.55}$$

2. Desired closed-loop characteristic Polynomial

In the previous section, we identify the desired closed-loop system as a second-order system determined by dominant poles. However, since the controlled object G_f is a third-order system, and the denominator of the PID controller G_C is first-order, then the closed-loop system actually is fourth-order rather than the expected second-order system. Therefore, we design a actual fourth-order system consisting of two pairs of conjugate complex roots, with one pair the dominant poles and the other the non-dominant poles. Hence the desired fourth-order closed-loop characteristic polynomial is

$$\widehat{D}(s)=(s^2+2\xi\omega_n s+\omega_n^2)(s^2+m_1 s+m_0) \tag{8.56}$$

where ξ and ω_n are selected to meet the design specifications, and the roots of $s^2+2\xi\omega_n s+\omega_n^2=0$ are the dominant poles. Conversely, we want the roots of $s^2+m_1 s+m_0=0$ to be the non-dominant poles. The dominant poles should be on a vertical line in the complex plane at a distance $\xi\omega_n$ away from the imaginary axis. Let

$$m_1=2\alpha\xi\omega_n \tag{8.57}$$

Then the roots of $s^2+m_1 s+m_0=0$ is on the vertical line of $s=-\alpha\xi\omega_n$ in the complex plane. By selecting $\alpha>1$, we effectively move the roots to the left of the dominant poles. The larger we select α, the further the non-dominant poles lie to the left of the dominant poles. The reasonable value of α in the system is $\alpha=8$.

We can experiment with different values of α in MATLAB. If we select

$$m_0=\alpha^2\xi^2\omega_n^2$$

Then we obtain two real roots

$$s^2+m_1 s+m_0=(s+\alpha\xi\omega_n)^2=0 \tag{8.58}$$

Selecting $m_0=\alpha^2\xi^2\omega_n^2$ is not required but reasonable since we hope the effect of the non-dominant poles on the overall system can be fast and non-oscillatory.

The desired characteristic polynomial is then

$$\widehat{D}(s)=(s^2+2\xi\omega_n s+\omega_n^2)(s^2+2\alpha\xi\omega_n s+\alpha^2\xi^2\omega_n^2) \tag{8.59}$$

It is expanded to

$$s^4+[2\xi\omega_n(1+\alpha)]s^3+[\omega_n^2(1+\alpha\xi^2(\alpha+4)]s^2+[2\alpha\xi\omega_n^3(1+\xi^2\alpha)]s+\alpha^2\xi^2\omega_n^4=0 \tag{8.60}$$

3. Characteristic polynomial of PID control system

As shown in Figure 8.6.2, the closed-loop transfer function of the system is

$$\frac{C(s)}{R(s)}=\frac{G_c G_f}{1+G_c G_f}=\frac{\dfrac{K_D s^2+K_P s+K_I}{s+p}*\dfrac{1}{s(s^2+10s+37)}}{1+\dfrac{K_D s^2+K_P s+K_I}{s+p}*\dfrac{1}{s(s^2+10s+37)}} \tag{8.61}$$

$$\frac{C(s)}{R(s)}=\frac{K_D s^2+K_P s+K_I}{s^4+(10+p)s^3+(37+10p+K_D)s^2+(37p+K_P)s+K_I} \tag{8.62}$$

The characteristic polynomial of the PID control system is

$$D(s)=s^4+(10+p)s^3+(37+10p+K_D)s^2+(37p+K_P)s+K_I \tag{8.63}$$

4. Determination of PID controller coefficients

Let $\widehat{D}(s)=D(s)$. Then the corresponding coefficients of Equations 8.60 and 8.63 are equal. We obtain

$$\begin{aligned}&10+p=2\xi\omega_n(1+\alpha)\\&37+10p+K_D=\omega_n^2(1+\alpha\xi^2(\alpha+4)\\&37p+K_P=2\alpha\xi\omega_n^3(1+\xi^2\alpha)\\&K_I=\alpha^2\xi^2\omega_n^4\end{aligned} \tag{8.64}$$

The damping ratio $\xi=0.52$ and natural oscillation frequency $\omega_n=10.19\ \text{rad/sec}$, $\alpha=8$ calculated previously are taken into the Formula 8.64 respectively. We obtain the parameters of the

PID controller

$$
\begin{aligned}
p &= 84.82 \\
K_D &= 1882.5 \\
K_P &= 24\,323 \\
K_I &= 184\,410
\end{aligned}
\tag{8.65}
$$

Then the PID controller is

$$G_c(s) = \frac{K_D s^2 + K_P s + K_I}{s + p} = 1882.5 \frac{s^2 + 12.92s + 97.96}{s + 84.82} \tag{8.66}$$

8.6.3 System Simulation

The zero-pole distribution of closed-loop PID control system is shown in Figure 8.6.5.

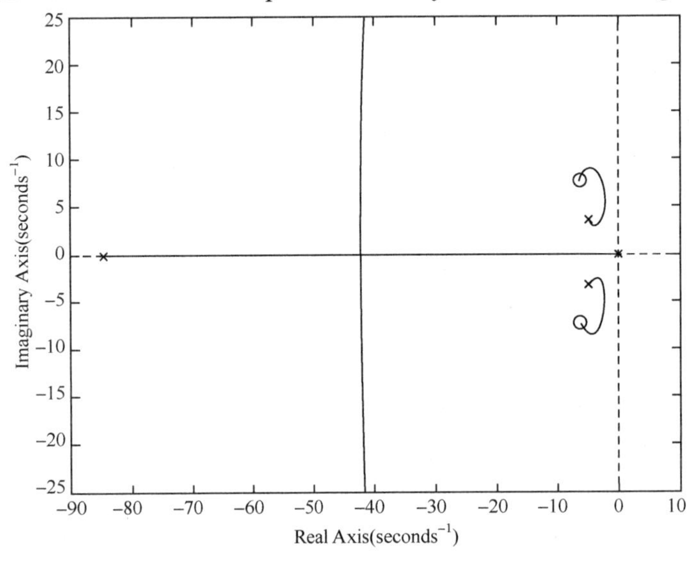

Figure 8.6.4 Root-Loci of Open-Loop PID Control Systems

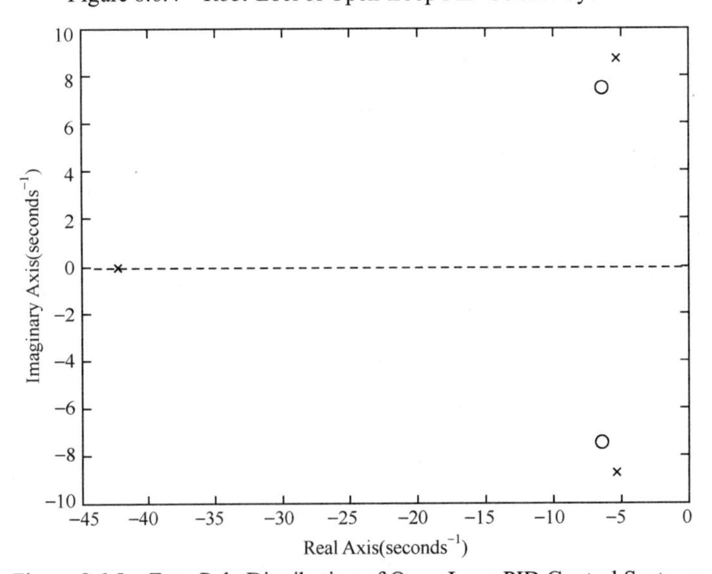

Figure 8.6.5 Zero-Pole Distribution of Open-Loop PID Control Systems

As can be seen from Figure 8.6.4, the four root loci of the open-loop system are in the left-half s plane, and the system is stable. As can be seen from Figure 8.6.5, the four poles of the closed-loop system are in the left-half s plane. A pair of dominant poles is close to the imaginary axis, at a distance from the imaginary axis of 5.2677. The other pair of non-dominant poles is the re-root, on the negative semi-axis, and at a distance from the imaginary axis of 42.1412. The ratio of the two poles to the imaginary axis is 8. Therefore, the performance of the system is mainly determined by the dominant pole close to the imaginary axis.

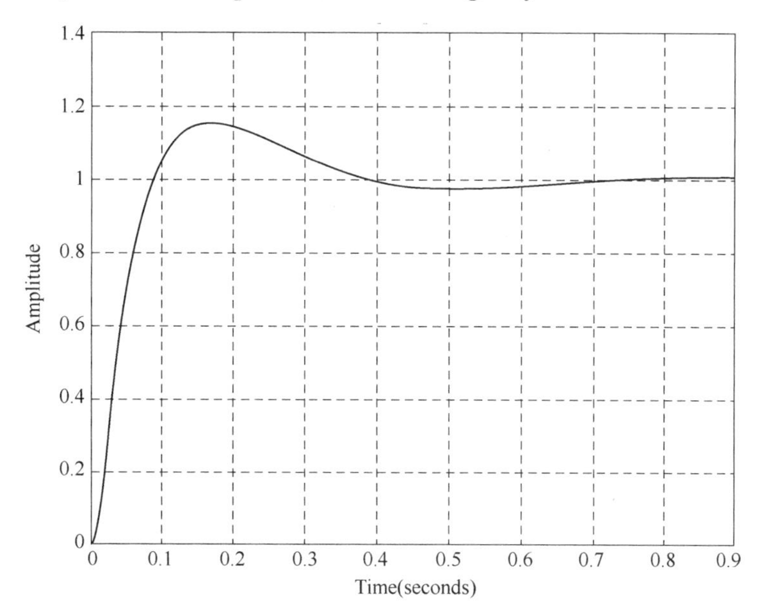

Figure 8.6.6 Step Response of PID Control Systems

As can be seen from Figure 8.6.6, the overshoot of the step response of the system is within 20%, and the adjustment time is less than 1 second.

```
% The MATLAB script used for Figure 8.6.4, Figure 8.6.5 and Figure 8.6.6
clc
clear
close all
POI=15;         %Input system overshoot indicator 2

wb=2*6.28;      %Input system closed-loop bandwidth indicator 1
af=8;           %The distance between the dominant pole and non-dominant pole from
                %the imaginary axis
a=10;b=37;
zeta=log(100/POI)/sqrt(pi^2+(log(100/POI))^2)  %Calculate system damping ratio according
                                               %to indicator 2
wn=wb/(-1.196*zeta+1.8508)                     %Calculate the natural vibration frequency of
                                               %the system according to the index 1
TsI=4/(wn*zeta)                                %Calculate the expected adjustment time
```

Figure 8.6.7 MATLAB Script and Its Results

```
s3=2*zeta*wn*(1+af);                    %The coefficient of s^3 in Formula 8.63
s2=wn^2*[1+af*zeta^2*(af+4)];           %The coefficient of s^2 in Formula 8.63
s1=2*af*zeta*wn^3*(1+zeta^2*af);        % The coefficient of s^1 in Formula 8.63
s0=af^2*zeta^2*wn^4;                    %The constant term in Equation 8.63
nf=[1];df=conv([1,0],[1,a,b]);          %Mathematical model in Equation 8.40, controlled
                                        %object model
p=s3-a;
KP=s1-b*p;
KD=s2-a*p-b;
KI=s0;
nc=[KD,KP,KI];dc=[0,1,p];               %controller Gc
[num,den]=series(nc,dc,nf,df);
[numc,denc]=cloop(num,den);             %Closed-loop model of PID controller
figure(1),rlocus(num,den);
figure(2),pzmap(numc,denc);
figure(3),step(numc,denc);
grid
[y,x,t]=step(numc,denc);
PO=100*(max(y) -1)                      % Calculate the overshoot of the step response
T=find(abs(y-1)>0.02);                  % Calculate the adjustment time of the step response
Ts=t(T(length(T)))
```

```
% Results of Command Window
zeta =

    0.5169


wn =

   10.1903


TsI =

    0.7594


PO =

   15.0127


Ts =

    0.5726
```

Figure 8.6.7 MATLAB Script and Its Results（continued）

```
ans =

              1882 s^2 + 2.432e04 s + 1.844e05
  -----------------------------------------------
  s^4 + 94.82 s^3 + 2768 s^2 + 2.746e04 s + 1.844e05

Continuous-time transfer function.


ans =

  -42.1412 + 0.0000i
  -42.1412 + 0.0000i
   -5.2677 + 8.7231i
   -5.2677 - 8.7231i
```

Figure 8.6.7 MATLAB Script and Its Results (continued)

From the results of Figure 8.6.7, we can see that the overshoot of the system step output is 15%, and the adjustment time is 0.57 seconds. The design specifications of system are as follows: overshoot is less than 15%, and the expected adjustment time is 0.76 seconds, which meets the design requirements compared with the expected value.

Since the time domain indicators are qualified, we go on to verify the frequency domain indicators.

The amplitude frequency characteristics of the closed-loop system are shown in Figure 8.6.8. The bandwidth is $\omega_b = 32.47\text{rad/sec} = 5.64\text{Hz}$, satisfying DS1, but larger than the design time $\omega_b = 2\text{Hz}$. Therefore, the higher the designed bandwidth is than expected, the less the adjustment time is than expected. In fact, the adjustment time is $T_s = 0.57\text{s}$ while the expected time is $T_s = 0.76\text{s}$.

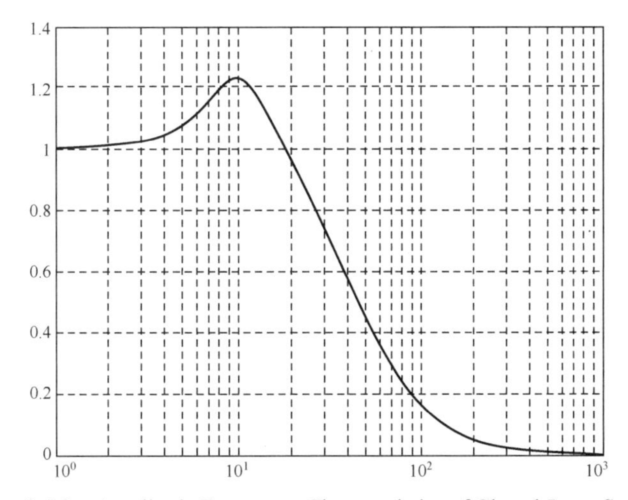

Figure 8.6.8 Amplitude Frequency Characteristics of Closed-Loop Systems

The peak is $M_r = 1.2301$, and what we expect is $M_r = 1.1299$. This slightly higher peak value of the design result basically meets the requirement. The procedure and results are shown in Figure 8.6.9.

```
% The MATLAB script used for Figure 8.6.8
clc
clear
close all
POI=15;     %Input of system overshoot indicator 2
wb=2*6.28; %Input of system closed-loop bandwidth indicator 1
af=8;           %The distance between the dominant pole and
                         %non-dominant pole from the imaginary axis
a=10;b=37;
zeta=log(100/POI)/sqrt(pi^2+(log(100/POI))^2);   %Calculate system damping ratio according to
                              % indicator 2
wn=wb/(-1.196*zeta+1.8508);   %Calculate the natural vibration frequency of the system
                                       % according to the index 1
TsI=4/(wn*zeta);                       %Calculate the expected adjustment time
s3=2*zeta*wn*(1+af);                   %The coefficient of s^3 in Formula 8.63
s2=wn^2*[1+af*zeta^2*(af+4)];   %The coefficient of s^2 in Formula 8.63
s1=2*af*zeta*wn^3*(1+zeta^2*af);   %The coefficient of s^1 in Formula 8.63
s0=af^2*zeta^2*wn^4;                   %The constant item in Equation 8.63
nf=[1];df=conv([1,0],[1,a,b]);   %Mathematical model in Equation 8.40, controlled object model
p=s3-a;
KP=s1-b*p;
KD=s2-a*p-b;
KI=s0;
nc=[KD,KP,KI];dc=[0,1,p];        %Controller Gc
[num,den]=series(nc,dc,nf,df);
[numc,denc]=cloop(num,den);   %Closed-loop model of PID controller

[y,x,t]=step(numc,denc);
wrI=wn*sqrt(1-zeta^2)
MrI=1/[2*zeta*sqrt(1-zeta^2)]
w=logspace(-1,2,400);
[mag,phase,w]=bode(numc,denc);
semilogx(w,mag);grid
[y,L]=max(mag);
wb_list=find(mag<0.707);
wb=w(wb_list(1));                        % Closed-loop system bandwidth
Mr=y,wr=w(L),wb                          %  Resonant amplitude peak of closed-loop system
```

Figure 8.6.9　MATLAB Program and Its Results

```
% Results of Command Window
wrI =

    8.7231


MrI =

    1.1299


Mr =

    1.2301


wr =

    10


 wb =

    32.4693
```

Figure 8.6.9 MATLAB Program and Its Results (continued)

Summary

1. The basic concept of PID controller is introduced.

2. Comparisons are made through examples between proportional and integral control as well as between proportional and differential control. Hence the following conclusions are drawn: proportional control is error control; integral control is error free control, but it may cause instability of the system; differential control can improve the stability of the system.

3. The example of PID control shows that the PID controller not only improves the static performance, but also improves the dynamic performance of the system.

4. Based on the concept of dominant pole, we designed a controller for the chess robot which meets the requirements of both bandwidth and overshoot specifications. Through simple analysis, we obtained an initial set of controller parameters. The design of the controller is proved to practically satisfy the design requirements. Slight modification would be needed to meet the exact design specifications.

生词注解

proportional	n. 比例 adj. 比例的，成比例的
integral	n. 积分 adj. 积分的，完整的
derivative	n. 导数，微分
deviation	n. 偏差
conjunction	n. 结合
characterized by	以……为特征的
isochronous	adj. 等时的，等步的
actuating	adj. 启动的 v. 驱使
damping	n. 阻尼，衰减 v. 抑制
implementation	n. 实现，履行
sensitivity	n. 敏感
inertia	n. 惯性，惰性
imaginary	adj. 虚构的
trajectories	n. 轨迹，轨道线
simplified	adj. 简化了的 v. 简化
calculation	n. 计算，估算
reproduce	vt. 复制；再现
predominantly	adv. 主要地
filters	n. 过滤器 v. 过滤
polynomial	n. 多项式 adj. 多项式的
amplitude	n. 振幅
signal components	信号分量
cutoff frequency	截止频率
approximation	n. 近似值，近似关系
resonance	n. 共振，共鸣

第 9 章 简单系统的 Simulink 建模实例

本章阐述的主要是 Simulink 的基础知识。以名为 simple_model 的简单模型为例子，说明了在 Simulink 中基本连线和函数模块的操作。接着详细说明了 RLC 模型的搭建，展示了建模的具体步骤，并实现对此模型的仿真。

9.1 Simulink 软件入门

9.1.1 启动 Simulink 软件

在启动 Simulink 软件之前，首先打开 MATLAB 专业软件环境，访问 MATLAB 文档以获取更多的信息。打开 Simulink 软件有以下两种方式：

- 在工具栏上单击图标 Simulink。
- 在 MATLAB 命令窗口中输入 Simulink。

随之出现 Simulink 的库浏览器，如图 9.1.1 所示，它显示了在 MATLAB 系统上安装的 Simulink 模块库的树形结构视图。从库浏览器中直接将模块复制（拖）到新建立的模型窗口中，可以建立新模型。

在 Simulink 库窗口中显示的图标都是已预装好的模块图标，可以从库浏览器中将模块拖到新建立的模型窗口中。

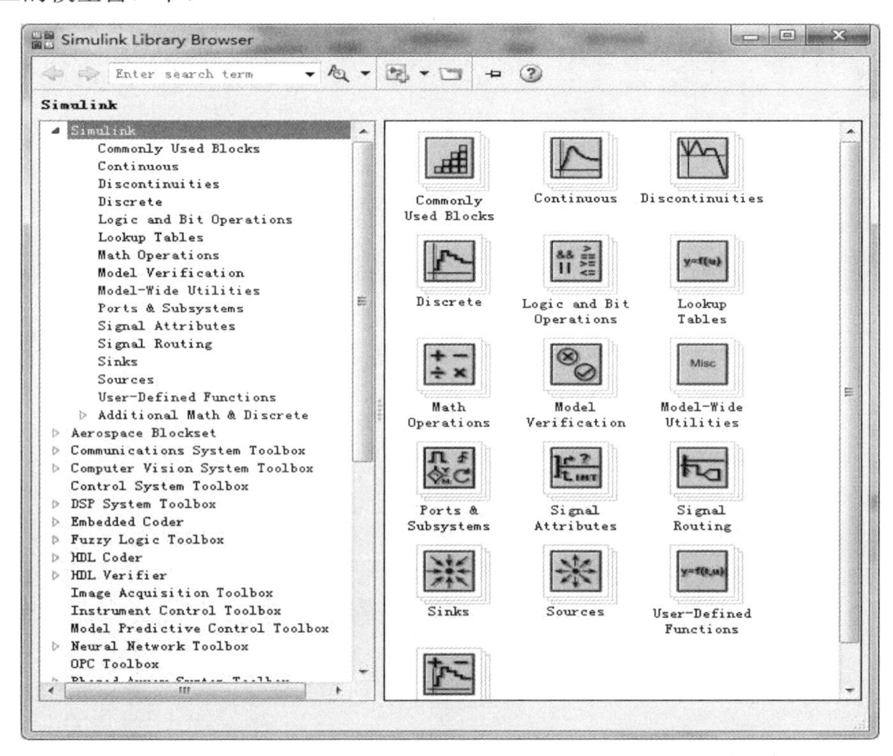

图 9.1.1 Simulink 库浏览器

9.1.2 Simulink的用户界面

9.1.2.1 Simulink 库浏览器

库浏览器显示的是安装在系统中的 Simulink 模块库，可以从模块库中复制模块到一个模型窗口中来建立模型（见图 9.1.2）。

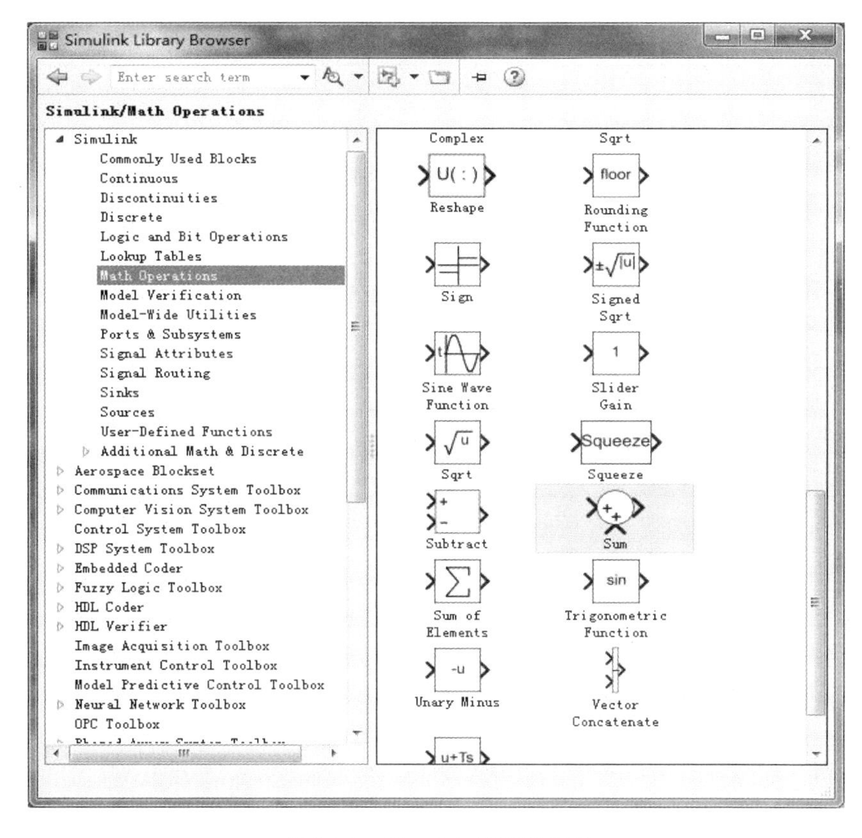

图 9.1.2　Simulink 的用户界面

9.1.2.2　使用库浏览器技巧

使用库浏览器时，应注意以下几点：

（1）可以在库浏览器的左侧选择库名称查看存储库中的模块，双击该库也可以查看。

（2）选择一个模块时，该模块的说明出现在浏览器底部。

（3）要了解一个模块的详细信息，先选中该模块，再选定**帮助→该模块的帮助**，即显示该模块的帮助页。

（4）右键单击模块，然后选择模块参数，可以查看该模块的参数。

（5）在模块搜索栏中，输入具体模块的名称，单击查找模块图标，可以搜索到这个具体模块。

9.1.2.3　标准模块库

Simulink 软件提供了 16 个标准模块库。表 9.1 简单说明了每个库。

表 9.1　16 个标准模块库

模　块　库	说　　明
常用模块组	包括许多常用模块集，如 Constant、In1、Out1、Scope 和 Sum 模块。在其他模块库中也包含了在这个库中的每个模块
连续系统模块	包括可对线性函数建模的模块，如微分单元 Derivative 和积分单元 Integrator 模块
非线性系统模块	包括非线性输入输出函数模块，如 Saturation 模块
离散系统模块	包括表示离散时间函数的模块库，如 Unit Delay 模块

续表

模　块　库	说　　明
逻辑与位操作模块	包括执行逻辑和位操作的模块，如 Logical Operator 和 Relational Operator 模块
查找表模块	包括使用查找表模块，该模块由输入决定输出，如 Cosine 和 Sine 模块
数学运算模块	包括执行数学和逻辑的函数模块，如 Gain，Product 和 Sum 模块
模型验证模块	包括能够创建自验证模型模块，如 Check Input Resolution 模块
模型扩充模块	包括有关模型的信息模块，如 Model Info 模块
端口与子系统模块	包括允许创建子系统的模块，如 In1、Out1 和 Subsystem 模块
信号属性模块	包括能修改信号属性的模块，如 Data Type Conversion 模块
信号线路模块	包括把信号从方框图的一个点传送到另一个点的模块，如 Mux 和 Switch 模块
显示和输出模块	包括显示和输出量的模块，如 Out1 和 Scope 模块
输入源模块	包括产生或导入系统输入量的模块，如 Constant、In1 和 Sine Wave 模块
用户自定义函数模块	包括允许自定义函数模块，如 Embedded MATLAB Function 模块
附加数学和离散模块	包括数学和离散函数模块两个附加库

9.1.3　Simulink 模型窗口

模型窗口包括该模型的模块框图，可以通过有条理地摆放模块的位置，并设置每个模块参数，然后用信号线把模块相连接来搭建自己的模型。

在模型窗口中也可以做以下操作：

（1）为模型设置配置参数，包括启动和停止的时间、算法类型的选择和数据导入/导出设置。

（2）启动和停止模型的仿真。

（3）保存模型。

（4）打印模块框图。

9.2　建立简单模型

本节以图 9.2.1 为例，详细介绍如何创建这个简单模型。

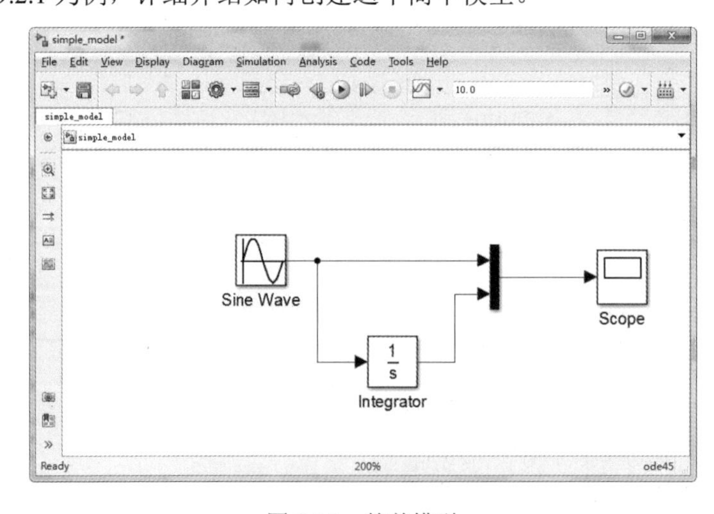

图 9.2.1　简单模型

9.2.1　创建一个新模型

在开始构建模型之前，必须启动 Simulink 并创建一个空白的模型窗口。

创建一个新模型：

（1）如果 Simulink 没有运行，则在 MATLAB 命令窗口中输入 Simulink，打开 Simulink 库浏览器。

（2）选择 File→New→Model，在 Simulink 库浏览器中创建一个新的模型。

Simulink 软件会打开一个空白模型窗口，如图 9.2.2 所示。

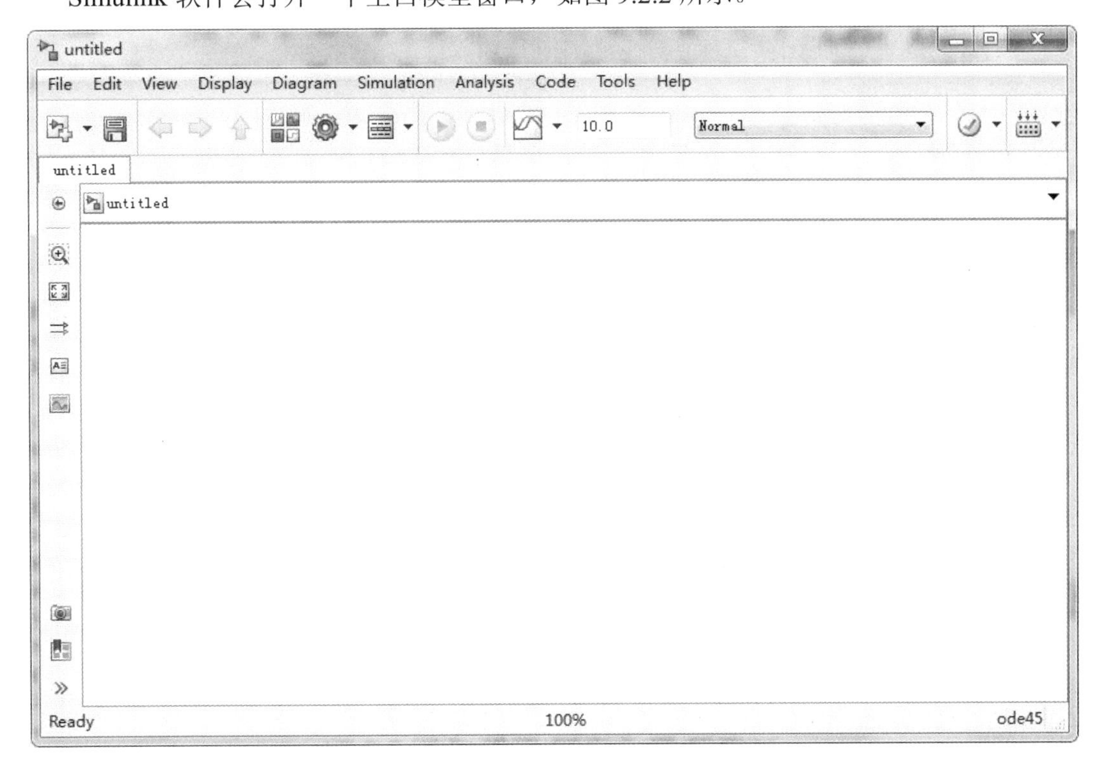

图 9.2.2　空白模型窗口

9.2.2　添加模块到空白模型窗口

构建一个模型，首先从 Simulink 库浏览器中拖动模块到空白模型窗口。创建如图 9.2.1 所示的简单模型，需要以下 4 个模块：

（1）正弦波模块——在模型中产生输入信号。

（2）积分器模块——处理输入信号。

（3）示波器模块——在模型中把信号可视化。

（4）合并模块——把多路输入信号以及经过处理的信号合并到一个示波器中。

添加上述模块到空白模型窗口：

（1）选择 Simulink 库浏览器中的 Sources 库（输入源模块库）。输入源模块库显示在 Simulink 库浏览器中，如图 9.2.3 所示。

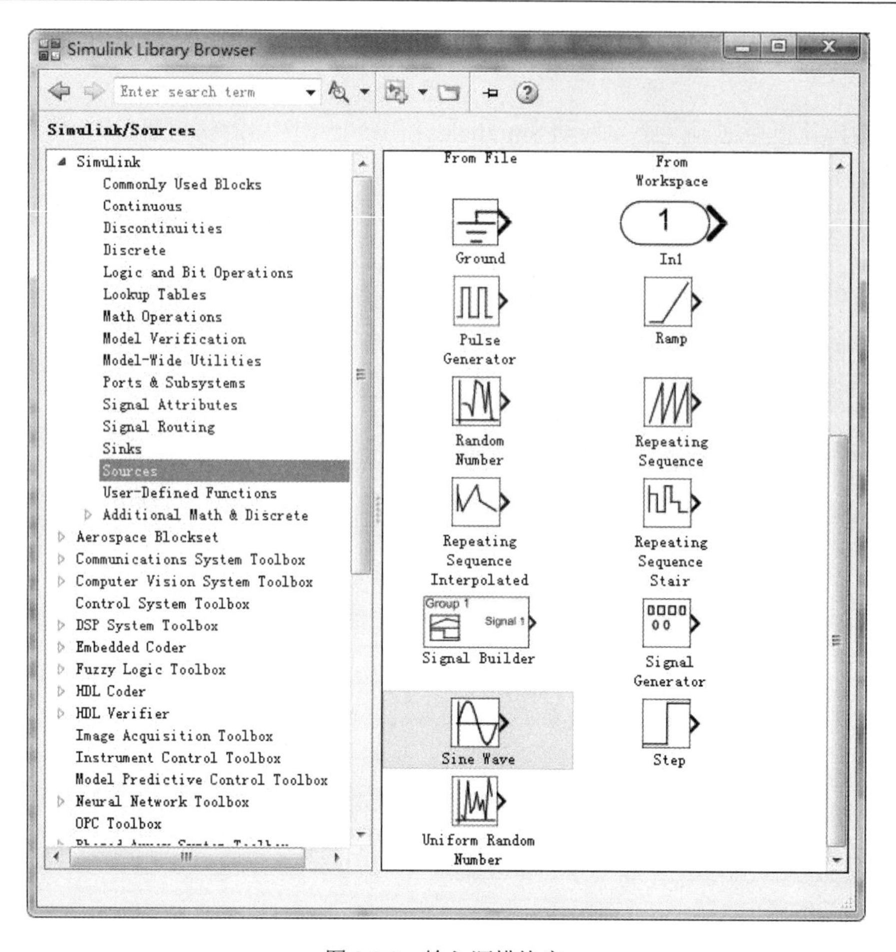

图 9.2.3　输入源模块库

（2）选择 Simulink 库浏览器中的正弦波模块 Sine Wave，将其拖动到模型窗口中，就会显示正弦波模块，如图 9.2.4 所示。

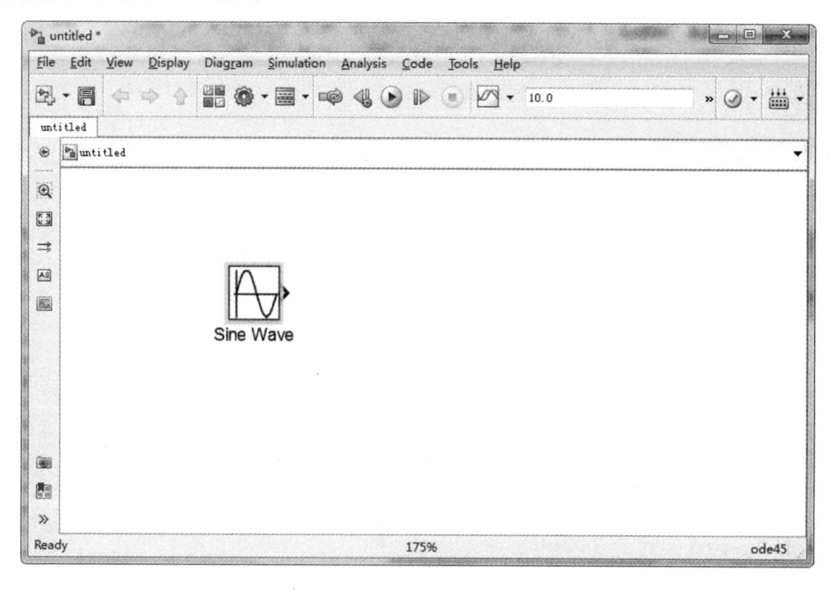

图 9.2.4　正弦波模块

（3）选择 Simulink 库浏览器中的 Sinks（显示和输出模块）。

（4）在 Sinks 中选择示波器模块，将其拖动到模型窗口中，就会显示示波器模块。

（5）在 Simulink 库浏览器中选择 Continuous（连续系统模块）。

（6）从连续系统模块中选择积分器模块 Integrator，将其拖动到模型窗口中，就会显示积分器模块。

（7）从库浏览器中选择 Signal Routing（信号线路模块）。

（8）从 Sinks 中选择合并模块 Mux，将其拖动到模型窗口中，就会显示合并模块。

9.2.3　在模型窗口中移动模块

在连接模块之前，把各个模块进行合理的排列，然后按照系统的信号流程将各系统模块尽可能简单地连接起来。

在模型窗口中移动模块，有以下两种方式：

- 拖动模块。
- 选择模块，然后按键盘上的箭头键。

在模型窗口中排列模块，使之如图 9.2.5 所示。

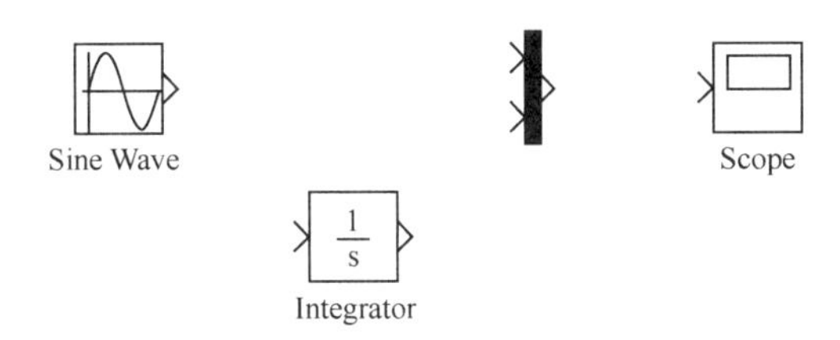

图 9.2.5　排列好的模块

9.2.4　连接模型窗口中的模块

把模块添加到模型窗口之后，必须将各种功能模块进行连接，以表示模型内信号的连接。

注意，每个模块的一侧或者两侧都有尖括号，这些尖括号表示输入输出端口。

（1）指向模块的符号>是输入端口。

（2）从模块指出的符号>是输出端口，如图 9.2.6 所示。

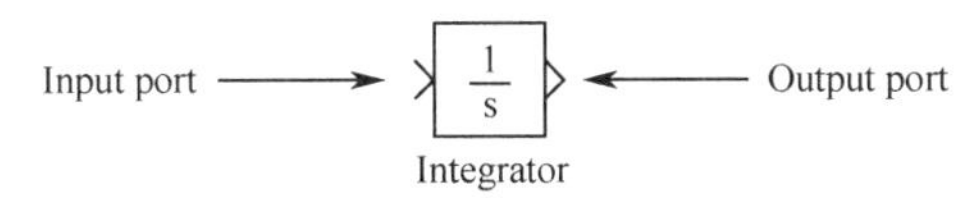

图 9.2.6　尖括号表示输入输出端口

以下介绍如何通过画线从输出端口到输入端口把模块连接起来。

9.2.4.1　模块的连接

在模型窗口中，可以通过在输出端口和输入端口之间的连线进行模块连接。

两个模块的连接：

（1）将鼠标指针指向正弦波模块右侧的输出端口，注意当鼠标指针指向端口时光标变成十字（+）形状，如图 9.2.7 所示。

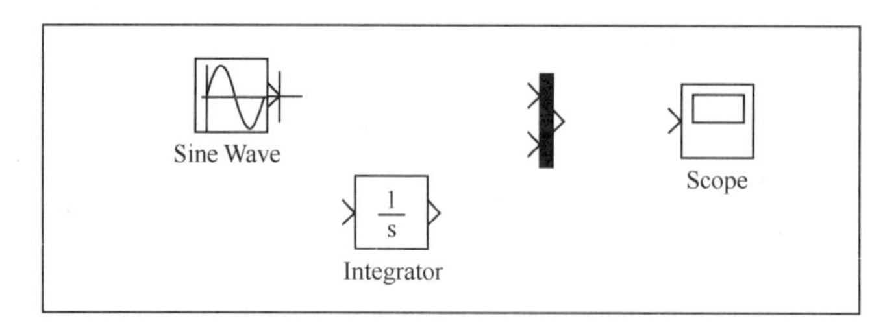

图 9.2.7　十字（+）形状

（2）拖动线从正弦波模块输出端口到合并模块顶端的输入端口相连接，请注意，按住鼠标按钮时线变为虚线，在接近合并模块输入端口时光标变成双线十字线，如图 9.2.8 所示。

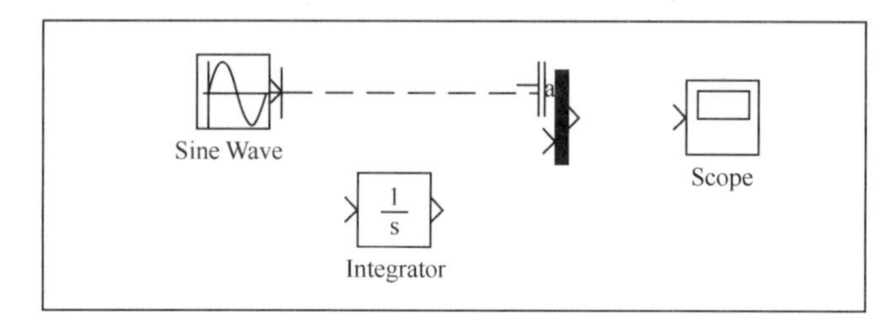

图 9.2.8　双线十字线

（3）然后松开鼠标，线就与合并模块输入端口相连接，用箭头表示信号流的方向，如图 9.2.9 所示。

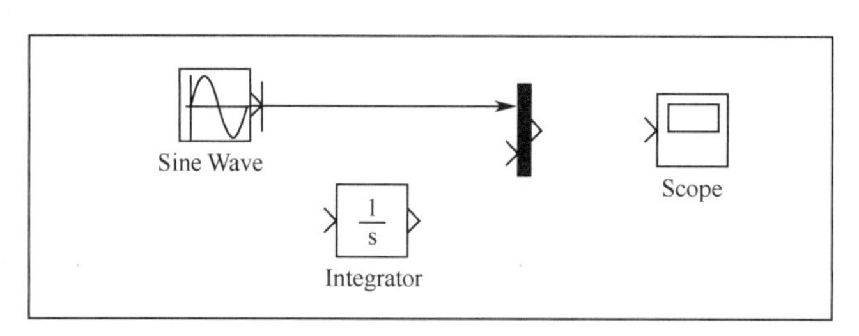

图 9.2.9　连接模块的箭头方向

（4）从积分器模块输出端口拖动一条线与合并模块底端输入端口相连接。

（5）软件自动连接模块：选择合并模块，然后按 Ctrl 键+单击示波器模块。该软件会自动绘制模块之间的连接线。连接好的模型如图 9.2.10 所示。

注意：连接多模块或复杂的模型时，按 Ctrl 键+单击快捷键是非常有用的。

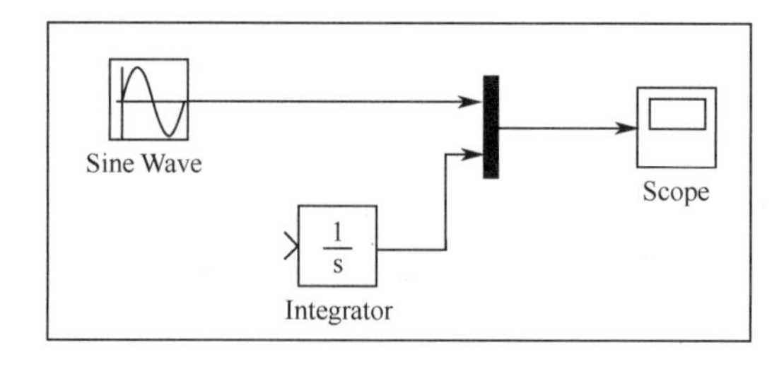

图 9.2.10　连接好的模型

9.2.4.2　画一个分支线

图 9.2.10 模型几乎是完整的，但缺少一个连接。为了完成该模型，必须将正弦波模块连接到积分器模块。

这个最后的连接线与其他三条已有连线稍微不同，表现在输出端口连接到输入端口有所不同。由于正弦波模块的输出端口已经有一个连接，必须把这个已有连接再连接到积分器模块的输入端口。这个新的连接线称为分支线，其传递的信号与从正弦波模块到合并模块的信号是相同的。

完成现有线路的连接：

（1）将鼠标指针放在正弦波模块和合并模块之间的连线上，如图 9.2.11 所示。

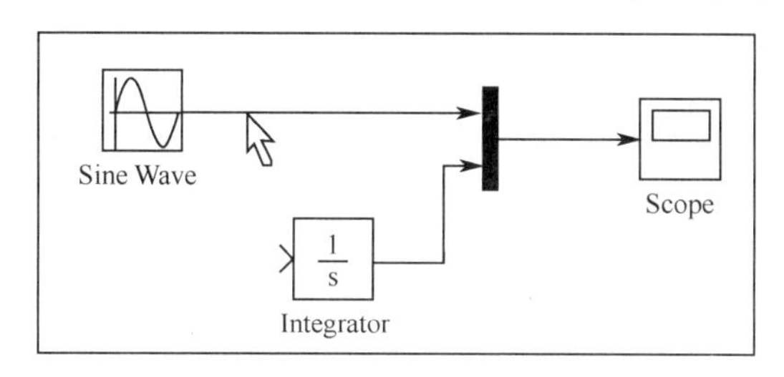

图 9.2.11　鼠标指针在正弦波模块和合并模块之间的连线上

（2）按住 Ctrl 键，然后拖动线到积分器模块输入端口，如图 9.2.12 所示。

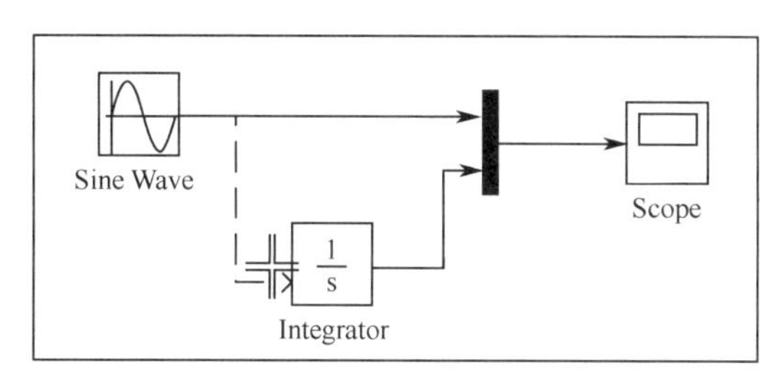

图 9.2.12　按 Ctrl 键拖动线到积分器模块输入端口

（3）在起点和积分器模块的输入端口之间连线，完成该模型。如图 9.2.13 所示。

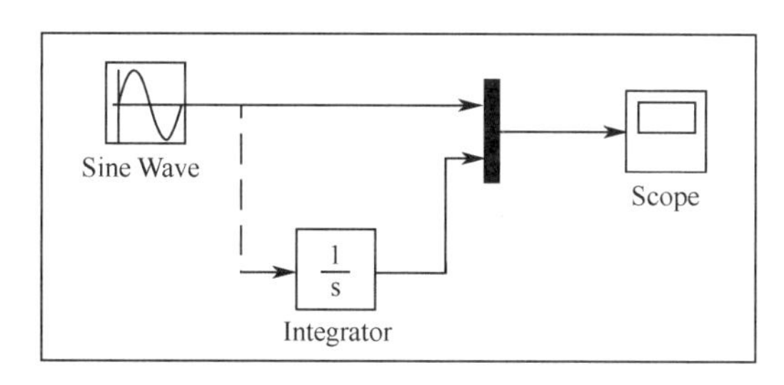

图 9.2.13　已完成的模型

9.2.5　保存模型

完成模型后，将其保存以备将来使用。

保存模型步骤：

（1）在模型窗口中选择 File→Save。

（2）指定要保存模型的位置。

（3）在文件名称中输入 simple_model。

（4）单击 Save，该软件保存模型的文件名为 simple_model. slx。

9.3 RLC 电路模型仿真

选择一个简单的 RLC 电路，如图 9.3.1 所示。其中，V_{in}为输入电压，L 为电感（H），R 为电阻（Ω），C 为电容（F），V_{out} 为输出电压，i 为通过电阻的电流，i_1 为通过电感的电流，i_2 为通过电容的电流。

图 9.3.1 RLC 电路

使用基尔霍夫电流定律，可以得到

$$i = i_1 + i_2 \tag{9.1}$$

$$\frac{1}{L}\int_0^t V_L \mathrm{d}t + C\frac{\mathrm{d}V_C}{\mathrm{d}t} = \frac{V_{in} - V_C}{R} \tag{9.2}$$

其中，R=2Ω，L=6H，C=1/3F。

将常数代入，观察电路有$V_C = V_L = V_{OUT}$，求上面的积分方程，得

$$\frac{\mathrm{d}^2V_C}{3\mathrm{d}t^2} + \frac{\mathrm{d}V_C}{2\mathrm{d}t} + \frac{1}{6}V_C = 0 \tag{9.3}$$

用框图模型表示微分方程（9.3），在 Simulink 中绘制方框图，如图 9.3.2 所示。

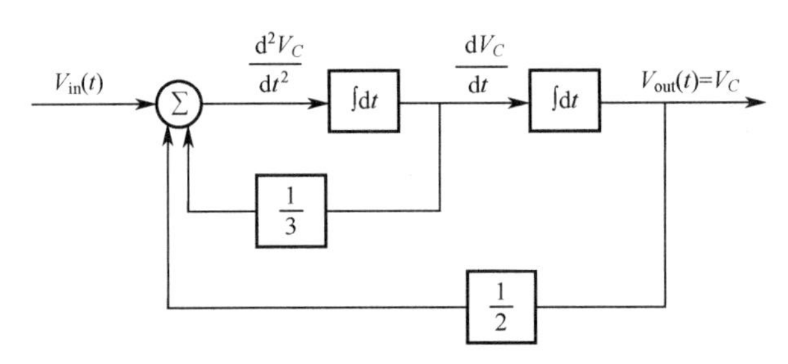

图 9.3.2 方程（9.3）的框图

用 Simulink 创建微分方程（9.3）模型，有以下步骤：

（1）在 Simulink 库浏览器中，单击图顶部标题栏上最左边的一个空白页面图标。将出现无标题的新模型窗口，如图 9.3.3 所示。

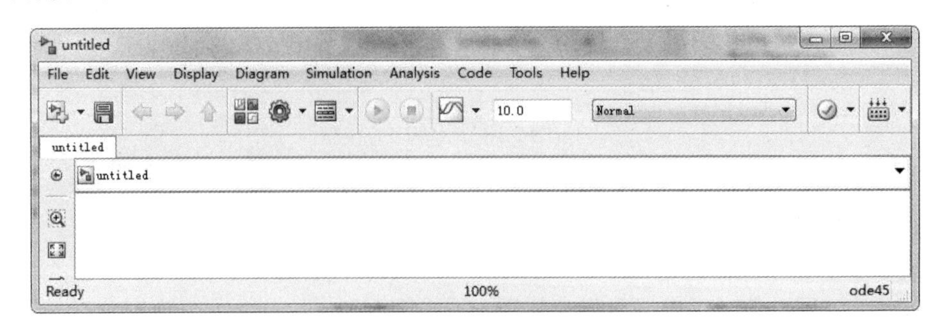

图 9.3.3 在 Simulink 中无标题的模型窗口

图 9.3.3 窗口是模型窗口，输入模块组成模块框图，我们将这个模型另存为名为 RLC 的模型文件。Simulink 将自动添加.slx 扩展名，新模型窗口将显示为 RLC，所有依此保存的文件，均以该方式显示。如图 9.3.4 所示。

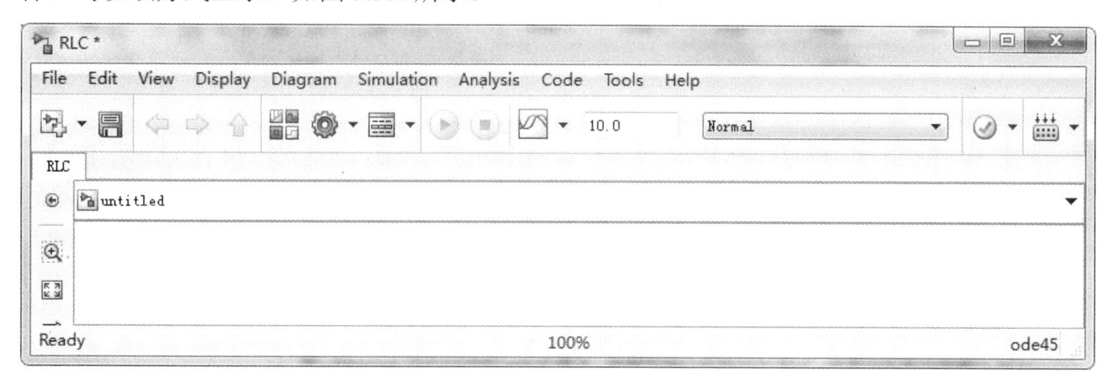

图 9.3.4　RLC.slx 文件模型窗口

（2）RLC 模型窗口和 Simulink 库浏览器均可见，单击 Sources 模块出现左侧列表，单击右侧向下滑动，会看到 Step（单位阶跃函数模块），如图 9.3.5 所示。

选择该模块，将其拖动到 RLC 模型窗口中，Step 模块会出现，如图 9.3.5 所示。在 RLC 模型窗口使用 File 下拉菜单保存文件 RLC（如图 9.3.5 右侧所示）。

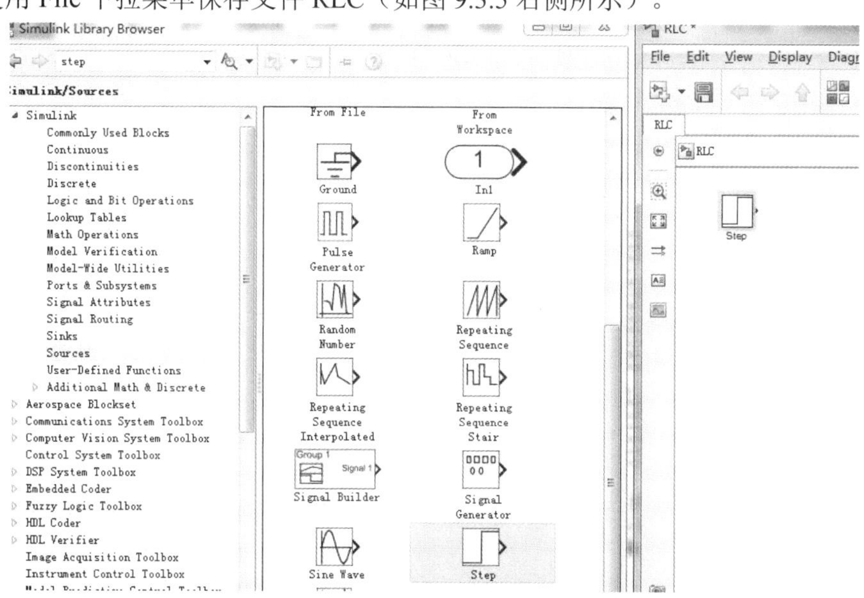

图 9.3.5　将 Step 单位阶跃函数拖到 RLC 模型窗口

（3）参照图 9.3.2，还需要连接一个放大器模块、增益模块和单位阶跃函数模块，在 Simulink 中，增益模块在常用模块组（常用模块组是 Simulink 库浏览器第一项）下，如图 9.3.5 所示。如果 RLC 模型窗口不可见，则单击库浏览器顶部菜单栏中的白页图标，然后模型出现。

（4）接下来需要添加有 3 个输入的加法器模块。加法器模块在 Simulink 库浏览器右下的数学运算模块中。用鼠标选中并拖到 RLC 模型窗口。双击，在出现的功能模块参数窗口上指定 3 个输入。然后，将单位阶跃函数模块的输出连接到加法器模块的第一个输入上，如图 9.3.6 所示。

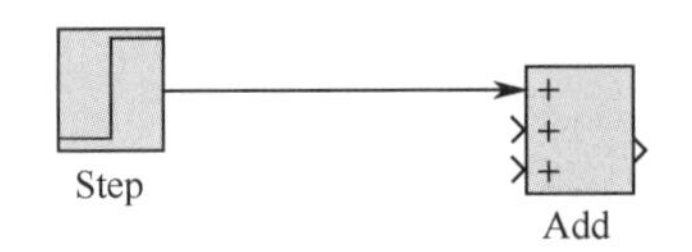

图 9.3.6 已增加了单位阶跃函数模块的 RLC 文件

（5）从 Simulink 库浏览器的常用模块中，选择积分器模块，把它拖到 RLC 模型窗口，将其连接到加法器模块的输出端口上。重复此步骤，添加第二个积分器模块。单击第一个积分器模块下文本"积分器"，命名为 Integrator1。同样，把第二个积分器模块命名为 Integrator2，如图 9.3.7 所示。

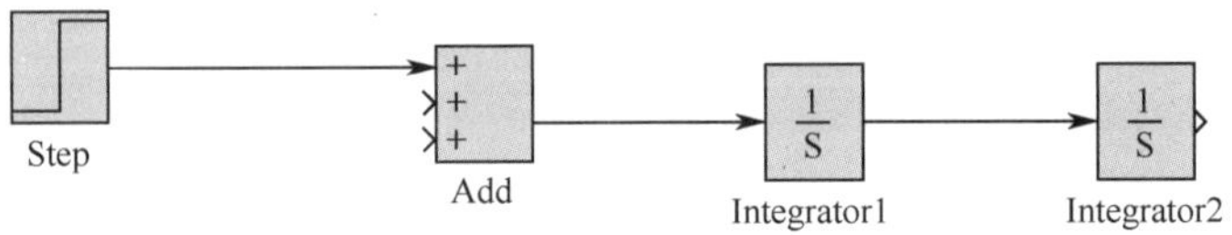

图 9.3.7 增加了两个积分器模块的 RLC 文件

（6）要完成框图，需继续添加示波器模块，该模块在 Simulink 库浏览器的常用模块组中，单击增益模块，将其复制并粘贴两次。在格式下拉菜单中使用 Flip Block 翻转模块命令来翻转已粘贴板增益模块，然后标注两个增益模块为 Gain1 和 Gain2。最后，双击增益模块打开功能模块参数窗口，将 Gain1 参数改为 1/3，Gain2 参数改为 1 /2，如图 9.3.8 所示。

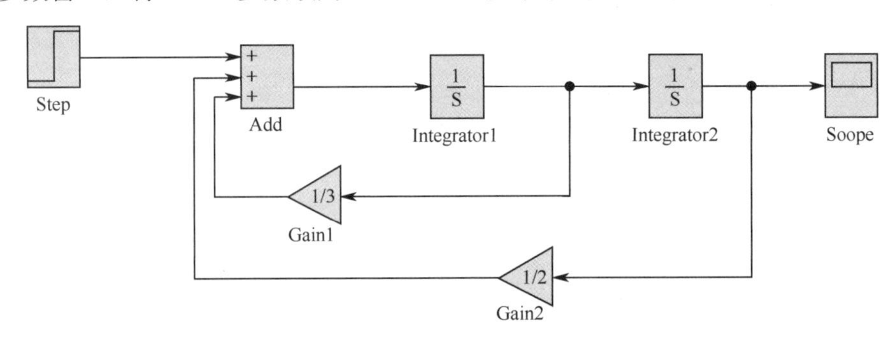

图 9.3.8 完整的 RLC 框图

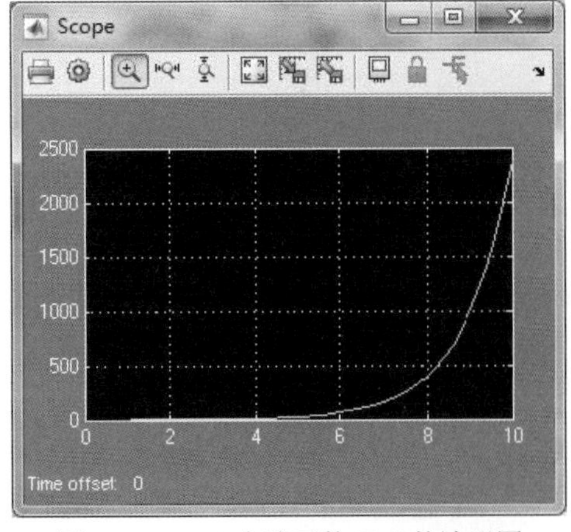

图 9.3.9 RLC 电路函数 $V_C(t)$的波形图

（7）双击积分器模块，将积分器模块的初始条件改为 0。还需指定仿真时间，可从 Simulation（仿真）下拉菜单的 Configuration Parameters（配置参数栏）将仿真时间设置为 10 秒。然后开始仿真，从模型下拉菜单中或者单击图标 ▶ 启动仿真。

（8）要观察输出波形，双击示波器模块，波形图如图 9.3.9 所示。

本章小结

（1）简单介绍了 Simulink 模块库；

（2）通过把具体 Sin 函数信号及其积分后信号共同显示在示波器中，介绍了连线与函数模块的操作；

（3）以 RLC 模型的搭建过程，对模型搭建进行详细讲解。

Chapter 9
Simulink Modeling Examples for Simple Systems

This chapter describes the basics of Simulink. By taking a simple model called *simple_model* as an example, the operation of wiring and function modules in Simulink is illustrated. Then the construction of the RLC model is explained in detail, the specific steps of modeling are demonstrated, and the simulation of this model is realized.

9.1 Introduction to Simulink Software

9.1.1 Starting Simulink Software

To start the Simulink software, you must first start the MATLAB technical computing environment. Consult your MATLAB documentation for more information. You can then start the Simulink software in two ways:

On the toolbar, click the *Simulink* icon.

Enter the "*simulink*" command at the MATLAB prompt.

The Library Browser appears, as shown in Figure 9.1.1. It displays a tree diagram of the Simulink block libraries installed on your system. You can build new models by copying blocks from the Library Browser into a model window.

The Simulink library window displays icons representing the pre-installed block libraries. You can create your models by copying blocks from the library into a model window.

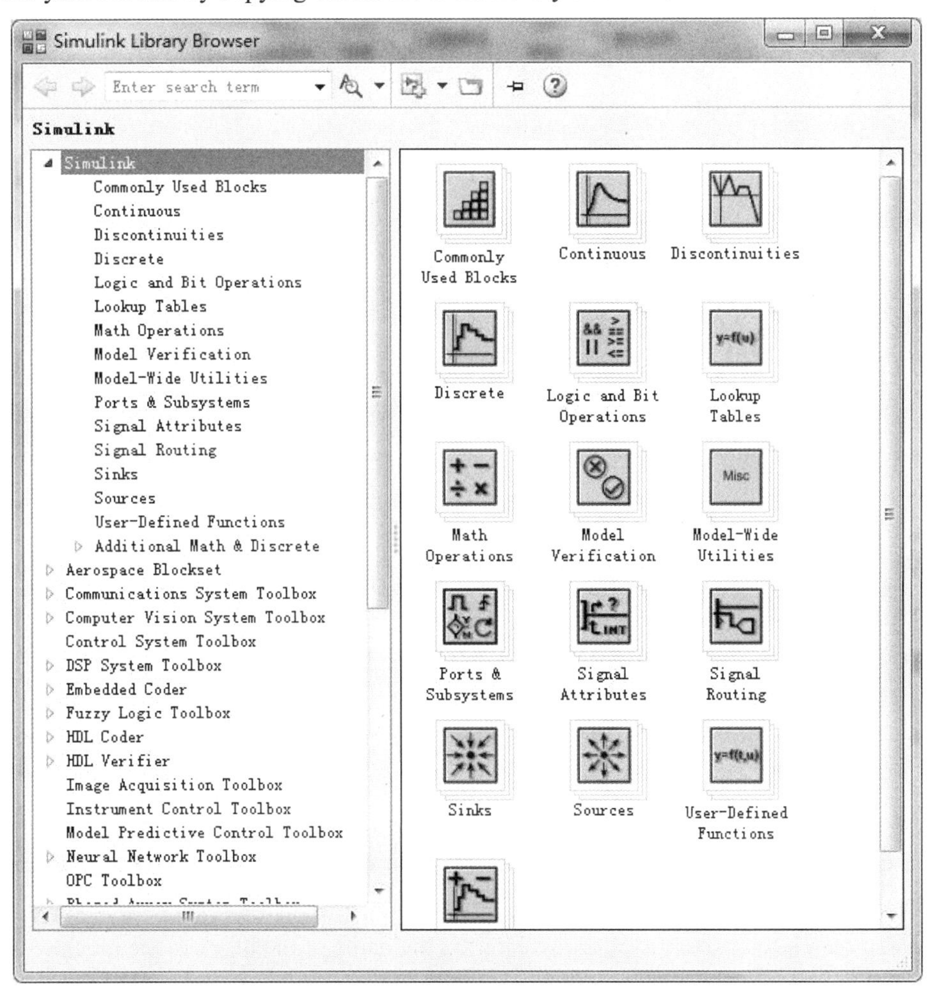

Figure 9.1.1 Simulink Library Browser

9.1.2 Simulink User Interface

9.1.2.1 Simulink Library Browser

The Library Browser displays the Simulink block libraries installed on your system. You can build new models by copying blocks from a library into a model window. See Figure 9.1.2.

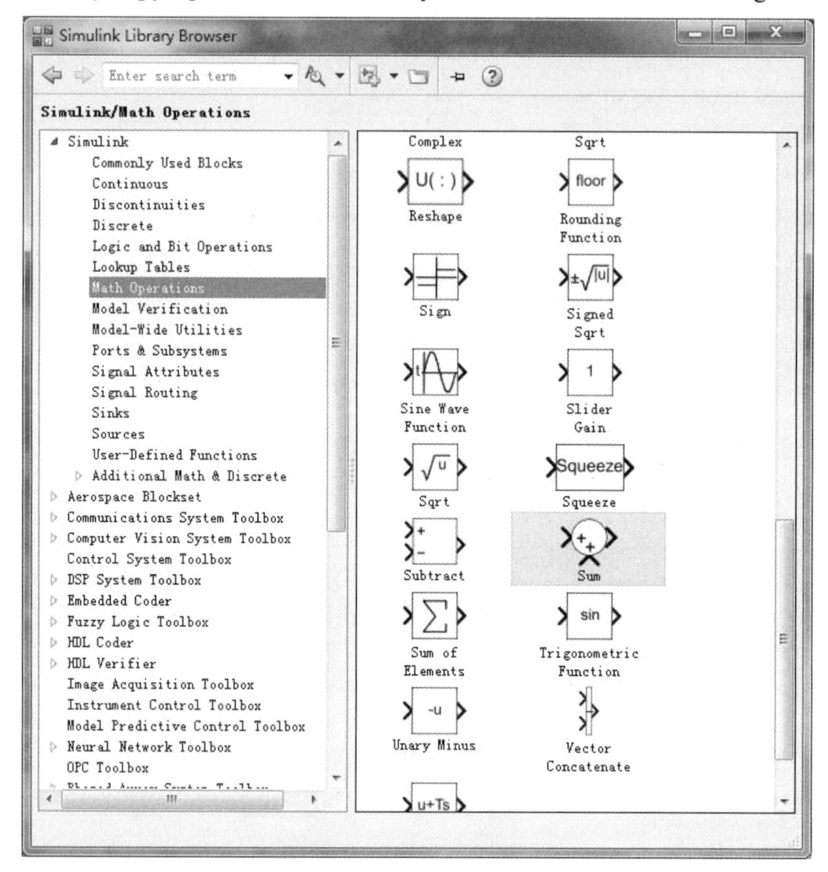

Figure 9.1.2 Simulink User Interface

9.1.2.2 Tips for Using the Library Browser

When using the Library Browser, please note the following points:

1. You can view the blocks in a library by selecting the library name on the left side of the Library Browser, or by double-clicking the library.

2. When you select a block, a description of that block appears at the bottom of the browser.

3. For more information on a block, select the block, then select Help>Help on the Selected Block to display the help page for the block.

4. You can view the parameters for a block by right-clicking the block, then selecting Block Parameters.

5. You can search for a specific block by entering the name of the block in the block search field, then clicking the Find block icon.

9.1.2.3 Standard Block Libraries

Simulink software provides 16 standard block libraries. Table 9.1 describes them each.

Table 9.1 16 Standard Block Libraries

Standard Block Libraries	Description
Commonly Used Block	Contains a group of the most commonly used blocks, such as the Constant, In1, Out1, Scope, and Sum blocks. Each of the blocks in this library are also included in other libraries.
Continuous	Contains blocks that model linear functions, such as the Derivative and Integrator blocks.
Discontinuities	Contains blocks with outputs that are discontinuous functions of their inputs, such as the Saturation block.
Discrete	Contains blocks that represent discrete time functions, such as the Unit Delay block.
Logic and Bit Operations	Contains blocks that perform logic or bit operations, such as the Logical Operator and Relational Operator blocks.
Look-Up Tables	Contains blocks that use look-up tables to determine their outputs from their inputs, such as the Cosine and Sine blocks.
Math Operations	Contains blocks that perform mathematical and logical functions, such as the Gain, Product, and Sum blocks.
Model Verification	Contains blocks that enable you to create self-validating models, such as the Check Input Resolution block.
Model-Wide Utilities	Contains blocks that provide information about the model, such as the Model Info block.
Ports & Subsystems	Contains blocks that allow you to create subsystems, such as the In1, Out1, and Subsystem blocks.
Signal Attributes	Contains blocks that modify the attributes of signals, such as the Data Type Conversion block.
Signal Routing	Contains blocks that route signals from one point in a block diagram to another, such as the Mux and Switch blocks.
Sinks	Contains blocks that display or export output, such as the Out1 and Scope blocks.
Sources	Contains blocks that generate or import system inputs, such as the Constant, In1, and Sine Wave blocks.
User-Defined Functions	Contains blocks that allow you to define custom functions, such as the Embedded MATLAB® Function block.
Additional Math & Discrete	Contains two additional libraries for mathematical and discrete function blocks.

9.1.3 Simulink Model Window

Simulink model window contains the block diagram of the model. You can build models in the window by arranging blocks logically, setting the parameters for each block, and connecting the blocks with signal lines.

The model window also allows you to:

1. Set configuration parameters for the model, including the start-up and stop time, type of solver to use, and data import/export settings.
2. Start and stop simulation of the model.
3. Save the model.
4. Print the block diagram.

9.2 Creating a Simple Model

This section illustrates how to create a simple model by the example as shown in Figure 9.2.1.

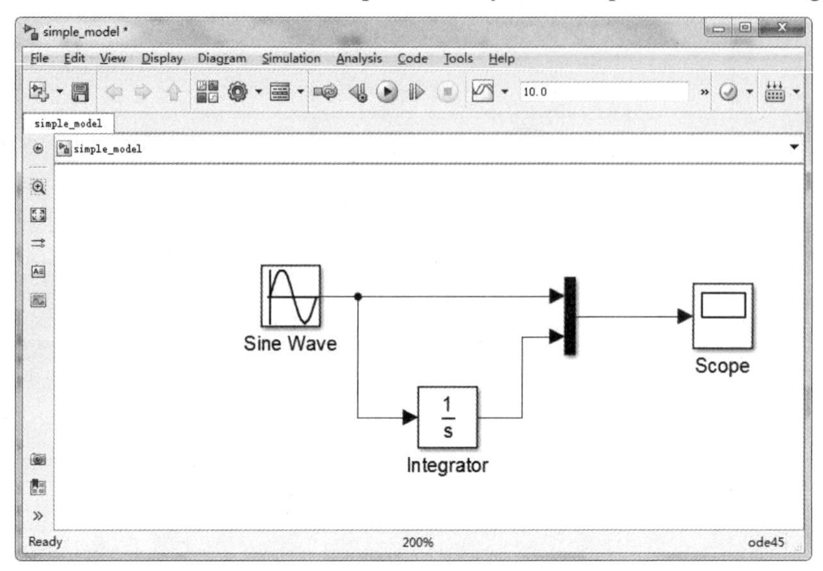

Figure 9.2.1 A Simple Model

9.2.1 Creating a New Model

Before you can begin building your model, you must start Simulink and create an empty model.

To create a new model:

1. If Simulink is not running, enter "*simulink*" in the MATLAB Command Window to open the Simulink Library Browser.

2. Select **File**>**New**>**Model** in the Simulink Library Browser to create a new model.

The software opens an empty model window as in Figure 9.2.2.

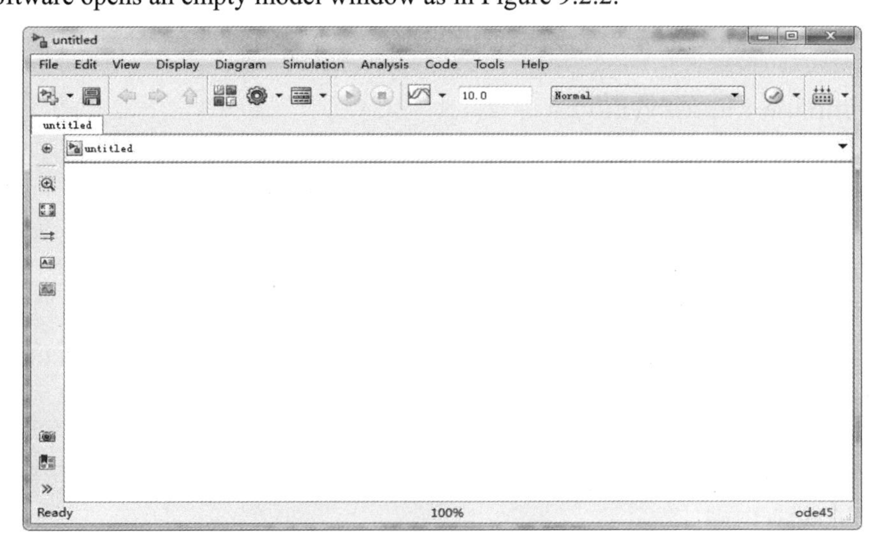

Figure 9.2.2 An Empty Model Window

9.2.2 Adding Blocks to Your Model

To construct a model, first copy blocks from the Simulink Library Browser to the model window. To create the Simple model in Figure 9.2.1, you need four blocks:

1. Sine Wave — To generate an input signal for the model.
2. Integrator — To process the input signal.
3. Scope — To visualize the signals in the model.
4. Mux — To multiplex the input signal and processed signal into a single scope.

To add blocks mentioned above to your model:

1. Select the Sources library in the Simulink Library Browser.

The Simulink Library Browser displays the Sources library as shown in Figure 9.2.3.

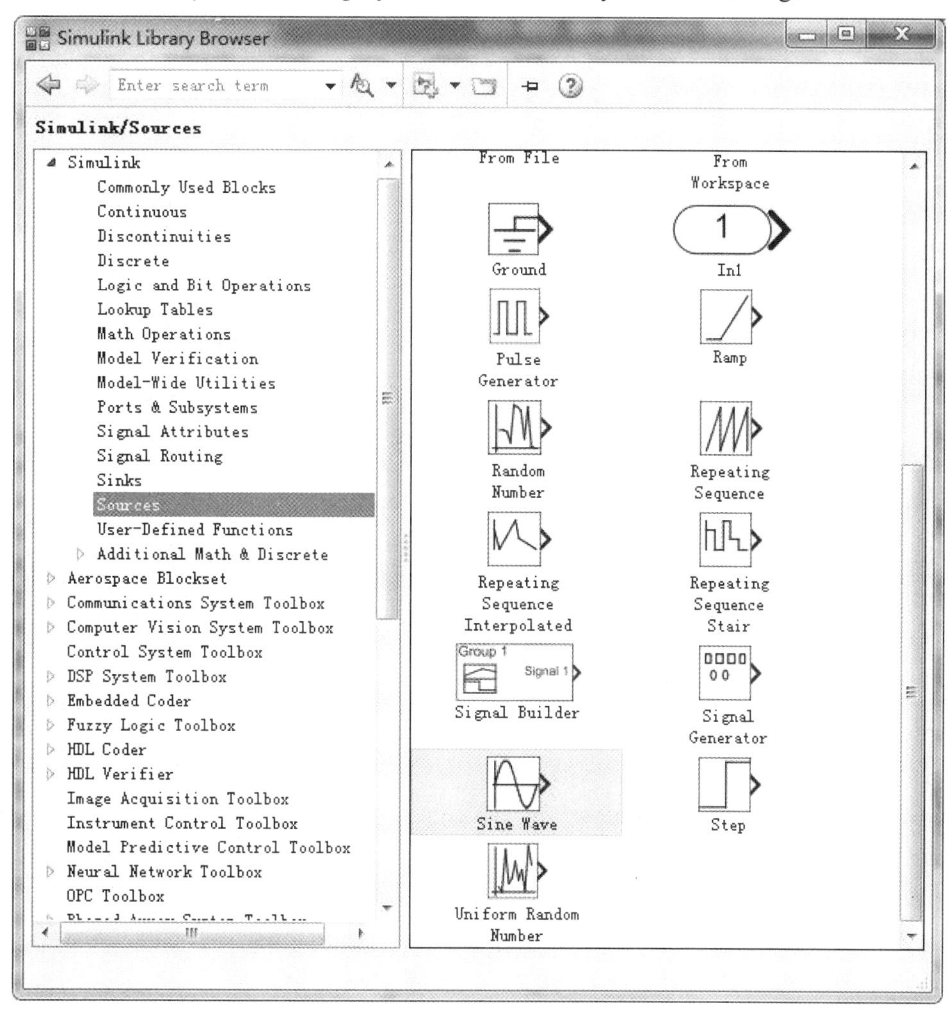

Figure 9.2.3 The Sources Library

2. Select the Sine Wave block in the Simulink Library Browser, then drag it to the model window. A copy of the Sine Wave block appears in the model window as shown in Figure 9.2.4.

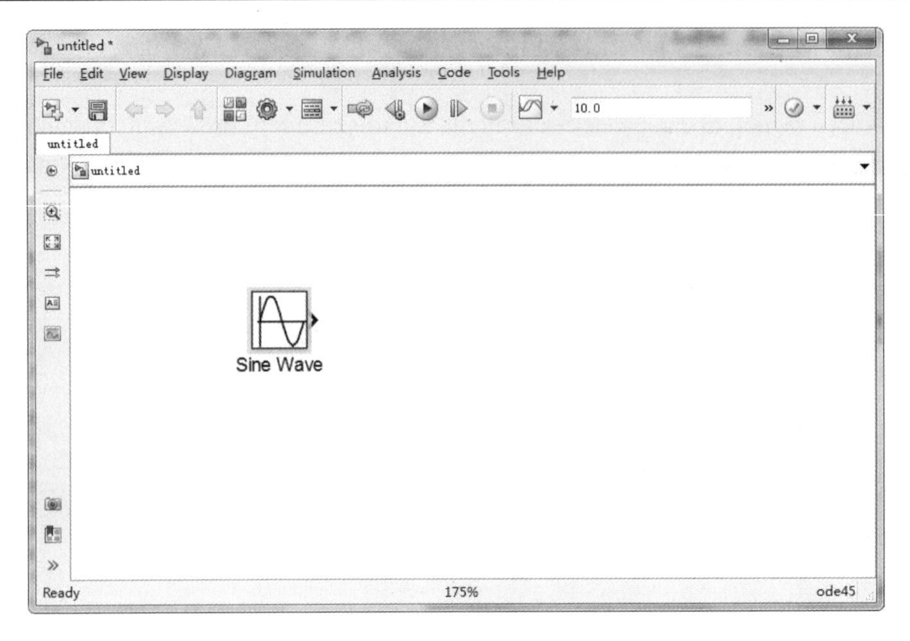

Figure 9.2.4 The Sine Wave Block

3. Select the Sinks library in the Simulink Library Browser.

4. Select the Scope block from the Sinks library, then drag it to the model window.

A Scope block appears in the model window.

5. Select the Continuous library in the Simulink Library Browser.

6. Select the Integrator block from the Continuous library, then drag it to the model window. An Integrator block appears in the model window.

7. Select the Signal Routing library in the Simulink Library Browser.

8. Select the Mux block from the Sinks library, then drag it to the model window. A Mux block appears in the model window.

9.2.3 Moving Blocks in the Model Window

Before connecting the blocks in your model, you should arrange them logically to make the signal connections as straightforward as possible.

To move a block in the model window, you can either

1. Drag the block.

2. Select the block, then press the arrow keys on the keyboard.

Arrange the blocks in the model to look like Figure 9.2.5 as follows.

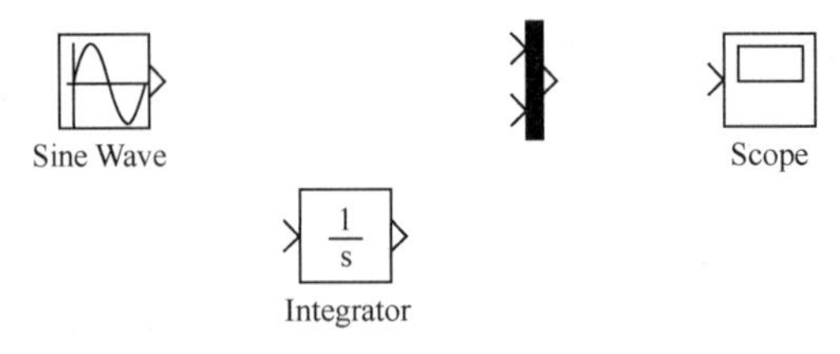

Figure 9.2.5 Well-Arranged Blocks

9.2.4 Connecting Blocks in the Model Window

After adding blocks to the model window, you must connect them to represent the signal connections within the model.

Note: each block has angle brackets on one or both sides. These angle brackets represent input and output ports (See Figure 9.2.6) :

1. The > symbol pointing into a block is an *input port.*
2. The > symbol pointing out of a block is an *output port.*

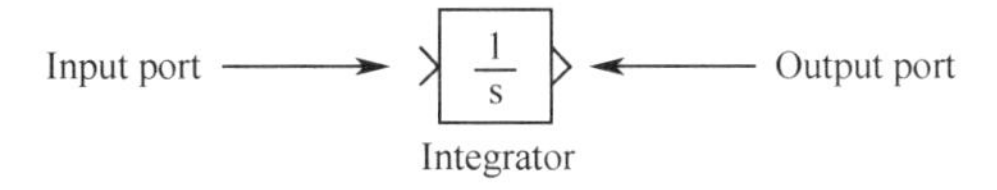

Figure 9.2.6 Angle Brackets Represent Input and Output Ports

Next, let's describe how to connect blocks by drawing lines from output ports to input ports.

9.2.4.1 Drawing Lines Between Blocks

You connect the blocks in your model by drawing lines between output ports and input ports. To draw a line between two blocks,

1. Point the mouse to the output port on the right side of the Sine Wave block.

Note that the pointer changes to a cross (+) while over the port (See Figure 9.2.7).

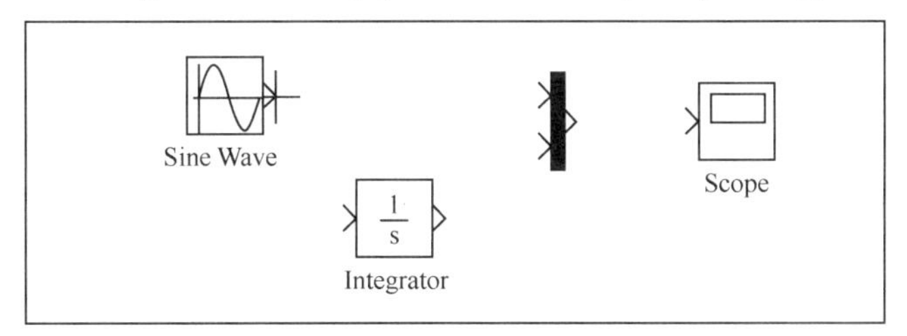

Figure 9.2.7 The Cross Cursor of the Port

2. Drag a line from the output port to the top input port of the Mux block.

Note: the line is dashed while you press the mouse button, and the pointer changes to a double-lined cross as it approaches the input port of the Mux block (See Figure 9.2.8) .

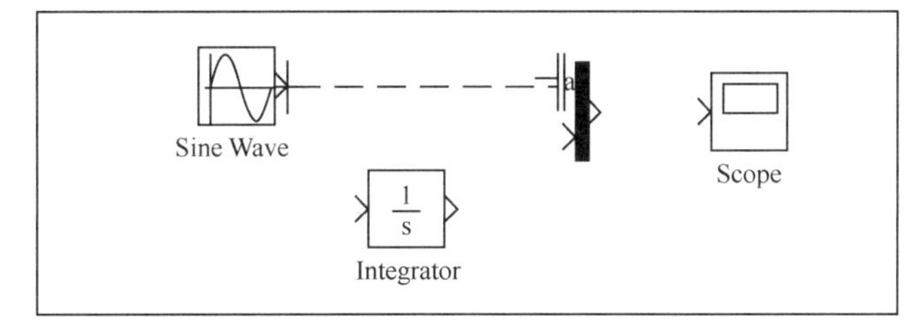

Figure 9.2.8 Double-Lined Cross

3. Release the mouse button over the output port and the line is connected to the input port of the Mux block. An arrow indicates the direction of signal flow (See Figure 9.2.9) .

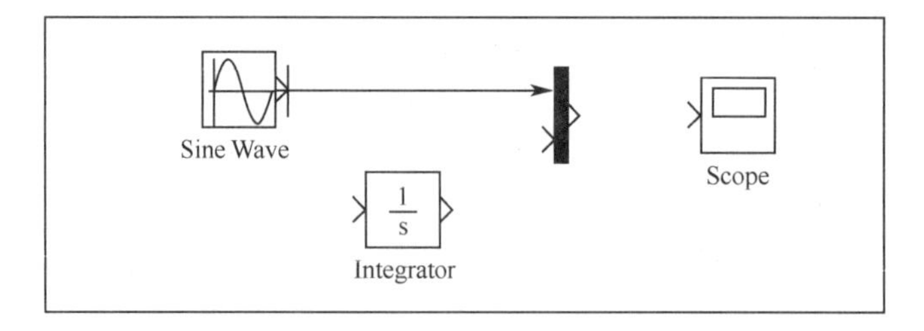

Figure 9.2.9 Arrow Direction of Software Connecting Blocks

4. Drag a line from the output port of the Integrator block to the input port at the bottom of the Mux block.

5. The self-connecting blocks between soft-wares: Select the Mux block, then Ctrl+click the Scope block. The software automatically draws the connection line between the blocks. The connected model is shown in Figure 9.2.10.

Note: The Ctrl+click shortcut is especially useful when you are connecting widely separated blocks, or when working with complex models.

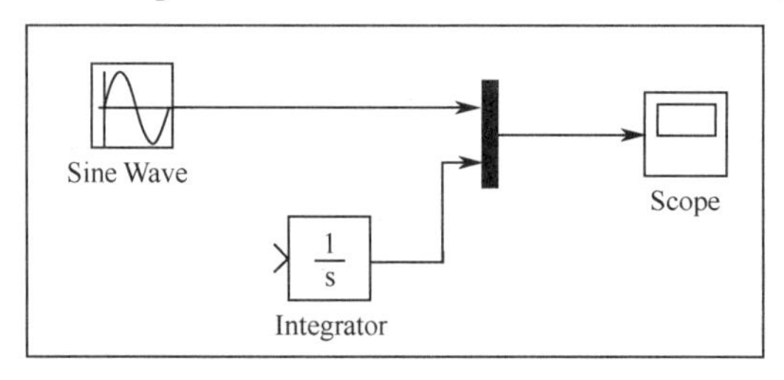

Figure 9.2.10 The Connected Model

9.2.4.2 Drawing a Branch Line

The model in Figure 9.2.10 is almost complete, but one connection is missing. To finish the model, you must connect the Sine Wave block to the Integrator block.

This final connection is somewhat different from the other three, which all connect output ports to input ports. Because the output port of the Sine Wave block already has a connection, you must connect this existing line to the input port of the Integrator block. The new line, called a branch line, carries the same signal that passes from the Sine Wave block to the Mux block.

To weld a connection to an existing line:

1. Position the mouse pointer on the line between the Sine Wave and the Mux block (See Figure 9.2.11).

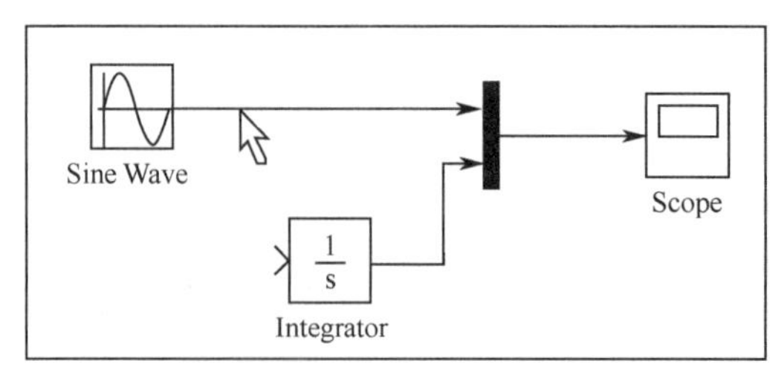

Figure 9.2.11 The Mouse Pointer *on the line* Between the Sine Wave and the Mux Block

2. Press and hold the Ctrl key, then drag a line to the Integrator block's input port (See Figure 9.2.12) .

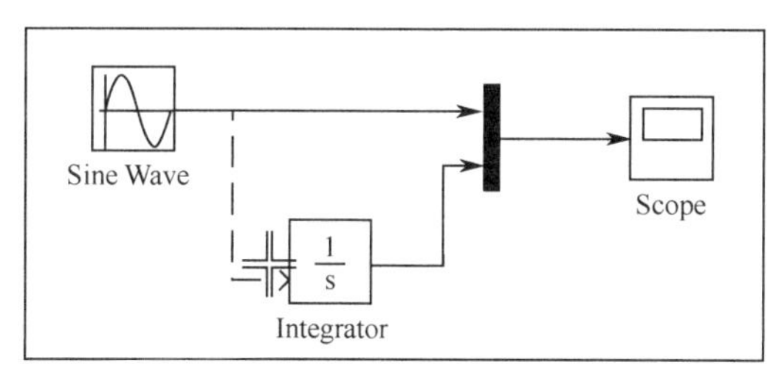

Figure 9.2.12 Ctrl and Drag a Line to the Integrator

3.The software draws a line between the starting point and the input port of the Integrator block. The model is now complete, similar to the figure shown in Figure 9.2.13.

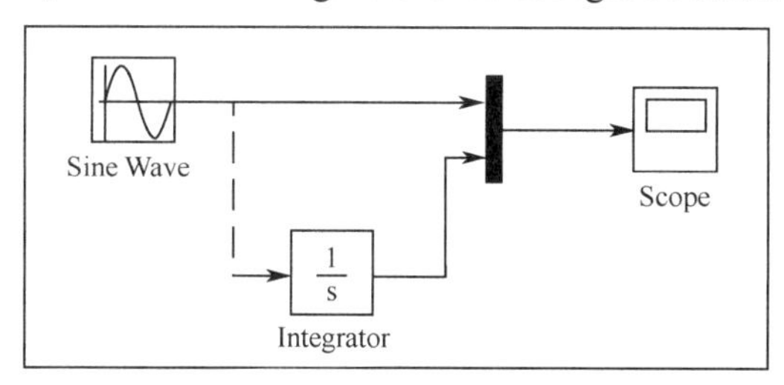

Figure 9.2.13 The Completed Model

9.2.5 Saving the Model

After completing the model, you should save it for future use.

To save the model:

1. Select File>Save in the model window.
2. Specify the location in which you want to save the model.
3. Enter *simple_model* in the File name field.
4. Click Save, and the software saves the model with the file name *simple_model.slx*.

9.3 RLC Circuit Model Simulation

Consider an RLC circuit shown in Figure 9.3.1, where

V_{in}=input voltage(volts),

L= inductance(H),

R=resistance(Ω),

C=capacitance(F),

V_{out}=output voltage(volts),

i=current across the resistance

i_1=current across the inductance

i_2=current across the capacitor

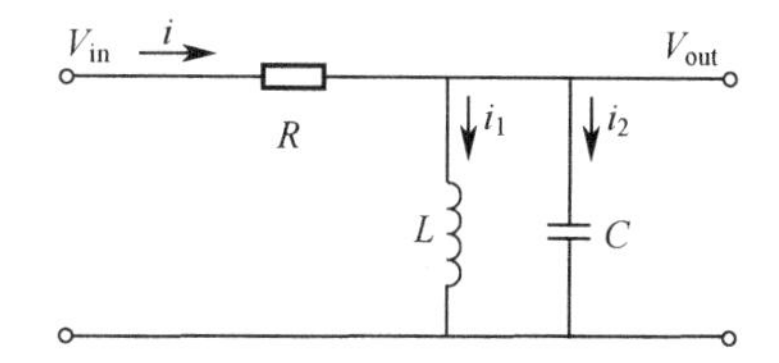

Figure 9.3.1 An RLC Circuit

By Kirchhoff's Current Law (KCL), where R=2Ω, L=6H, C=1/3F.

$$i = i_1 + i_2 \tag{9.1}$$

$$\frac{1}{L}\int_0^t V_L \mathrm{d}t + C\frac{\mathrm{d}V_C}{\mathrm{d}t} = \frac{V_{\mathrm{in}} - V_C}{R} \tag{9.2}$$

By substitution of the circuit constants, observing that $V_C = V_L = V_{\mathrm{OUT}}$ and differentiating the above integro-differential equation, we get

$$\frac{\mathrm{d}^2V_C}{3\mathrm{d}t^2} + \frac{\mathrm{d}V_C}{2\mathrm{d}t} + \frac{1}{6}V_C = 0 \tag{9.3}$$

A block diagram representing the equation (9.3) is shown by Simulink in Figure 9.3.2.

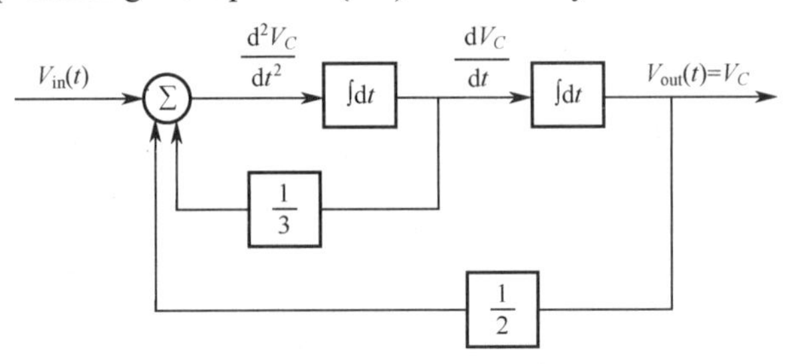

Figure 9.3.2 Block Diagram for Equation (9.3)

The steps to model the differential equation (9.3) using Simulink are as follows,

1. On the Simulink Library Browser, we click on the leftmost icon shown as a blank page on the top title bar. A new model window named untitled will appear as shown in Figure 9.3.3.

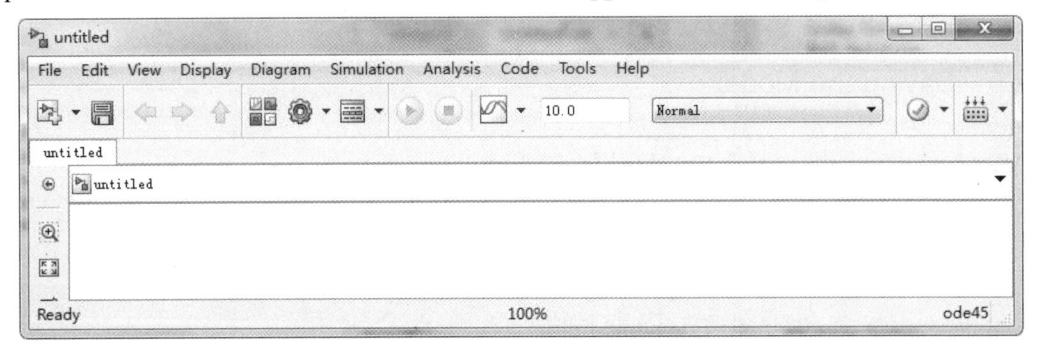

Figure 9.3.3 The Untitled Model Window in Simulink

The window of Figure 9.3.3 is the model window where we enter our blocks to form a block diagram. We save it as model file name RLC to which Simulink will add the extension *.slx*. The new model window will now be shown as RLC, and all saved files will appear this way. See Figure 9.3.4.

2. With the RLC model window and the Simulink Library Browser both visible, we click on the Sources to show the left side list, and click on the right side to scroll down until we see the unit step function(See Figure 9.3.5.).

Select and drag it into the RLC model window which now appears as shown in Figure 9.3.5. We save the file RLC using the File drop menu on the RLC model window (right side of Figure 9.3.5).

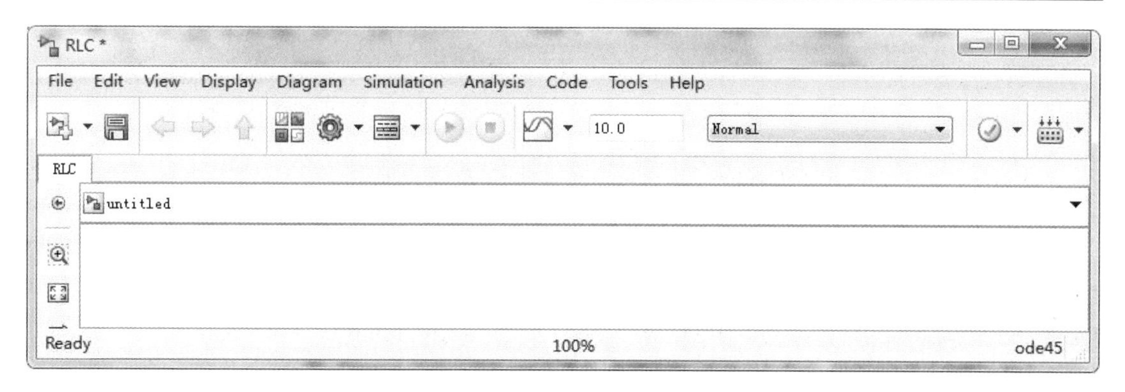

Figure 9.3.4 Model Window for RLC.slx File

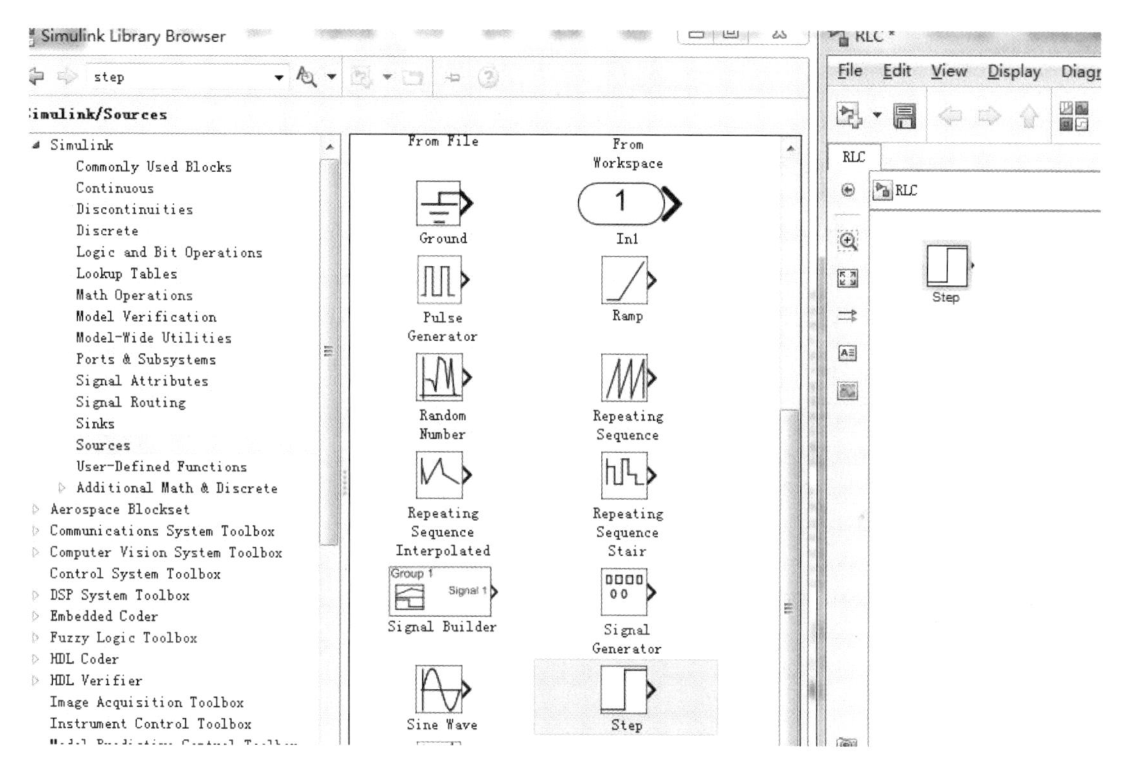

Figure 9.3.5 Dragging the Unit Step Function into File RLC

3. With reference to block diagram of Figure 9.3.2, we need to connect an amplifier with Gain to the unit step function block. The gain block in Simulink is under Commonly Used Blocks (first item under Simulink on the Simulink Library Browser). See Figure 9.3.5. If the RLC model window is no longer visible, it can be recalled by clicking on the white page icon on the top bar of the Simulink Library Browser.

4. Next, we need to add a three-input adder. The adder block appears on the right side of the Simulink Library Browser under Math Operations. Select it and drag it into the RLC model window. Then double-click it, and on the Function Block Parameters window which appears, we specify 3 inputs. We then connect the output of the step block to the first input of the adder block as shown in Figure 9.3.6.

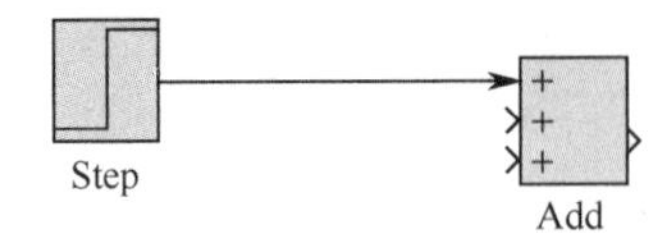

Figure 9.3.6 File RLC with Added Step Block

5. From the Commonly Used Blocks of the Simulink Library Browser, we choose the Integrator block, drag it into the RLC model window, and connect it to the output of the Add block. Then, repeat this step and add a second Integrator block. We click on the text "Integrator" under the first integrator block, and change it to Integrator 1. Then, we change the text Integrator1 under the second Integrator to Integrator2 as shown in Figure 9.3.7.

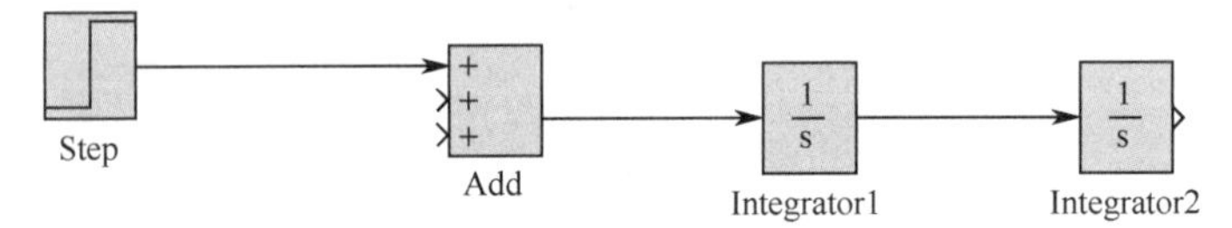

Figure 9.3.7 File RLC with the Addition of Two Integrators

6. To complete the block diagram, add the Scope block which is found in the Commonly Used Blocks on the Simulink Library Browser. We click on the Gain block, copy and paste it twice. We flip the pasted Gain blocks by using the Flip Block command from the Format drop menu, and label these as Gain1 and Gain2. Finally, we double-click on the Gain blocks and on the Function Block Parameters window, we change the gains from to 1/3 and 1/2 as shown in Figure 9.3.8.

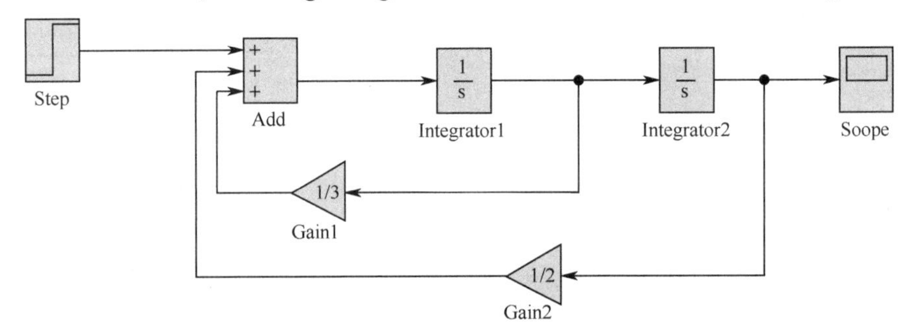

Figure 9.3.8 File RLC Complete Block Diagram

7. The initial conditions are entered double clicking the Integrator blocks and entering the values 0 for them. We also need to specify the simulation time. This is done by specifying the simulation time to be 10 seconds on the Configuration Parameters from the Simulation drop menu. We can start the simulation on Start from the Simulation drop menu or by clicking on the ▶ icon.

8. To see the output waveform, we double click on the Scope block, we obtain the waveform shown in Figure 9.3.9.

Summary

1. The Simulink Library Browser is briefly introduced.

2. By displaying the Sin function signal and its integrated signal in the scope, the operation of the wiring and function module is introduced.

3. By taking RLC modeling as an example, the modeling process is shown in detail.

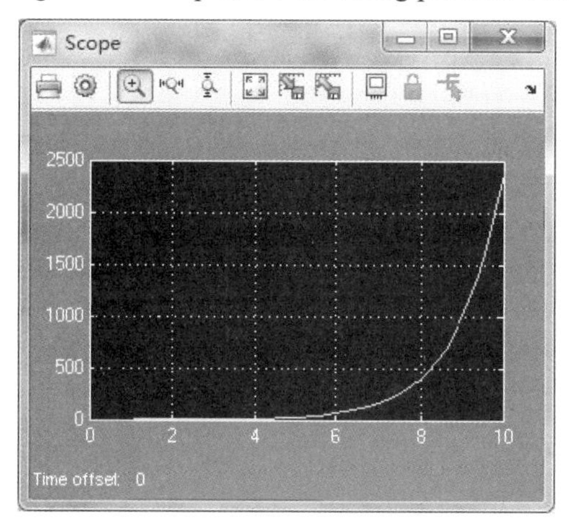

Figure 9.3.9 The Waveform for the Function $V_C(t)$ for RLC Circuit

生词注解

Simulink library browser	Simulink 库浏览器
display	vt. 显示；表现；陈列
documentation	n. 文件
logically	adv. 逻辑上
port	n. 端口
mux	abbr. 合并模块；多路复用器
substitution	n. 代替；[数] 置换；代替物
integrator	n. 积分器
configuration	n. 配置；结构；外形
autoscale	自动定标；自动比例

第 10 章 房屋加热系统建模实例

本章在第 9 章的基础上，介绍了如何创建子系统和子系统的封装技术。以房屋加热系统模型为例，详细说明了该系统中创建子系统的步骤，以及对子系统实现封装的过程。最后，对该系统进行了仿真。

10.1　创建和封装子系统

10.1.1　创建子系统

10.1.1.1　概述

子系统是一套模块组，该模块组可由某个独立模块（又称子系统模块）替代。当用户模型较大且复杂时，可以在该模型内把多个模块合并成相应的子系统来简化模型。采用子系统具有以下优点：

（1）减少了模型窗口中显示的模块数目。

（2）使功能相关的模块集中在一起。

（3）可以建立分层次模块图，图中子系统模块在一层，而组成该子系统的模块在另一层。

创建子系统有两种方式：

- 在模型中添加 Subsystem 模块，然后打开该模块，在打开的子系统窗口中添加其所包含的各种模块。
- 添加模块到子系统中，组合在一起的模块构成了子系统。

10.1.1.2　添加 Subsystem 模块创建子系统

创建子系统，在向子系统中添加所包含的模块之前，首先必须添加 Subsystem 模块到模型中，然后再添加 Subsystem 的构成模块。

（1）将 Ports & Subsystems 模块库中的 Subsystem 模块复制到模型窗口中。

（2）双击打开 Subsystem 模块。

子系统会在当前窗口或新的模型窗口中打开。在空白 Subsystem 窗口中组建子系统。子系统窗口中的 Inport 模块表示来自子系统的外部输入，Outport 模块表示子系统对外部的输出。

例如，在图 10.1.1 中，子系统包含一个 Relay1 模块，Inport 和 Outport 模块表示子系统的输入和输出。

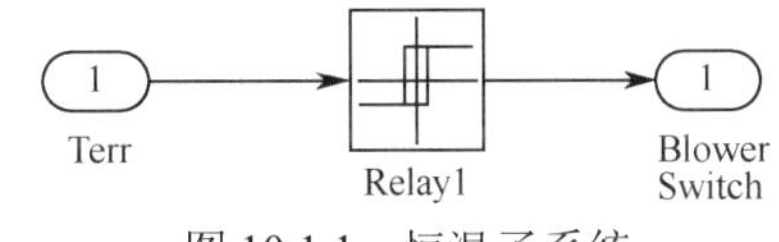

图 10.1.1　恒温子系统

10.1.1.3　组合已有模块创建子系统

如果模型中已经包含了用户想要转换为子系统的模块，那么可以把这些模块组合在一起构建子系统：

（1）选中子系统模块和想要包含在子系统内的连接线，不能一个个单独地选择模块来确定待创建的子系统区域，也不能通过使用 Select All 命令选择区域。

例如，图 10.1.2 是一个加热器模型，Sum、Product、Constant 和 Gain 模块都被选中在虚线框内。

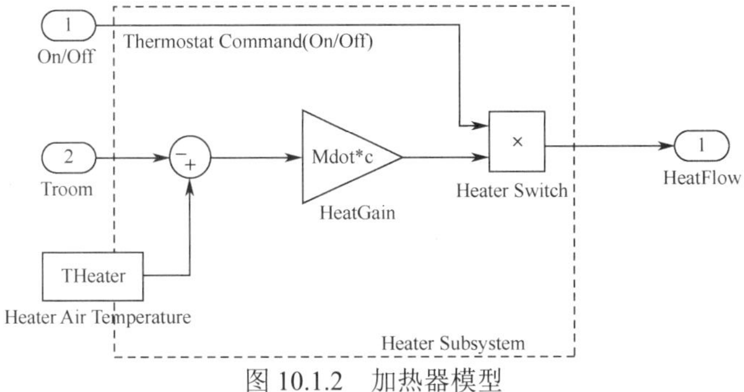

图 10.1.2　加热器模型

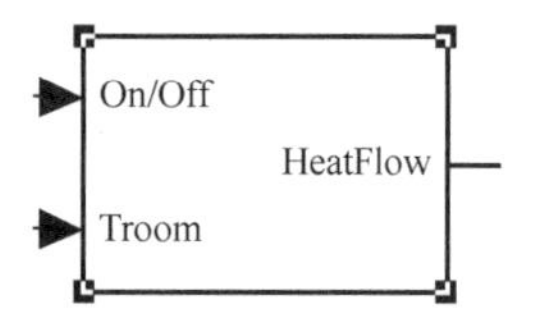

图 10.1.3　加热器子系统模块

（2）在 Edit 菜单中选择 Create Subsystem，选中的模块由一个 Subsystem 模块取代。选择 Create Subsystem 命令之后，图 10.1.3 显示了该模型。

注意：Simulink 软件添加 Inport 和 Outport 模块分别表示从外部输入、输出到外部。

10.1.2　封装 Subsystem

10.1.2.1　概述

封装就是创建子系统的自定义用户界面，隐藏了那些底层模块的原有用户界面，通过替换图标和所定义的参数对话框来实现封装。封装可用于 Subsystem 模块、Model 模块或 S-Function 模块中的任一模块。该模块可自由放置于用户自定义的库中。封装仅改变模块的使用界面，而不改变它的基本特征。

已封装模块的图标和参数对话框，可以实现与系统原有模块的图标和参数对话框一样的功能。当设置封装模块参数值时，新设置的参数值就可以实现对封装模块的图标和对话框进行动态调整，同时可以计算出在封装内所要用到的参数值。在子系统内，封装模块能根据封装的参数值相应动态调整子系统。

10.1.2.2　创建一个封装模块

1. 打开封装编辑器

Simulink Mask Editor 为封装一个模块提供所有功能。要调用封装模块的编辑器。

（1）图 10.1.4 是一个房屋模型。在虚线框内选中 Sum 模块、Integrator 模块和 Gain 模块，然后从 Edit 菜单中选择 Create Subsystem，则被选中的模块被一个子系统模块取代（见图 10.1.5）。

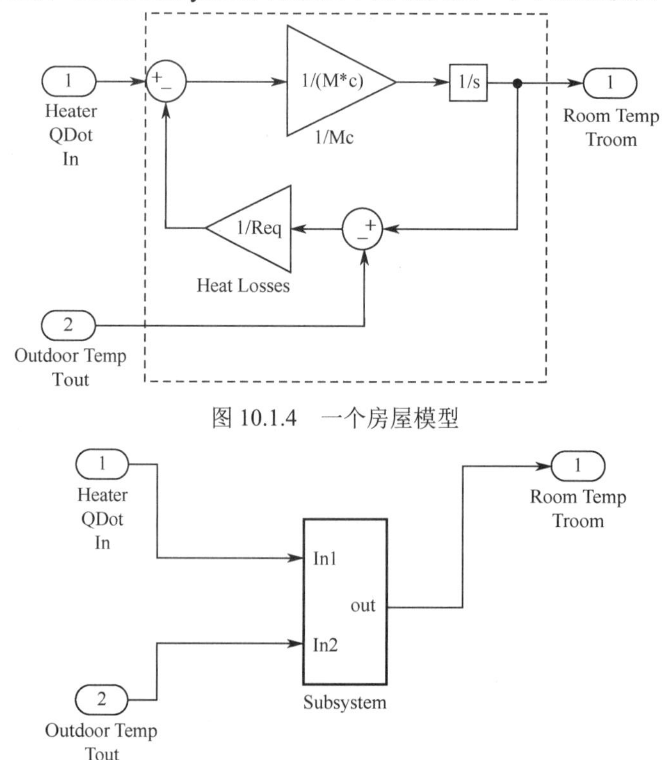

图 10.1.4　一个房屋模型

图 10.1.5　一个房屋子系统模型

（2）从 Edit 菜单中选择 Mask Subsystem，打开封装编辑器。封装编辑器如图 10.1.6 所示。

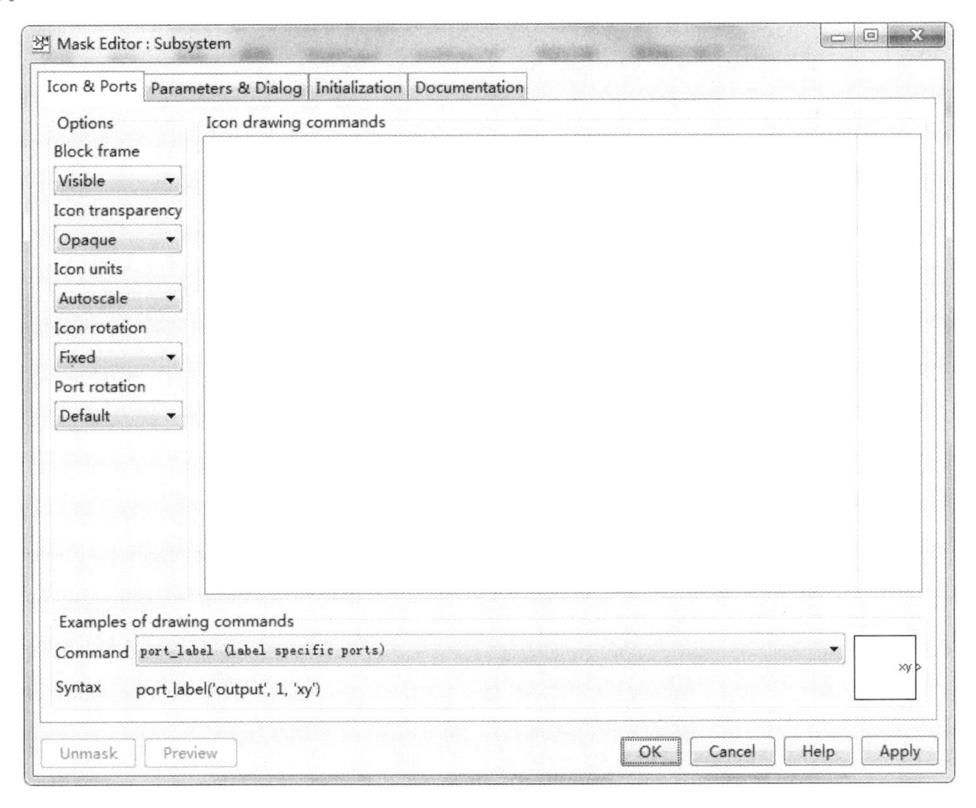

图 10.1.6　封装编辑器

只要最后结果正确，就可以按照任一顺序定义组件，也可以打开封装编辑器对已经封装的模块进行编辑。要保存对封装模块的改变，单击 Apply 或 OK 按钮修改模型，关闭编辑器。如果要永久保存更改，必须保存模型本身。如果不保存，那么，所有未保存的封装编辑的更改将全部丢失。

2. 定义封装图标

当模型中出现封装模块时，封装图标可以替换该封装模块的默认图标。封装图标可以包含描述性的文本、图形、图像、状态方程、一个或多个平面图，或者传递函数。在 Simulink 中使用命令来绘制封装图标。

使用编辑器的 Icon & Ports 窗格指定封装模块的图标。打开封装编辑器时，此窗格总是被默认选中。

指定封装图标的内容，在图标绘制命令文本框输入一个或多个描述封装图标的绘制命令，用这些命令就能画一个封装图标。如果绘图命令不能成功运行，图标将显示三个问号。

可以使用如下命令来绘制一个房屋封装图标。

```
image（imread（'fangwu.png'））    % Use fangwu icon for subsystem。
```

（注释：此处 fangwu.png 是一个图像文件。）

在下面的图标绘制命令窗格中输入如下内容：image（imread（'fangwu.png'）），如图 10.1.7 所示。

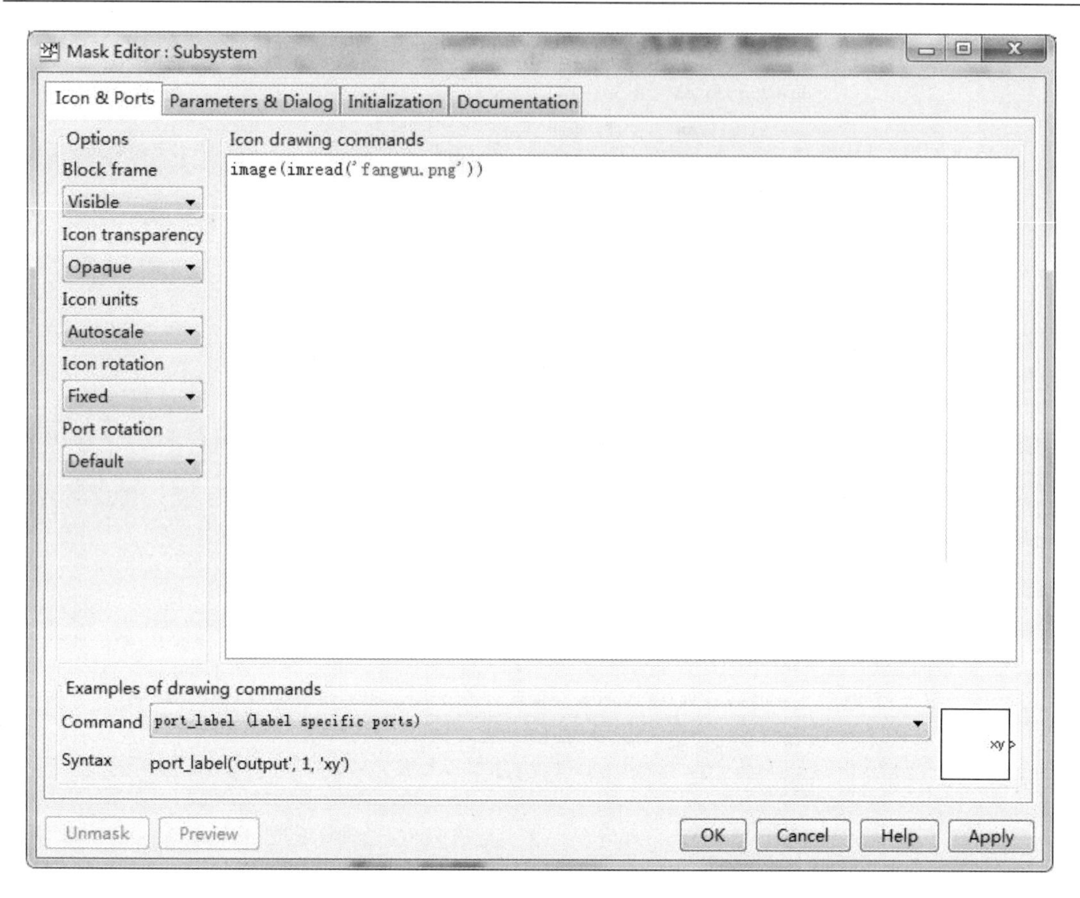

图 10.1.7　命令窗格

3. 单击 Apply

如果仿真软件不能运行命令，以此获得一个可显示的图标结果，则图标的内容是 3 个问号。单击 Apply，子模块的图标结果如图 10.1.8 所示。

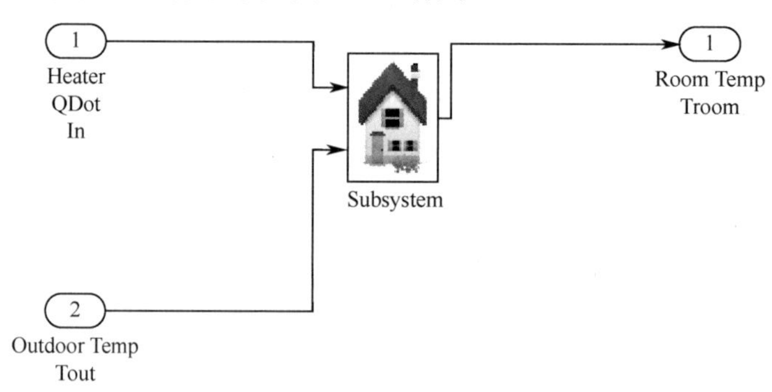

图 10.1.8　子模块的图标结果

可以通过图标的下拉选项来指定图标框的样式，透明度，绘制内容，旋转图标和端口，这样就完成了一个子系统的封装。

例如，房子和恒温子系统可显示自定义图标，来描述物理对象。双击转换子系统时，则显示自定义对话框，如图 10.1.9 所示。

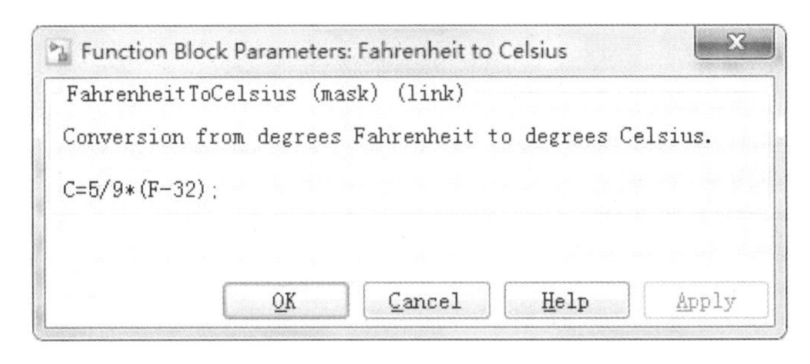

图 10.1.9　转换子系统的对话框

查看转换子系统底层模块，需要右键单击该子系统，然后选择 Look Under Mask，如图 10.1.10 所示。

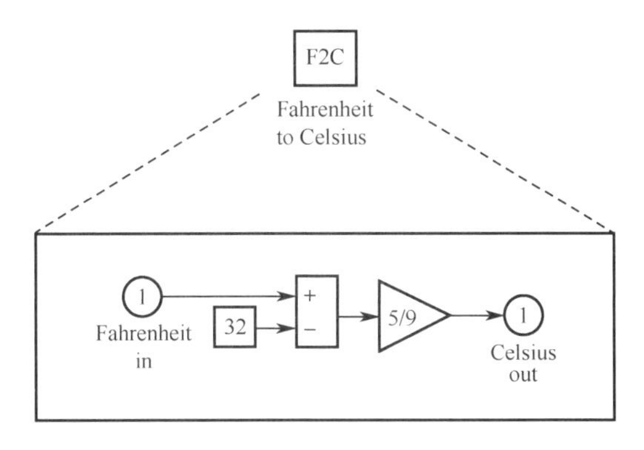

图 10.1.10　转换子系统的底层模块

10.2　房屋加热模型剖析

10.2.1　打开演示模型

本节描述的演示模型名为 sldemo_househeat，可以在 MATLAB 命令窗口中打开。

打开演示模型步骤：

（1）确保 MATLAB 已经打开。

（2）在 MATLAB 命令窗口中输入 sldemo_househeat。

启动软件并打开 sldemo_househeat 模型，如图 10.2.1 所示。

10.2.2　剖析模型

系统模拟室外环境、房子的热特性和取暖系统对房子的影响。它可以针对恒温器的设定和室外环境如何影响室内温度与供热费用进行仿真。

该模型包括多个在前一节创建简单模型中使用的相同模块。例如：

（1）Scope 模块，最右边显示仿真结果的示波器模块。

（2）Mux 模块，右下角最底端的合并模块，将室内外温度信号合并后传给示波器。

（3）Sine Wave 模块，底部左侧为模型提供了三个数据源之一的正弦波模块（标为每日温度变化）。

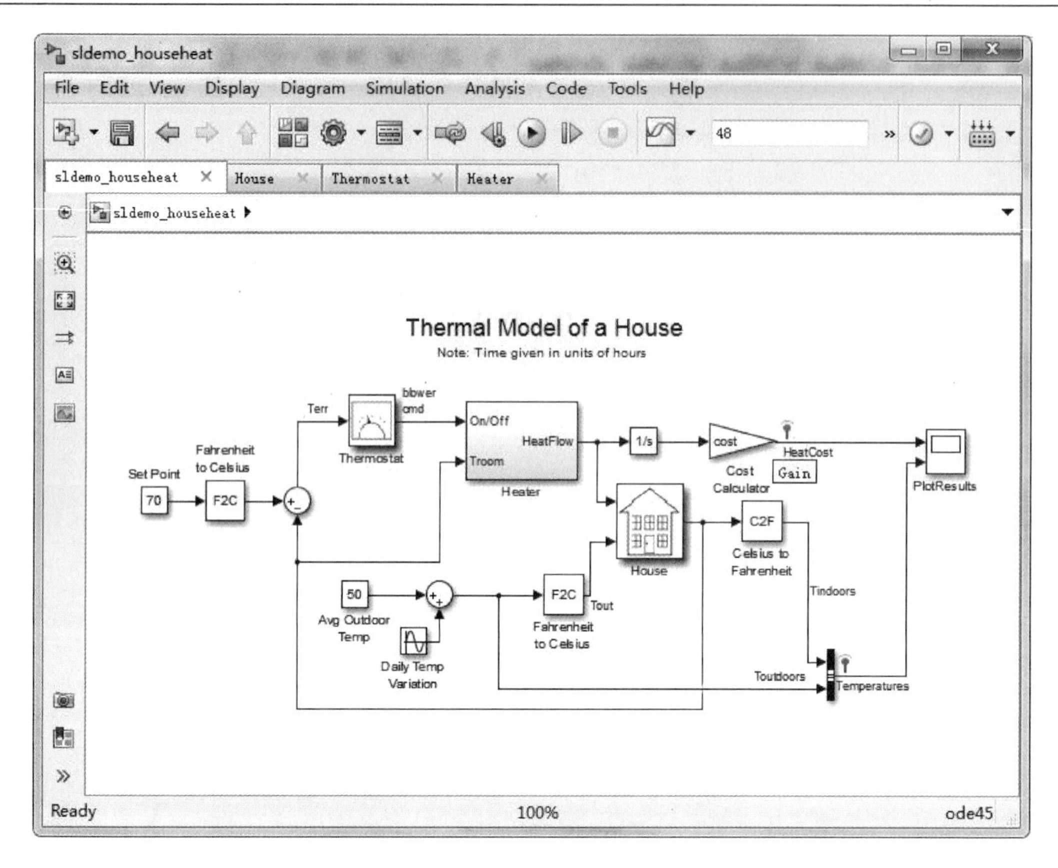

图 10.2.1 房子的热模型

模型中的恒温器系统设置为 70 华氏度。系统会用正弦波模拟室外温度的波动（振幅 15 华氏度，基底温度 50 华氏度）。

三个数据输入端（源）由两个常数模块（标设定点和平均室外温度），以及正弦波模块（标设每日温度变化）提供。表示图形结果的示波器模块是一个输出模块（sink）。

10.2.3 子系统模型的使用

该 sldemo_househeat 模型采用了子系统模型的概念来简化模型图，创建了可再使用的子模块组件，并自定义子模块外观。子系统是一组分层模块，它由独立的子系统模块封装而成。

演示模型使用以下子系统（如图 10.2.2 所示）：

（1）Thermostat（恒温器）

（2）Heater（加热器）

（3）House（房屋）

（4）Fahrenheit to Celsius（华氏温度转换成摄氏温度）

（5）Celsius to Fahrenheit（摄氏温度转换成华氏温度）

子系统可能很复杂且包含许多模块，如果这些模块没有采用子系统封装，模型图可能看上去较为混乱。例如，双击 House 子系统，将其打开。如图 10.2.3 所示。

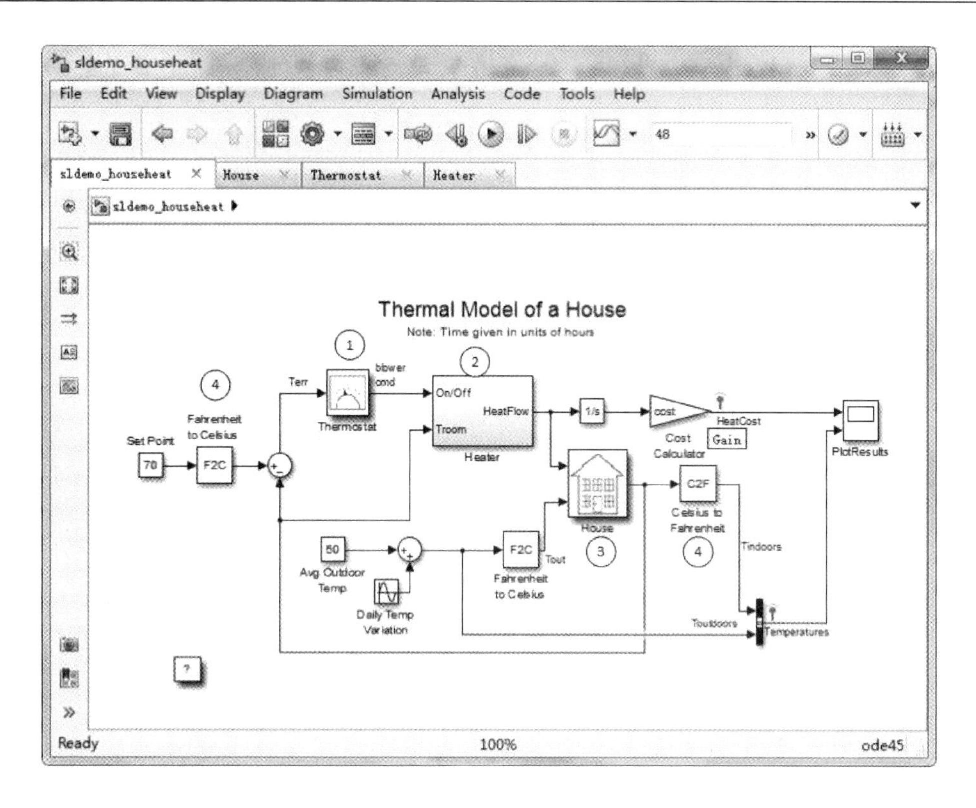

图 10.2.2　5 个子系统

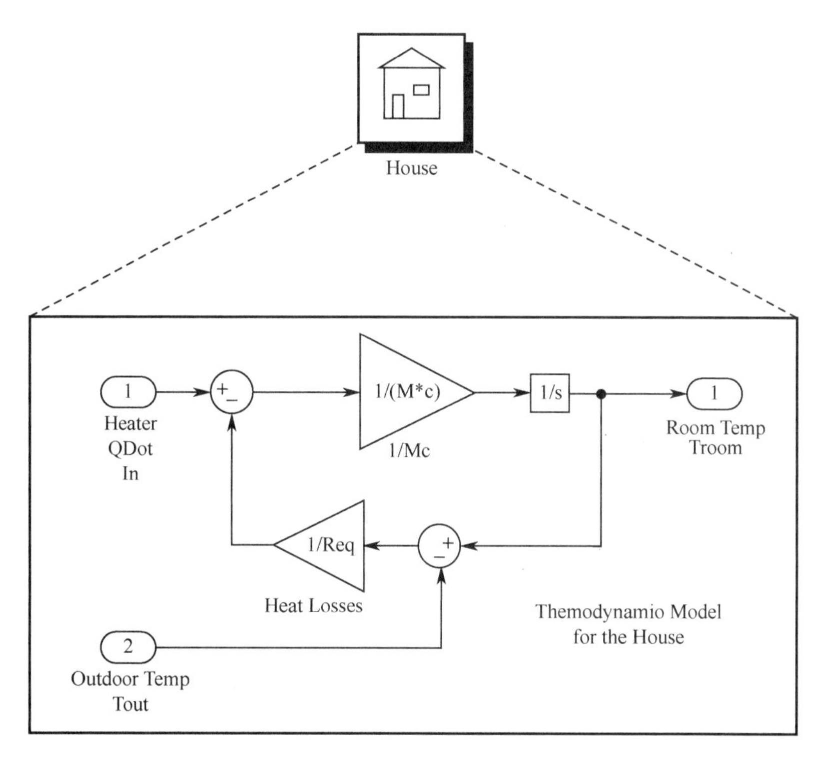

图 10.2.3　房子子系统

可以看到，该子系统把热流量和外部温度作为输入，并利用这两个输入来计算当前室温。虽然可以把这些模块直接放在主模型窗口，但是，把它们组合成子系统，有助于简化主模型框图。

子系统也可以很简单，仅包含几个模块。例如，双击恒温器子系统将其打开（见图 10.2.4）。

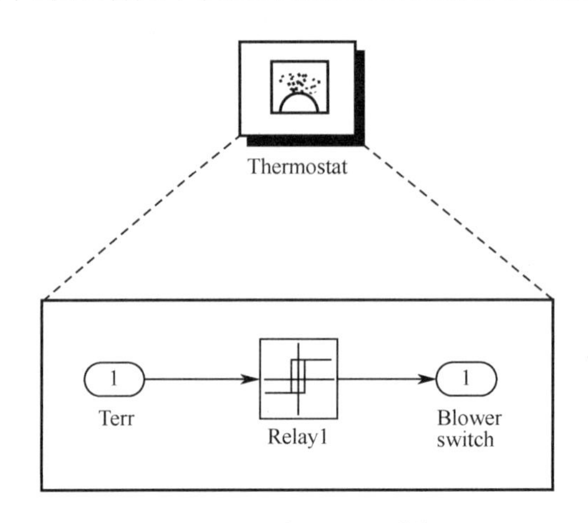

图 10.2.4 恒温器子系统

这个子系统模拟恒温器的操作，决定加热系统何时开和关。该子系统由一个继电器模块组成，但在逻辑上表示该模型图中的恒温器。

子系统也可重复使用，实际操作中，搭建一次，可供多次使用。例如，某模型中包含名为华氏转摄氏的完全相同的两个子系统（F2C）。该子系统（F2C）实现的是把内部和外部的温度从华氏度转换到摄氏度。

创建子系统，你可以将多个相关的模块装到一个子系统模块中。

创建子系统的步骤如下：

（1）在模型中选择恒温器模块。

（2）在模型中按住 Shift 键并单击该加热器模块，将其选中。

（3）在模型窗口中，选择编辑→创建子系统。

（4）该软件创建一个包含恒温器和加热器模块的子系统模块。

（5）选择编辑→撤销创建子系统，将模型返回到初始的状态。

10.3 房屋热力系统模型

10.3.1 概述

本模型展示如何使用 Simulink 创建一个房屋热力系统模型，该模型仿真了室外环境、房屋的热特性及加热系统。现在打开模型（如图 10.3.1 所示）。

图 10.3.2 中的脚本文件将模型工作空间的数据进行了初始化。要想修改模型，可以直接在工作空间对模型参数直接编辑，或者编辑 m 文件，还可以在工作空间重装载这个模型。要查看模型工作空间，在 Simulink 编辑器中选择 View→Model Explorer。

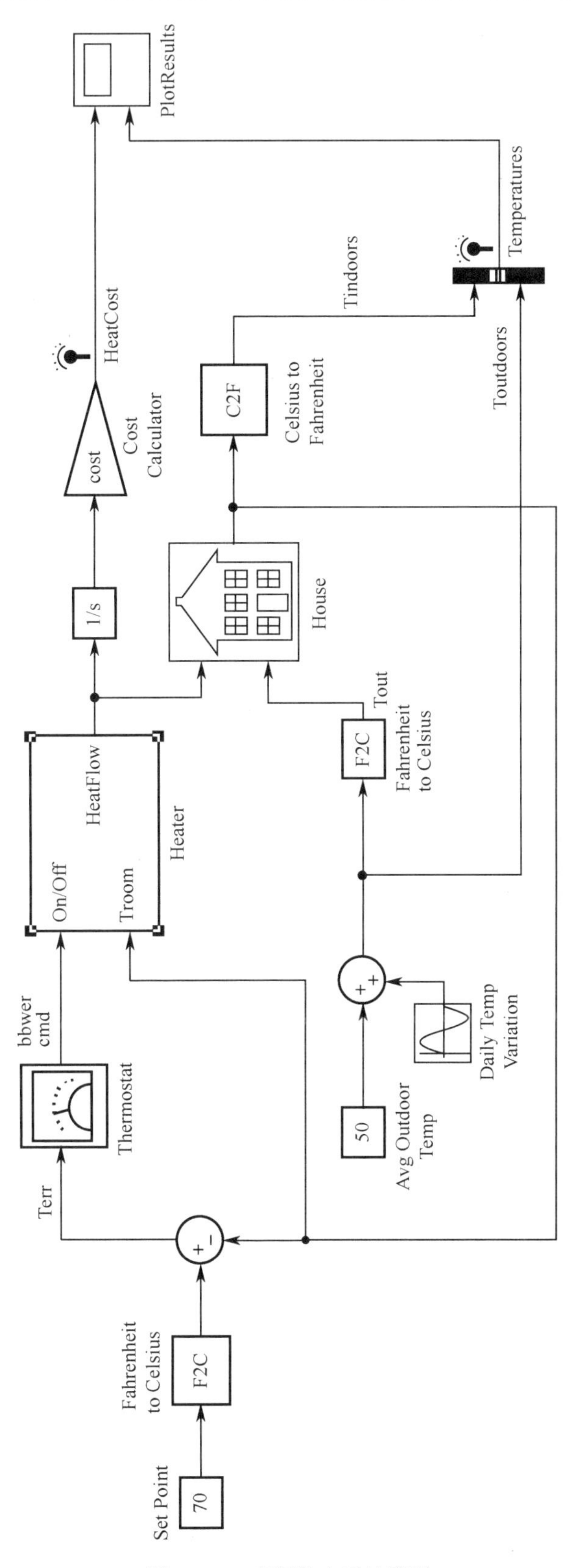

图 10.3.1　房屋热力系统模型

```
r2d = 180/pi;
lenHouse = 30;
widHouse = 10;
htHouse = 4;
pitRoof = 40/r2d;
numWindows = 6;
htWindows=1;
widWindows = 1;
windowArea = numWindows*htWindows*widWindows;
wallArea = 2*lenHouse*htHouse + 2*widHouse*htHouse +…
           2*(1/cos(pitRoof/2))*widHouse*lenHouse +…
           tan(pitRoof)*widHouse - windowArea;
kWall = 0.038*3600;
LWall = .2;
RWall = LWall/(kWall*wallArea);
kWindow = 0.78*3600;
LWindow = .01;
RWindow = LWindow/(kWindow*windowArea);
Req = RWall*RWindow/(RWall + RWindow);
c = 1005.4;
THeater = 50;
Mdot = 3600;
densAir = 1.2250;
M = (lenHouse*widHouse*htHouse+tan(pitRoof)*widHouse*lenHouse)*densAir;
cost = 0.09/3.6e6;
TinIC = 20;
```

图 10.3.2 脚本文件

该模型计算了普通住宅采暖费。打开模型时，它会加载图 10.3.2 中脚本文件的房屋信息。M 文件执行以下操作：

（1）定义房屋的几何结构；

（2）指定房屋材料的热性能；

（3）计算房屋的热阻；

（4）提供加热器的特征（热空气的温度、流速）；

（5）定义用电的费用（0.09$/kW·h）；

（6）指定初始室温（20 摄氏度= 68 华氏度）。

注意，时间以小时为单位。某些数量，如空气流速，用每小时来表达（而不是每秒）。

10.3.2 房屋加热模型的构成

Constant（设定值）模块

Constant 模块是一个常数模块，它的目录在 Commonly Used Blocks 之下，其设定值是必须保持的室温，默认为 70 华氏度。温度是按照华氏度给出的，但最终会转换为摄氏度来计算。

Fahrenheit to Celsius 模块

Fahrenheit to Celsius 模块利用转换公式将华氏的输入温度换算为摄氏的输出温度。该模块的目录在 simulink_extras/Transformations 下面，转换公式为

$$T_C = \left(\frac{5}{9}\right)(T_s - 32)$$

然而，Celsius to Fahrenheit 模块是相反的

$$T_C = \left(\frac{9}{5}\right)(T_C + 32)$$

其中，T_C 指摄氏温度，T_F 指华氏温度。

Sum（求和）模块

Sum 模块可以对输入信号进行加法和减法，该模块可以加上或者减去标量、矢量或者矩阵的输入。可以用 List of signs 参数确定模块的操作。加（+）、减（-）和间隔符（|）显示待输入信号的具体操作。

（1）如果有两个或多个输入，则加、减符号的数量必须等于输入的数量。例如，“+ - +”需要三个输入，从第一输入（顶部）减去第二输入（中间），再加第三输入（底部）。

（2）所有非标量输入必须有相同的维度。标量输入可以扩展为和其他输入具有相同的维度。

（3）间隔符在模块图标端口之间创建额外空间。

（4）对于圆形 Sum 求和模块，第一个输入端口最接近十二点的位置，输入端口在模块中逆时针方向排列。同样地，其他输入端口也按照逆时针方向排列。

（5）如果仅需要对所有的输入进行相加，那么将提供与输入数量一致的参数值，而非加号字符。

（6）如果只需要一个输入端口，那么，可通过指定一个“+”或一个“-”来实现对一个输入信号的叠加操作。

Integrator（积分器）模块

Integrator 模块会按照当前的时间步长对输入信号进行积分的输出。Integrator Limited（积分器限定）模块和 Integrator 模块是相同的，不同的是该模块的输出受到上、下饱和界限的限制。

下面的等式表示模块输出 y 是输入 u 的函数，并且初始条件为 y_0，其中，y 和 u 是当前仿真时间 t 的矢量函数。

$$y(t) = \int_{t_0}^{t} u(t)\mathrm{d}t + y_0$$

Gain（增益）模块（费用计算）

Gain 模块是输入乘以一个恒定值（增益）。输入和增益可以分别是标量、矢量或矩阵。

在 Gain 参数中可以指定增益值。Multiplication 参数可以设定向量点积或者矩阵乘法。对于矩阵乘法，该参数可以显示被乘数的次序。

增益值由双精度数据类型，通过最近取整或饱和的方式，转换为在模块封装离线状态下指定的数据类型。把输入（值）乘以增益（值），其结果会通过指定的舍入/溢出方式而转换为输出数据类型。

Scope（示波器）模块（PlotResults）

Scope 模块显示了与仿真时间有关的输入。它可以有多个轴（每个端口有一个），并且所有轴都具有独立 y 轴的公共时间范围。Scope 模块允许你调整时间和输入值的范围。可以移动和调整示波器的窗口，同时还可以在仿真过程中修改示波器的参数值。

开始仿真时，示波器窗口还没打开，但是数据已写入到示波器中，因此，如果在仿真之后打开示波器，该示波器的输入信号则会显示。

示波器提供了工具栏按钮，能够放大数据，把所有输入数据显示在示波器上，在多次仿真之间会保存轴设置，并限制数据显示，把数据保存到工作空间。图 10.3.3 显示，当打开 Scope 模块时，示波器窗口的显示情况，图中列出了工具条的各个按钮。

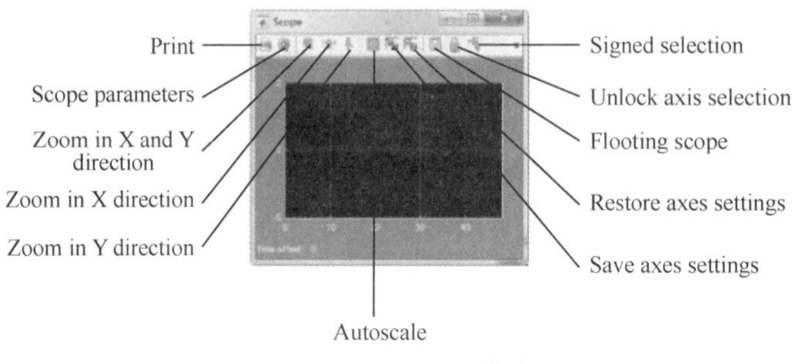

图 10.3.3 Scope 模块

Sine Wave（正弦波）模块

Sine Wave 模块提供正弦波，该模块可以在时间域或离散域下运行，它与数学运算库（Math Operations）中的正弦波函数（Sine Wave Function）模块一样。如果在模块对话框中的 Time 参数中选择 Use external source，会得到正弦波函数模块。

Bus Creator（总线创建器）模块

Bus Creator 模块把一组信号合并在一个总线上。把一组信号与 Bus Creator 模块相连，把 Numbe of inputs 的参数设置为该组输入信号的个数。模块会显示数值所确定的端口数。然后，把待分组的信号依次连接到相应的输入端口。

可以把任何类型的信号连接到输入端口，包括其他的总线信号。若要解绑信号，把该模块的输出端口连接到 Bus Selector 端口即可。

10.3.3 运行仿真并观察结果

先打开“PlotResults”示波器来观察结果。取暖成本和室内外温度会显示在示波器上。室外温度是正弦变化，而室内温度保持在设定值的华氏 5 度以内（见图 10.3.4）。

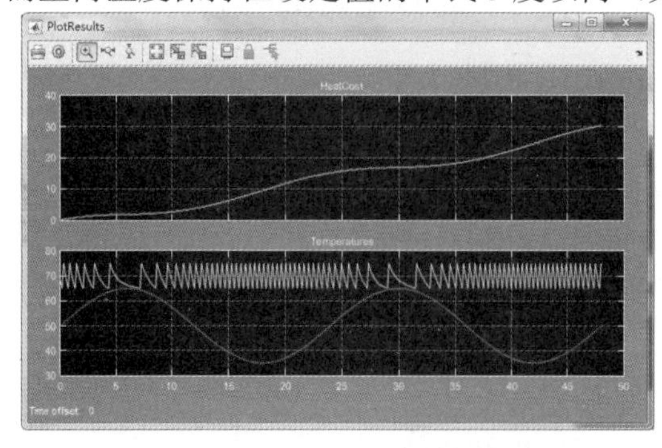

图 10.3.4 仿真结果（时间轴是小时）

根据这个模型，给房子供暖两天会花费大约 30 美元。也可尝试改变参数，并观察系统的反应。

10.3.4　注释

设计此模型仅仅是为了计算取暖费用。如果外部空气的温度高于室内温度，则室内温度会超过所期望的“设定值”。

可以修改模型把空调器包含在内，将其作为一台修正的取暖器用。要做到这一点，请添加下列参数到脚本文件：

（1）冷空气输出；

（2）来自空调器的气流温度；

（3）空调效率。

此外，还需要调整恒温器以便控制空调器和加热器。

本章小结

（1）本章主要介绍如何创建子系统及其封装技术；

（2）通过房屋加热系统模型例子，详细介绍了该系统中的创建子系统步骤，以及把子系统封装的过程；

（3）最后编程实现系统的仿真运行。

Chapter 10
Modeling the House Heating System

This chapter, on the basis of chapter 9, describes how to create subsystems and their masking techniques. Taking the house heating system model as an example, the steps of creating subsystems in the system and the process of masking the subsystems are described in detail. Finally, the system is simulated.

10.1 Creating and Masking Subsystems

10.1.1 Creating Subsystems

10.1.1.1 Introduction

A subsystem is a set of blocks that have been replaced by a single block called a Subsystem block. As your model increases in size and complexity, you can simplify it by grouping blocks into subsystems. Using subsystems has these advantages:

1. It helps reduce the number of blocks displayed in your model window.

2. It allows you to keep functionally related blocks together.

3. It enables you to establish a hierarchical block diagram, where a Subsystem block is on one layer and the blocks that make up the subsystem are on another.

You can create a subsystem in two ways:

1. Add a Subsystem block to your model, then open that block and add the blocks it contains to the subsystem window.

2. Add the blocks that make up the subsystem, then group those blocks into a subsystem.

10.1.1.2 Creating a Subsystem by Adding the Subsystem Block

To create a subsystem before adding the blocks it contains, add a subsystem block to the model, then add the blocks that make up the subsystem:

1. Copy the subsystem block from the Ports & Subsystems library into your model.

2. Open the subsystem block by double-clicking it.

The subsystem is opened in the current or a new model window. In the empty Subsystem window, create the subsystem. Use Inport blocks to represent input from outside the subsystem and Outport blocks to represent external output.

For example, in Figure 10.1.1, the subsystem shown includes a Relay1 block and Inport and Outport blocks to represent input to and output from the subsystem.

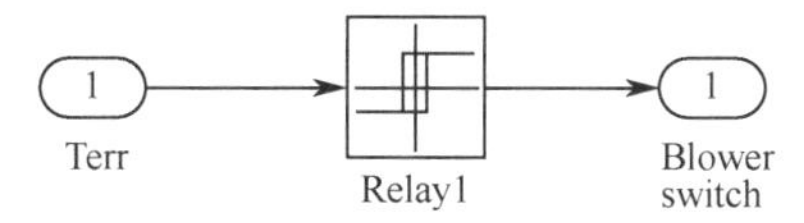

Figure 10.1.1 Thermostat Subsystem

10.1.1.3 Creating a Subsystem by Grouping Existing Blocks

If your model already contains the blocks you want to convert to a subsystem, you can create the subsystem by grouping those blocks:

1. Enclose the blocks and connecting lines that you want to include in the subsystem within a bounding box. You cannot specify the blocks to be grouped by selecting them individually or by using the Select All command.

For example, Figure 10.1.2 is a heater model, the Sum, Product, Constant and Gain blocks are selected within a bounding box.

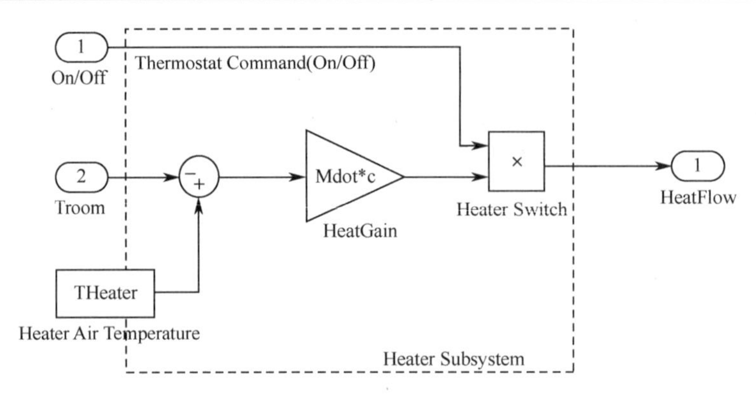

Figure 10.1.2 Heater Subsystem

2. Choose Create Subsystem from the Edit menu. The selected blocks are replaced with a Subsystem block. This Figure 10.1.3 shows the model after you choose the Create Subsystem command.

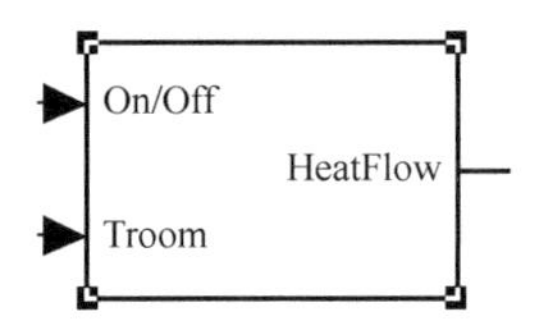

Figure 10.1.3 The Heater Subsystem Block

Notice that the Simulink software adds Inport and Outport blocks to represent input from and output to blocks outside the subsystem.

10.1.2 Masking Subsystems

10.1.2.1 Introduction

A mask is a custom user interface for a Simulink block. The mask hides the native user interface of the underlying block, substituting an icon and a parameters dialog box defined by the mask. You can apply a mask to any Subsystem block, Model block, or S-Function block. The block can optionally reside in a user-defined library, masking a block changes only the block's user interface, not its underlying characteristics.

A mask's icon and parameters dialog box can provide any capability that a block's native icon and dialog box can provide. When you set mask parameter values, the mask can use the values to dynamically change the mask's icon and dialog box, and to calculate values to be used under the mask. A mask on a subsystem can dynamically change the subsystem to reflect mask parameter values.

10.1.2.2 Creating a Block Mask

1. Opening the Mask Editor

The Simulink Mask Editor provides all capabilities necessary for masking a block. To invoke the editor on an unmasked block:

1. Figure 10.1.4 is a model of a house. Box to select the Sum on the dotted line module, the Integrator module and Gain module, then choose Create Subsystem from the Edit menu. The

selected blocks are replaced with a Subsystem block(see Figure 10.1.5).

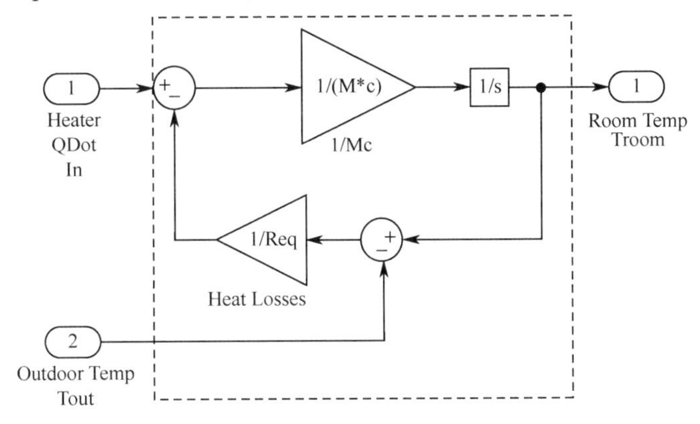

Figure 10.1.4 A House Model

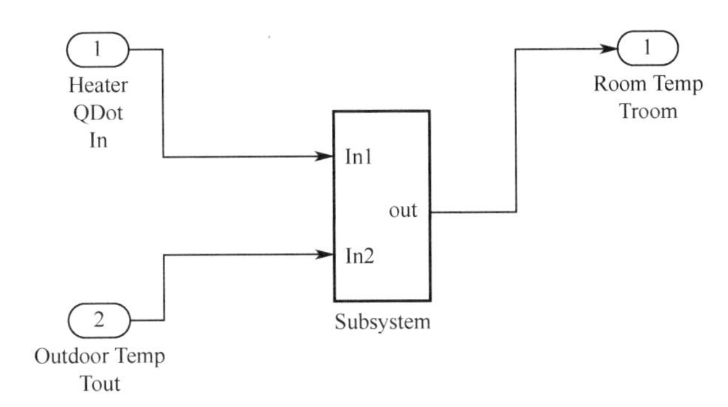

Figure 10.1.5 A House Subsystem Block

2. Then choose Mask Subsystem from the Edit menu. The Mask Editor will be opened. The Mask Editor opens as shown in Figure 10.1.6.

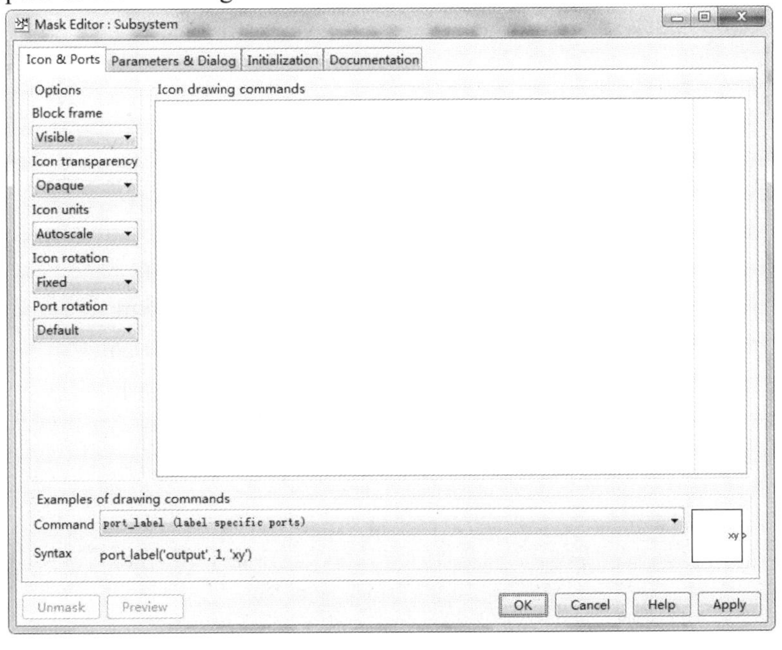

Figure 10.1.6 The Mask Editor

You can define mask components in any order, as long as the final result is correct. You can also open the Mask Editor on block that is already masked. Click Apply, or OK to make changes within the model and close the editor. To permanently save changes, you must save the model itself. If you close the model without saving it, all unsaved Mask Editor changes are lost.

2. Defining the Mask Icon

A mask icon replaces a masked block's standard icon when the masked block appears in a model. Mask icons can contain descriptive text, graphics, images, state equations, one or more plots, or a transfer function. Simulink uses commands that you supply to draw the mask icon.

Use the Mask Editor's Icon & Ports pane to specify the icon of a masked block. This pane is always selected by default when you open the Mask Editor.

To specify the contents of a mask icon, enter one or more of the commands described in Mask Icon Drawing Commands in the Icon Drawing commands text box. You can use only these commands to draw a mask icon. If any drawing command cannot successfully execute, the icon displays three question marks.

You can use the command to draw a fangwu mask icon in MATLAB.

```
image(imread('fangwu.png'))    % Use fangwu icon for subsystem。
```

(Note: Here fangwu.png is an image file.)

Type the following command into the Icon Drawing commands pane:

image(imread('fangwu.png')) , see Figure 10.1.7.

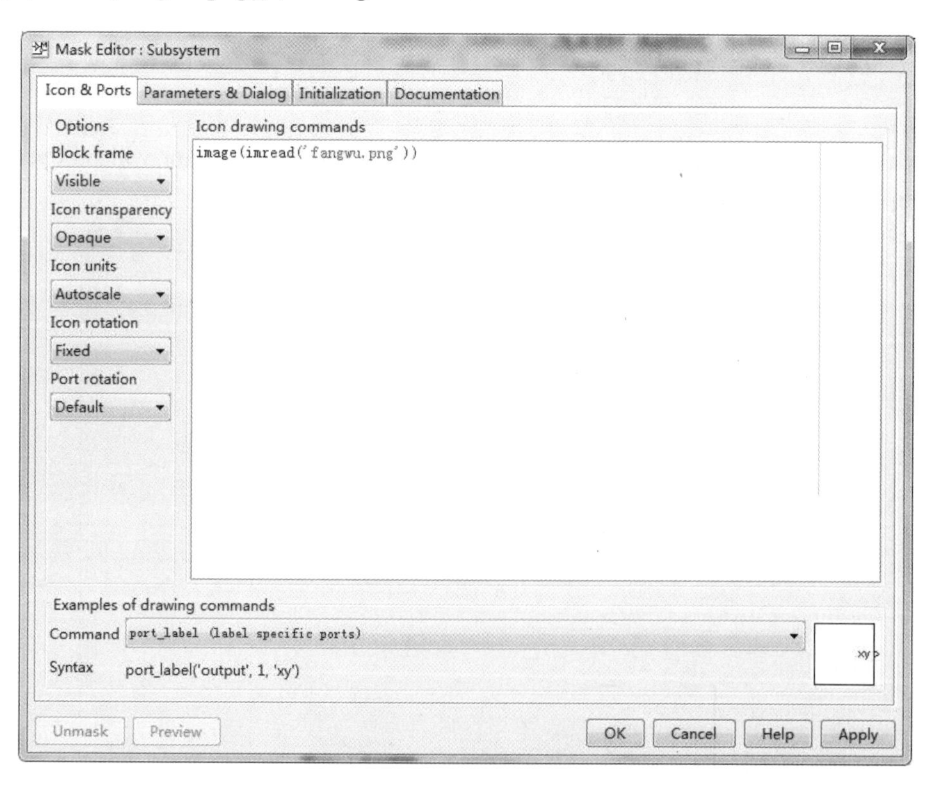

Figure 10.1.7 Commands Pane

3. Click Apply

If Simulink cannot evaluate all commands in the Drawing Commands pane to obtain a displayable result, the content of the icon is three question marks. Click Apply, and the submodule icon results in Figure 10.1.8.

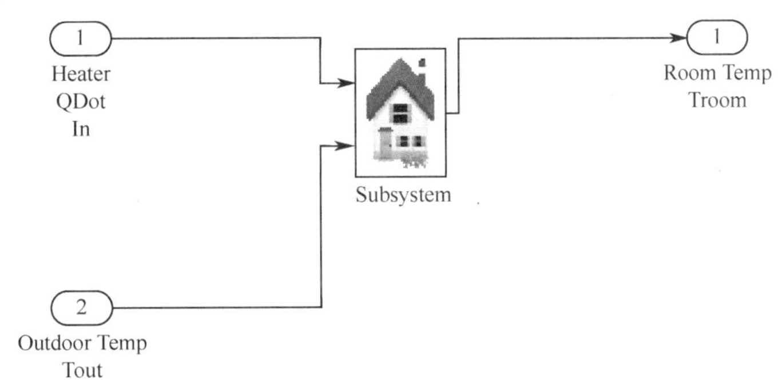

Figure 10.1.8 Icon Result of the Subsystem

The pull downs under Icon Options allow you to specify the icon frame visibility, icon transparency, drawing context, icon rotation, and port rotation. And the subsystem is masked completely.

For example, the House and Thermostat subsystems display custom icons that depict physical objects, while the conversion subsystems display custom dialog boxes when you double-click them(as shown in Figure 10.1.9).

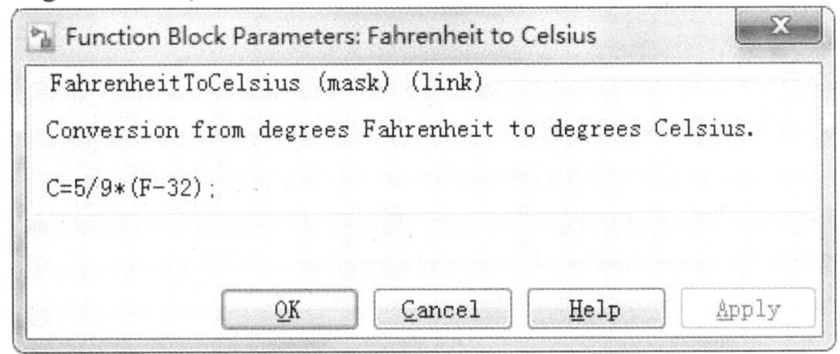

Figure 10.1.9 The Conversion Subsystems

To view the underlying blocks in the conversion subsystem, right-click the subsystem block, then select Look Under Mask (see Figure 10.1.10).

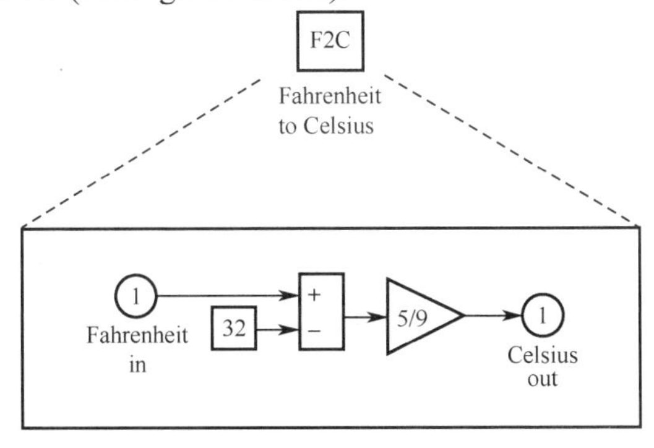

Figure 10.1.10 The Underlying Blocks in the Conversion Subsystem

10.2 Anatomy of the House Heating Model

10.2.1 Opening the Demo Model

The demo model described in this section is called sldemo_househeat. You can open it from the MATLAB Command Window.

To open the demo model:

1. Ensure that MATLAB is open.
2. Enter sldemo_househeat in the MATLAB Command Window.

The software starts and opens the sldemo_househeat model. See Figure 10.2.1.

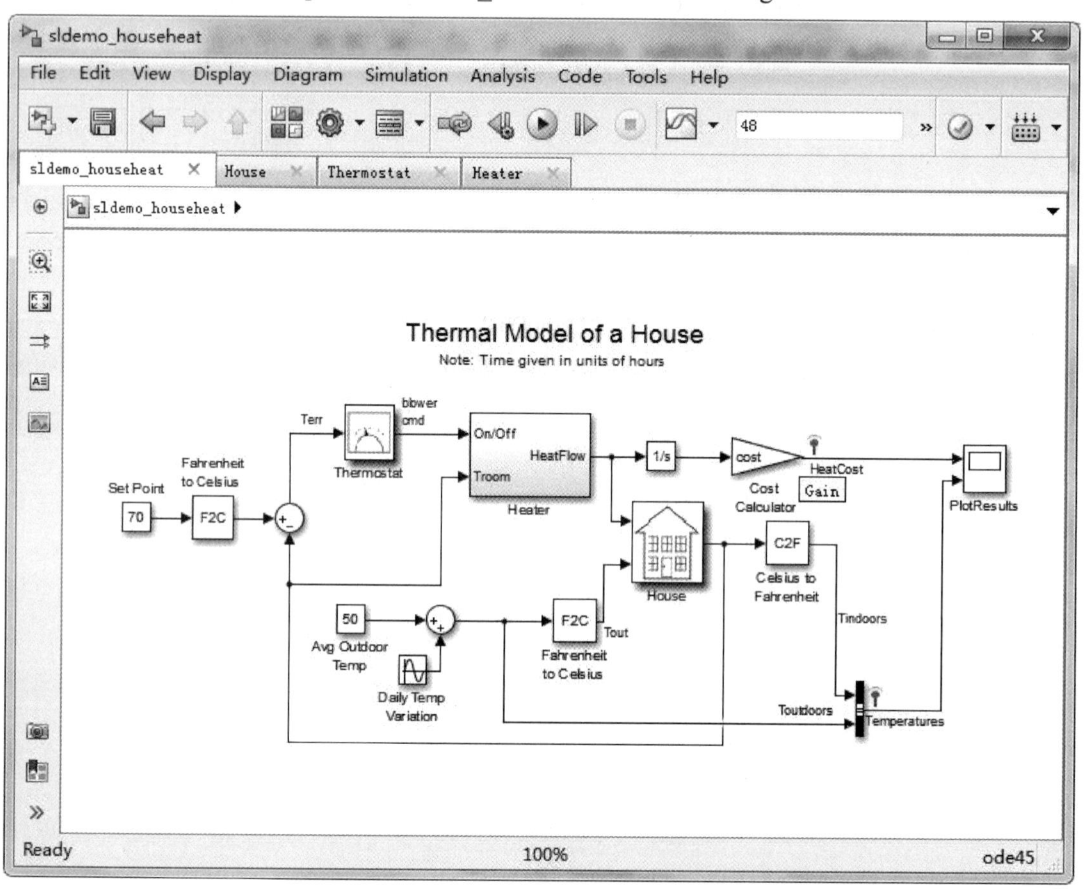

Figure 10.2.1 Thermal Model of a House

10.2.2 Anatomy of the Model

The system models the outdoor environment, the thermal characteristics of the house, and the house heating system. It allows you to simulate how the thermostat setting and outdoor environment affect the indoor temperature and cumulative heating costs.

The model includes many of the same blocks you used to create the simple model in Creating a Simulink Model. These include:

1. A Scope block (labeled PlotResults) on the far right displays the simulation results.

2. A Mux block at the bottom right combines the indoor and outdoor temperature signals for the Scope.

3. A Sine Wave block (labeled Daily Temp Variation) at the bottom left provides one of three data sources for the model.

In the demo, the thermostat is set to 70 degrees Fahrenheit. The system models fluctuations in outdoor temperature by applying a sine wave with amplitude of 15 degrees to a base temperature of 50 degrees.

The three data inputs (sources) are provided by two Constant blocks (labeled Set Point and Avg Outdoor Temp), and the Sine Wave block (labeled Daily Temp Variation). The Scope block labeled PlotResults is the one output (sink).

10.2.3 Using Subsystems

The sldemo_househeat demo model uses *subsystems* to simplify the appearance of the block diagram, create reusable components, and customize the appearance of blocks. A subsystem is a hierarchical grouping of blocks encapsulated by a single Subsystem block.

The demo model uses the following subsystems(see Figure 10.2.2)

1. Thermostat
2. Heater
3. House
4. Fahrenheit to Celsius
5. Celsius to Fahrenheit

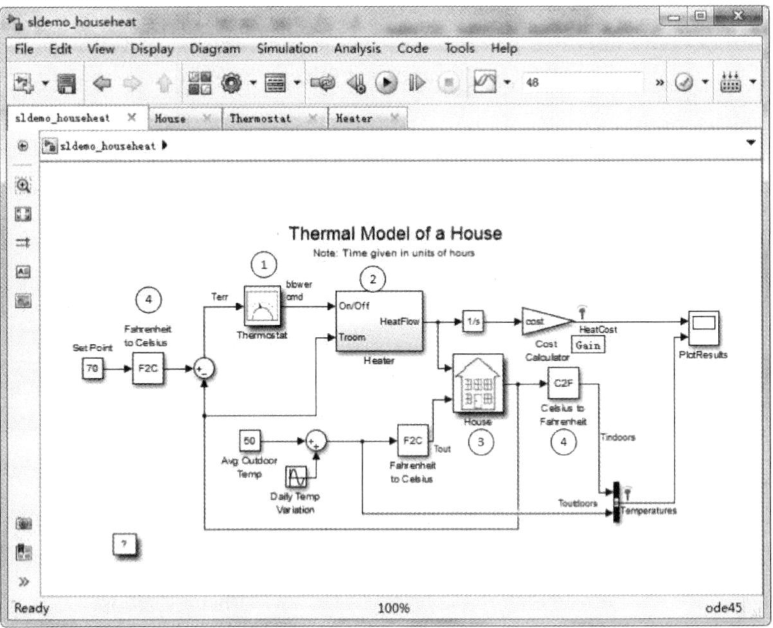

Figure 10.2.2 Five Subsystems

Subsystems can be complex and contain many blocks that might otherwise clutter a diagram. For example, double-click the House subsystem to open it (see Figure 10.2.3).

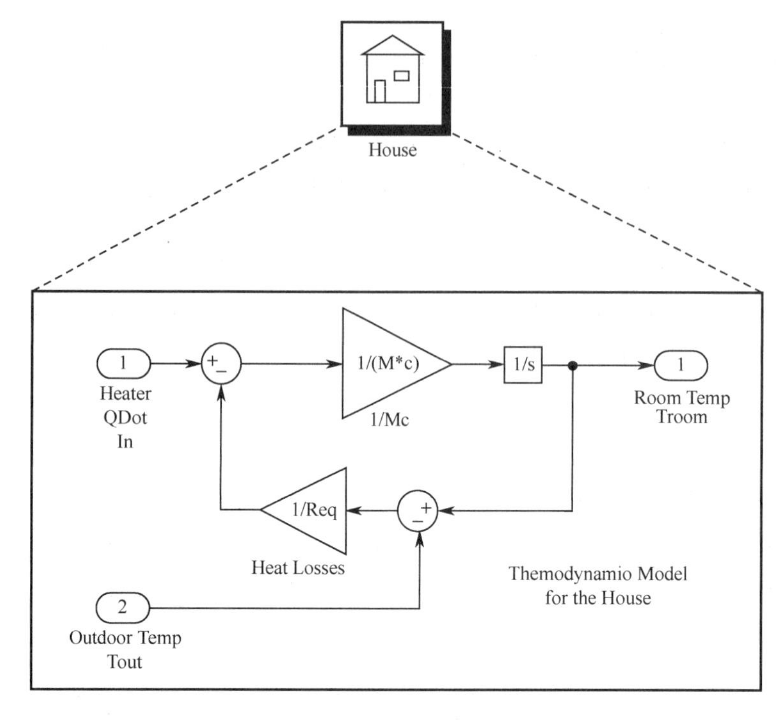

Figure 10.2.3 House Subsystem

You can see that the subsystem receives heat flow and external temperature as inputs, which it uses to compute the current room temperature. You could leave each of these blocks in the main model window, but combining them as a subsystem helps simplify the block diagram.

Subsystems can also be simple and contain only a few blocks. For example, double-click the Thermostat subsystem to open it (see Figure 10.2.4).

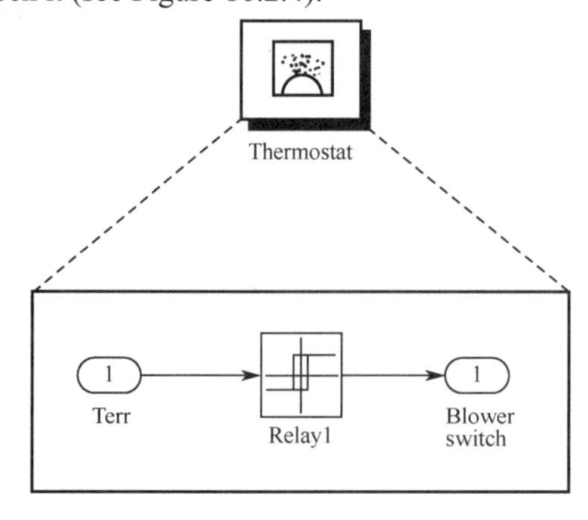

Figure 10.2.4 Contents of Thermostat Subsystem

This subsystem models the operation of a thermostat, determining when the heating system is on or off. It contains only one Relay block, but logically represents the thermostat in the block diagram.

Subsystems are also reusable, enabling you to implement an algorithm once and use it multiple times. For example, the model contains two instances of identical subsystems named Fahrenheit to Celsius(F2C). These subsystems(F2C) convert the inside and outside temperatures from degrees Fahrenheit to degrees Celsius.

Creating a subsystem allows you to group multiple related blocks into one subsystem block.

To create a subsystem:

1. Select the Thermostat block in the demo model.
2. Shift-click the Heater block in the demo model to select it.
3. Select Edit>Create Subsystem in the model window.
4. The software creates a Subsystem block containing the Thermostat and Heater blocks.
5. Select Edit>Undo Create Subsystem to return the model to its original configuration.

10.3 Thermal Model of a House

10.3.1 Introduction

This demo illustrates how you can use Simulink to create the thermal model of a house. This system models the outdoor environment, the thermal characteristics of the house, and the house heating system. Then open the Thermal Model of a House(shown in Figure 10.3.1).

Figure 10.3.2 the script file initializes the data for the model workspace.To make changes, you can edit the model workspace directly or edit the m-file and re-load the model workspace. To view the model workspace, select View > Model Explorer from the Simulink editor.

This model calculates heating costs for a generic house. When the model is opened,the model loads the house information of the script file in Figure 10.3.2. The M-file does the following:

1. Defines the house geometry.
2. Specifies the thermal properties of house materials.
3. Calculates the thermal resistance of the house.
4. Provides the heater characteristics (temperature of the hot air, flow-rate).
5. Defines the cost of electricity (0.09$/kW·h).
6. Specifies the initial room temperature (20deg. Celsius = 68deg. Fahrenheit).

Note: Time is given in units of hours. Certain quantities, like air flow-rate, are expressed per hour (not per second).

10.3.2 The House Heating Model Components

Constant(Set Point)

"Set Point" is a constant block.This block directory under the Commonly Used Blocks specifies the temperature that must be maintained indoors. It is 70 degrees Fahrenheit by default. Temperatures are given in Fahrenheit, but then are converted to Celsius to perform the calculations.

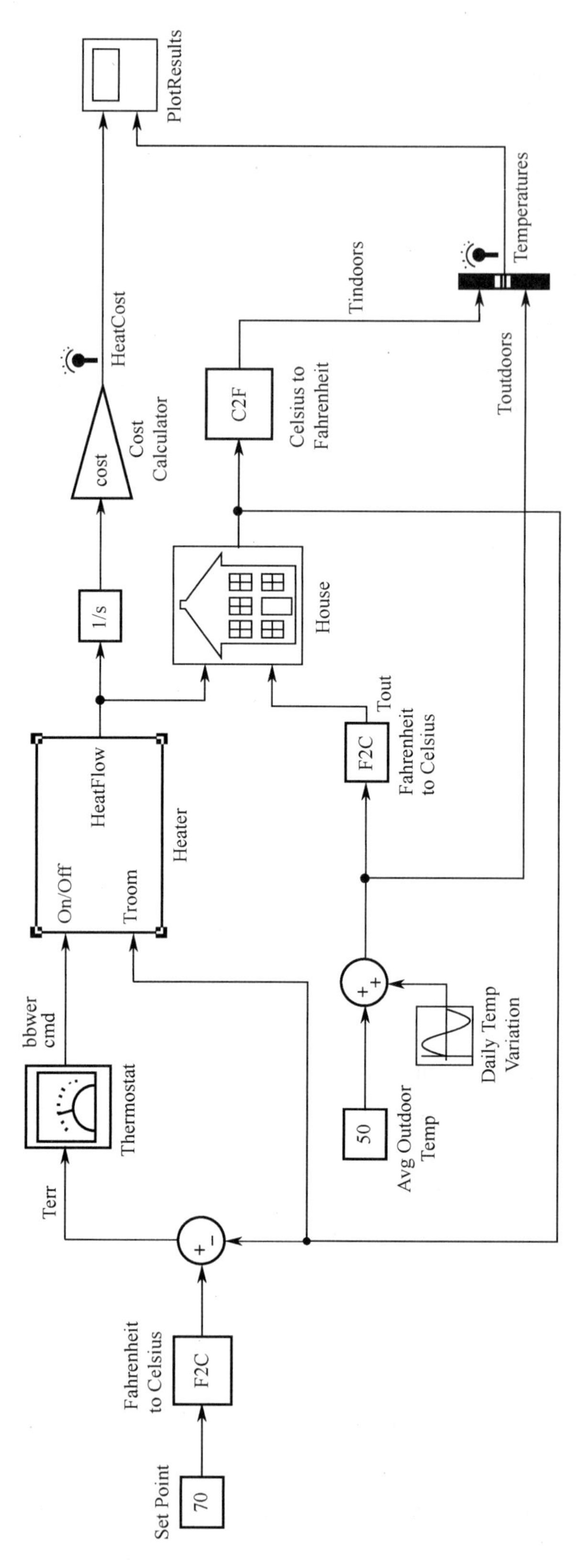

Figure 10.3.1 The House Heating Model

```
r2d = 180/pi;
lenHouse = 30;
widHouse = 10;
htHouse = 4;
pitRoof = 40/r2d;
numWindows = 6;
htWindows=1;
widWindows = 1;
windowArea = numWindows*htWindows*widWindows;
wallArea = 2*lenHouse*htHouse + 2*widHouse*htHouse + ...
           2*(1/cos(pitRoof/2))*widHouse*lenHouse + ...
           tan(pitRoof)*widHouse - windowArea;
kWall = 0.038*3600;
LWall = .2;
RWall = LWall/(kWall*wallArea);
kWindow = 0.78*3600;
LWindow = .01;
RWindow = LWindow/(kWindow*windowArea);
Req = RWall*RWindow/(RWall + RWindow);
c = 1005.4;
THeater = 50;
Mdot = 3600;
densAir = 1.2250;
M = (lenHouse*widHouse*htHouse+tan(pitRoof)*widHouse*lenHouse)*densAir;
cost = 0.09/3.6e6;
TinIC = 20;
```

Figure 10.3.2 Script File

Fahrenheit to Celsius

The Fahrenheit to Celsius block computes the output temperature in degrees Celsius from the input temperature in degrees Fahrenheit using the conversion factor. This block directory is under the simulink_extras/Transformations.

$$T_C = \left(\frac{5}{9}\right)(T_s - 32)$$

However, the Celsius to Fahrenheit block is inverse.

$$T_C = \left(\frac{9}{5}\right)(T_C + 32)$$

where T_F is the temperature in degrees Fahrenheit and T_C is the temperature in degrees Celsius.

Sum block

The Sum block performs addition or subtraction on its inputs. This block can add or subtract scalar, vector, or matrix inputs.You specify the operations of the block with the **List of signs** parameter. Plus (+), minus (−), and spacer (|) characters indicate the operations to be performed on the inputs:

1. If there are two or more inputs, then the number of + and − characters must equal the number

of inputs. For example, "+−+" requires three inputs and configures the block to subtract the second (middle) input from the first (top) input, and then add the third (bottom) input.

2. All nonscalar inputs must have the same dimensions. Scalar inputs will be expanded to have the same dimensions as the other inputs.

3. A spacer character creates extra space between ports on the block's icon.

4. For a round Sum block, the first input port is the port closest to the 12 o'clock position going in a counterclockwise direction around the block. Similarly, other input ports appear in counterclockwise order around the block.

5. If only addition of all inputs is required, then a numeric parameter value equal to the number of inputs can be supplied instead of "+" characters.

6. If only one input port is required, a single "+" or "−" collapses the element via the specified operation.

Integrator block

The Integrator block outputs the integral of its input at the current time step. The Integrator Limited block is identical to the Integrator block with the exception that the output of the block is limited based on the upper and lower saturation limits.

The following equation represents the output of the block y as a function of its input u and an initial condition y_o, where y and u are vector functions of the current simulation time t.

$$y(t) = \int_{t_0}^{t} u(t)\mathrm{d}t + y_0$$

Gain block (cost calculator)

The Gain block multiplies the input by a constant value (gain). The input and the gain can each be a scalar, vector, or matrix.

You specify the value of the gain in the Gain parameter. The Multiplication parameter lets you specify element-wise or matrix multiplication. For matrix multiplication, this parameter also lets you indicate the order of the multiplication.

The gain is converted from doubles to the data specified in the block mask offline using round-to-nearest and saturation. The input and gain are then multiplied, and the result is converted to the output data type using the specified rounding and overflow modes.

Scope (PlotResults)

The Scope block displays its input with respect to simulation time. The Scope block can have multiple axes (one per port) and all axes have a common time range with independent y-axes. The Scope block allows you to adjust the amount of time and the range of input values displayed. You can move and resize the Scope window and you can modify the Scope's parameter values during the simulation.

When you start a simulation the Scope windows are not opened, but data is written to connected Scopes. As a result, if you open a Scope after a simulation, the Scope's input signal or signals will be displayed.

The Scope provides toolbar buttons that enable you to zoom in on displayed data, display all the data input to the Scope, preserve axis settings from one simulation to the next, limit data displayed, and save data to the workspace. The toolbar buttons are labeled in Figure 10.3.3,which shows the Scope window as it appears when you open a Scope block.

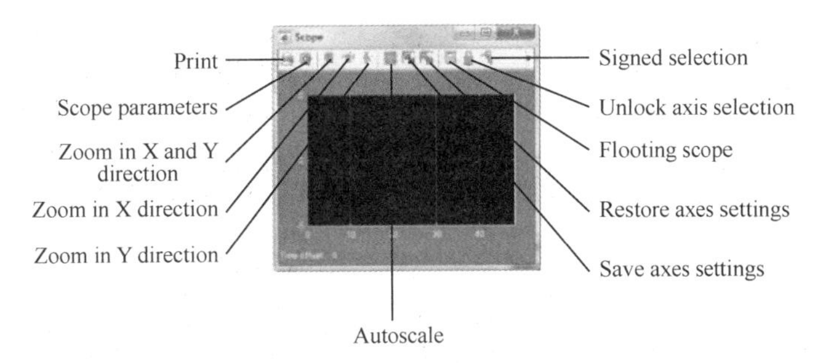

Figure 10.3.3 Scope Block

Sine Wave block

The Sine Wave block provides a sinusoid. The block can operate in time-based or sample-based mode. This block is the same as the Sine Wave Function block that appears in the Math Operations library. If you select Use external source for the Time parameter in the block dialog box, you get the Sine Wave Function block.

Bus Creator block

The Bus Creator block combines a set of signals into a bus. To bundle a group of signals with a Bus Creator block, set the block's Number of inputs parameter to the number of signals in the group. The block displays the number of ports that you specify. Connect the signals to be grouped to the resulting input ports.

You can connect any type of signal to the inputs, including other bus signals. To ungroup the signals, connect the block's output port to a Bus Selector port.

10.3.3 Running the Simulation and Visualizing the Results

Open the "PlotResults" scope to visualize the results. The heat cost and indoor versus outdoor temperatures are plotted on the scope. The temperature outdoor varies sinusoidally, whereas the indoors temperature is maintained within 5 degrees Fahrenheit of "Set Point"(see Figure 10.3.4).

According to this model, it would cost around $30 to heat the house for two days. Try varying the parameters and observe the system response.

10.3.4 Remarks

This particular model is designed to calculate the heating costs only. If the temperature of the outside air is higher than the room temperature, the room temperature will exceed the desired "Set Point".

You can modify this model to include an air conditioner. You can implement the air conditioner as a modified heater. To do this, add parameters like the following to script:

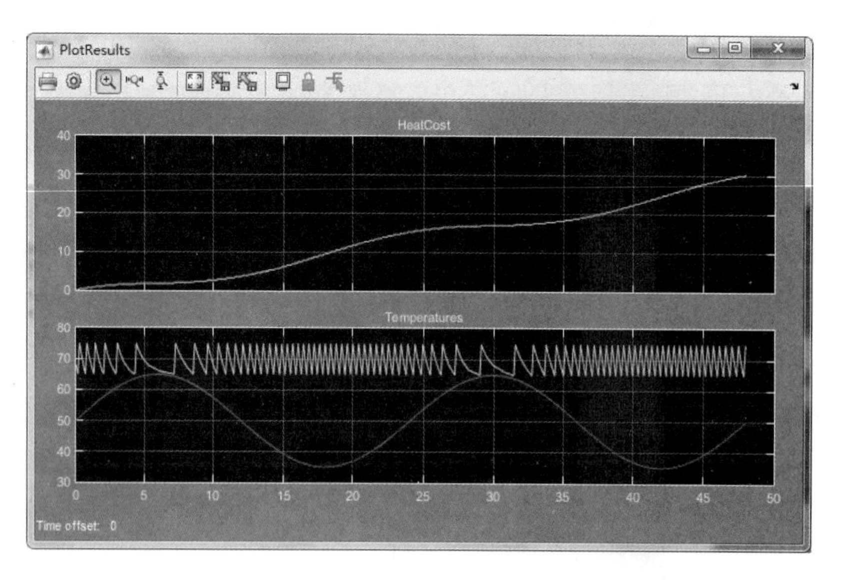

Figure 10.3.4 Simulation Results (Time Axis Labeled in hours)

1. Cold air output.
2. Temperature of the stream from the air conditioner.
3. Air conditioner efficiency.

You would also need to modify the thermostat to control both the air conditioner and the heater.

Summary

1. How to create and mask subsystems is mainly introduced in this chapter.

2. The steps to create subsystems and the process of masking substems are illustrated through the building thermal model of a house.

3. The system simulation is implemented by programming in simulink.

生词注解

creating a block mask	创建一个封装块
plots	多个平面图
subsystem	n. 子系统
dialog	n. 对话
specify	v. 指定，详细说明
encapsulated	adj. 密封的
implement	n. 工具，器具
	v. 实施
algorithm	n. 算法
sinusoidally	正弦地